Rapid Methods for Analysis of Biological Materials in the Environment

NATO ASI Series

Advanced Science Institute Series

A Series presenting the results of activities sponsored by the NATO Science Committee, which aims at the dissemination of advanced scientific and technological knowledge, with a view to strengthening links between scientific communities.

The Series is published by an international board of publishers in conjunction with the NATO Scientific Affairs Division

A. Life Sciences **B. Physics**	Plenum Publishing Corporation London and New York
C. Mathematical and Physical Sciences **D. Behavioural and Social Sciences** **E. Applied Sciences**	Kluwer Academic Publishers Dordrecht, Boston and London
F. Computer and Systems Sciences **G. Ecological Sciences** **H. Cell Biology** **I. Global Environment Change**	Springer-Verlag Berlin, Heidelberg, New York, London, Paris and Tokyo

PARTNERSHIP SUB-SERIES

1. Disarmament Technologies	Kluwer Academic Publishers
2. Environment	Springer-Verlag / Kluwer Academic Publishers
3. High Technology	Kluwer Academic Publishers
4. Science and Technology Policy	Kluwer Academic Publishers
5. Computer Networking	Kluwer Academic Publishers

The Partnerschip Sub-Series incorporates activities undertaken in collaboration with NATO's Cooperation Partners, the countries of the CIS and Central and Eastern Europe, in Priority Areas of concern to those countries.

NATO-PCO-DATA BASE

The electronic index to the NATO ASI Series provides full bibliographical references (with keywords and/or abstracts) to about 50,000 contributions from international scientists published in all sections of the NATO ASI Series. Access to the NATO-PCO-DATA-BASE is possible via a CD-ROM "NATO Science and Technology Disk" with user-friendly retrieval software in English, French, and German (©WTV GmbH and DATAWARE Technologies, Inc. 1989). The CD-ROM contains the AGARD Aerospace Database.

The CD-ROM can be ordered through any member of the Board of Publishers or through NATO-PCO, Overijse, Belgium.

Serie 1: Disarmament Technologies – Vol. 30

Rapid Methods for Analysis of Biological Materials in the Environment

edited by

Peter J. Stopa
U.S. Army Edgewood Chemical and
Biological Center,
APG, MD, U.S.A.

and

Michael A. Bartoszcze
Veterinary Research Center,
Military Institute of Hygiene and Epidemiology,
Pulawy, Poland

Kluwer Academic Publishers

Dordrecht / Boston / London

Published in cooperation with NATO Scientific Affairs Division

Proceedings of the NATO Advanced Research Workshop on
Rapid Methods for Monitoring the Environment for Biological Hazards
Warsaw, Poland
19–22 May 1997

A C.I.P. Catalogue record for this book is available from the Library of Congress.

ISBN 0-7923-6304-3

Published by Kluwer Academic Publishers,
P.O. Box 17, 3300 AA Dordrecht, The Netherlands.

Sold and distributed in North, Central and South America
by Kluwer Academic Publishers,
101 Philip Drive, Norwell, MA 02061, U.S.A.

In all other countries, sold and distributed
by Kluwer Academic Publishers,
P.O. Box 322, 3300 AH Dordrecht, The Netherlands.

Printed on acid-free paper

Printed in the Netherlands.

DEDICATION

This book is dedicated to the memory of my father, Peter A. Stopa,
and all those others,
whose hearts were on both sides of the Atlantic.

TABLE OF CONTENTS

PREFACE

The work described in this report was presented at the NATO advanced Research Workshop (ARW) entitled "Rapid Methods for Monitoring the Environment for Biological Hazards". This workshop was held in Warsaw, Poland in May 1997. This ARW was funded by the NATO Science Committee, the Polish Ministry of Defense, the US Army European Research Office, and the US Edgewood Chemical and Biological Center. In addition, the organizers would also like to thank the many corporate sponsors who contributed to the success of this workshop.

ACKNOWLEDGMENTS

The authors and the organizing committee wish to thank the NATO Science Committee and the other contributors for their generous support and encouragement that made this fruitful event possible. In addition, the organizers would like to thank the following individuals for their participation and contribution to the NATO Advanced Research Workshop: Dr. Henryk Arciuch (WIHE), Dr. Michal Bartoszcze (WIHE), Ms. Agata Bielawska (WIHE), Dr. Camille Boulet (DRES), Dr. Wayne Bryden (JHU-APL), Dr. J. Bzdega (WIHE), Professor Krzysztof Chomiczewski (WIHE), Dr. Adrian Clark (CBD), Mr. Philip Coon (ERDEC), Dr. Malcolm Dando (U of Bradford), Dr. Hazel Davey (UWA), Dr. Vito G. Del Vecchio (U of Scranton), Dr. Mildred Donlon (DARPA), Dr. Norm Dovichi (U of Alberta), Dr. Heinrich Egghart (USAERO), Dr. Henri Garrigue (CEB), Dr. Royston Goodacre (UWA), Dr. Sergey Grinshpun (U of Cincinnati), Dr. Bruce Harper (DPG), Mr. Jeff Harvey (Bio-Rad Labs), Dr. Eric Henchal (USAMRIID), Dr. James Ho (DRES), Dr. Sergei Ignatov (ORC), Dr. Vladimir Ivanov (UNU), Dr. Graziella Jula (REIE), Dr. Arsny Kaprylants (MSU), Mrs. Zuzanna Krocova (PMI), Dr. Ann Kusterbeck (NRL), Dr. Michaela Lazarescu (MWFEP), Dr. Andrzej Lidacki (WIHE), Mr. Larry Loomis (NHD), Mrs. Jadwiga Matras (WIHE), Dr. Henry Meier (USAMRICD), Dr. Beatta Mizak (PRIWET), Mr. Hank Mottl (Dycor), Dr. Michaela Nicolae (REIE), Dr. Christian Otterbein (FRGMMA), Mr. Larry Paul (TSI), Dr. W. Palec (PRIWET), Mrs. Dorothea Paterno (ERDEC), Mr. Carl Peterson (SCP), Dr. Vern Seligy (HC), Dr. Elena Stabnikova (UNU), Dr. Elzbieta Trafny (WIHE), Dr. Gerard Trouiller (CEB), Dr. Elena Vorobyova (MSU), Dr. B. Wojton (PRIWET), Dr. A. Zakharov (RSI), Dr. Nettie Zegers (TNO).

In addition, the editors wish to thank Mr. Preston "Konrad" Smith and Ms. Karen Lopez-Vado without whose help this manuscript would not have been possible.

INTRODUCTION

Mr. Peter J. Stopa of the U.S. Army Edgewood Chemical and Biological Center, Aberdeen Proving Ground, MD, USA, and COL Michael Bartoszcze, Commander of the Veterinary Research Center of the Military Institute of Hygiene and Epidemiology, Pulawy Poland, convened the first NATO Advanced Research Workshop on "Rapid Methods for Monitoring the Environment for Biological Hazards". The meeting was held on 19 May 1997, in Warsaw, Poland, at the Staszica Palace, which houses the Polish Academy of Sciences. The purpose of the workshop was to discuss issues regarding monitoring of the environment for biological materials in a variety of contexts, including biological warfare; treaty inspections for the Biological Weapons Convention (BWC); food safety; and water monitoring. Mr. Stopa's opening remarks reflected that there was a "changing of the guard with respect to biological materials. People are becoming more concerned about the impact of biological materials on environmental matters."

The opening session included participation from the Polish Ministry of Sanitation and the US Embassy in Poland, both of which stressed the importance of the meeting. Director Ryszard Krol, who is the Chief of the Department for NATO Cooperation within the Polish Ministry of National Defense, also participated. Other opening session talks included a talk by Dr. Heinrich Egghart of the U.S. Army European Research Office, whose office also contributed to the meeting. Dr. Millie Donlon of the Defense Advanced Research Projects Agency (USA) also presented an overview of DARPA and discussed some future technologies that were being developed to address the biological detection problem. At the request of Mrs. Schulte of NATO, Dr. Malcolm Dando was invited to give a talk on BWC issues. COL Michael Bartoszcze, the co-chair, then gave a presentation on "Some Problems Concerning Biological Threats." In his remarks, he supported the formation of an international team of experts to respond to biological incidents, such as disease outbreaks, as was proposed by others.

The afternoon session was an overview of what is known about the biology of the environment. One of the highlights was the presentation by Dr. Zakharov, on the approaches to study biological environments in preparation for future missions to Mars.

The next sessions reviewed current methodology and short-term, future improvements. Luminescence techniques, immunoassay test kits and sensors, current nucleic acid analysis techniques, and mass spectometry were discussed.

Wednesday morning was dedicated to aerobiology techniques and included an overview of sampling and particle analysis methodologies by university, government, and industrial experts. Dr. Bruce Harper of Dugway Proving Ground and Mrs. Zuzanna Krocova of the Czech Republic then tied the technology aspects together. They discussed how these techniques are used to measure bacterial aerosols for the testing of biological detection equipment and the evaluation of the efficacy of vaccines against aerosol challenges.

The afternoon session focused on emerging technologies. Flow cytometry was the main topic of discussion. The talks included both designs and applications of this new technique to microbiological analyses. The last two papers of the session discussed techniques and methods on how to automate and send data electronically from field sensor suites.

Dr. Beatta Mizak of the National Veterinary Institute, Pulawy, Poland, expertly chaired Thursday morning's session on DNA Probe techniques. The presentations discussed current and emerging techniques in this area. Dr. Garrigue of France tied it all together with a paper on the problems associated with the use of these techniques to measure DNA in soil samples.

The last session of the workshop dealt with applications of these techniques to current problems. The paper by Dr. Vern Seligy of Health Canada discussed some potential dangers on the use of bio-pesticides. Most of the technologies described during the workshop were used in some fashion in this study and clearly showed the importance of monitoring the environment.

This workshop was an overall success. The organizers managed to bring together experts in the field of biological analysis and have a fruitful conference from scientific, cultural, and social aspects. The participants presented their perceptions on how biological materials impact the environment; their work on the state of the art of biological analysis; projected future improvements; and how this work impacts on the environmental analysis of biological materials.

In addition to the scientific content of the meeting, the hosts prepared a social and cultural program that was rich in content. Participants were treated to tours of some of the historic sites in and around Warsaw, and were also treated to the culinary aspects of Polish culture. They came away with a better appreciation and understanding of Poland. In short, the goal of the NATO science program was met – this conference helped to foster scientific and cultural exchange among the scientists of NATO and Cooperation Partner countries.

LIST OF AUTHORS

Dr. Michael Bartoszcze
Military Institute of Hygiene and Epidemiology
2 Lubelska Street
Pulawy 24-100
Poland

Miss Agata Bielawska
Military Institute of Hygiene and Epidemiology
2 Lubelska Street
Pulawy 24-100
Poland

Dr. Camille Boulet
Defense Research Establishment Suffield
P.O. Box 4000 Stn Main
Medicine Hat, Alberta T1A 8K6
Canada

Dr. Wayne Bryden
Johns Hopkins Applied Physics Laboratory
Laurel, MD 20723-6099
USA

Dr. Malcolm Dando
University of Bradford
Department of Peace Studies
West Yorkshire BD7 1DP
UK

Dr. H.M. Davey
Institute of Biological Sciences
Unversity of Wales
Aberystwyth SY23 4RB
Wales, U.K.

Dr. Vito G. Del Vecchio
Institute of Molecular Biology and Medicine
University of Scranton
Scranton PA 18510
USA

Dr. Henri Garrigue
Biology Department
Centres d"Etudes du Bouchet
91710 Vert-le Petit BP3
France

Dr. Royston Goodacre
Institute of Biological Sciences
University of Wales
Aberswtwyth, Ceredigion, S72 3DA
Wales, U.K.

Sergey Grinshpun
Aerosol Research and Exposure Assessment Laboratory
Department of Environmental Health
University of Cincinnati
P.O. Box 670056
Cincinnati, OH 45267-0056
USA

E.A. Henchal
USAMRIID
1425 Porter Street
Fort Detrick, MD 21702-5011
USA

Dr. James Ho
National Defence Canada
Defence Research Establishment Suffield
P.O. Box 4000 Stn Main
Medicine Hat, Alberta T1A 8K6
Canada

S.G. Ignatov
State Research Center for Applied Microbiology
Moscow Region
Obolensk 142279
Russia

Vladimir Ivanov
60 Vladimizskaya Str.
Biological Faculty, Shevchenko's Univ.
Kiev 252017
Ukraine

Dr. A.S. Kaprelyants
Bakh Institute of Biochemistry
Russian Academy of Sciences
Leninskii Prospekt 33
117071 Moscow
Russia

Zuzana Krocova
Institute for Immunology
Purkyne Military Medical Academy
Hradec Kralove
561 66 Techonin
Czech Republic

Anne W. Kusterbeck
Center for Bio/Molecular Science and Engineering
Biosensors and Biomaterials Laboratory
Code 6910, Naval Research Laboratory
4555 Overlook Avenue SW
Washington DC 20375-5348
USA

Dr. Andrzej Lidacki
Poland
Military Institute of Hygiene and Epidemiology
2 Lubelska ul.
Pulawy 24-100

Professor Jerzy Mierzejewski
Military Institute of Hygiene and Epidemiology
2 Lubelska ul.
Pulawy 24-100
Poland

Dr. Beata Mizak
Department of Carnivora and Fur Animal
National Veterinary Research Institute
Pulaway 24-100
Poland

M.T.C. Offermans
Division of Immunological and Infectious Diseases
TNO Prevention and Health
P.O. Box 2215
2301 CE Leiden
The Netherlands

Wieslawa Palec
Veterinary Research Center
Military Institute of Hygiene and Epidemiology
Pulawy
Poland

Dr. Guy Patras
France
Centres d"Etudes du Bouchet
Vert-le Petit B. P. 3 91710

Dr. V.L. Seligy
Department of Health
Environmental Health Centre, 0803A Tunney's Pasture
Ottawa K1A0L2
Canada

A. Peter Snyder
USASBCCOM, ECBC
5183 Blackhawk Road
APG, MD 21010-5424
USA

Elena Stabnikova
68 Vladimirskaya Str
Ukraine State University of Food Technologies
Kiev 252017
Ukraine

Dr. Peter J. Stopa
USASBCCOM, ECBC
5183 Blackhawk Road
AMSSB-RRT-DD E3160
APG, MD 21010-5424

Elena Vorobyova
Soil Biology Department
Moscow State University
Moscow
Russia

Alexander V. Zakharov
Space Research Institute
Profsoyuznaya 84/32
Moscow 117810
Russia

SOME PROBLEMS CONCERNING BIOLOGICAL THREATS

M. BARTOSZCZE AND J. MIERZEJEWSKI
The Military Institute of Hygiene and Epidemiology, 24-100 Pulawy, Lubelska 2, Poland

Biological threats can be divided into two groups:

Natural: Concerns those diseases or outbreaks of diseases which occur naturally in the territory of a given country as well as those diseases which can be imported and those newly emerging diseases which can occur as a result of mutations, natural selection and other mechanisms. In diseases of this type there is no clear evidence of human participation in the introduction or creation of these threats.

Man-made: Concerns those diseases, which are introduced to the environment for military reasons, through the actions of terrorist groups or industrial accidents. As in the case of military or terrorist acts, the targets are usually clearly indicated.

Until recently, common opinions held that infectious disease were now under control due to the effect of modern medical methods such as vaccines or antibiotics. It is hardly an exaggeration to say that opinion was almost canonized with successes such as the eradication of smallpox on a global scale.

Unfortunately, since this initial wave of euphoria there has been a period of relative laxness bordering on negligence. Our intoxication with our successes against some diseases, such as smallpox, has allowed previously known, and what we considered conventional diseases to appear with renewed strength (tuberculosis, malaria and new strains of cholera). In this same period, new threats such as HIV, Ebola, Hepatitis C, Legionella diseases, Lyme disease and others like the cases of visceral Leishmanasis which appeared in the wake of the Gulf War. It is extremely difficult to combat the effects of these new diseases, as there are no effective treatments available at this time. Prevention has also proved very difficult. Another problem has been the growing resistance shown by a number of pathogens to antibiotics which has and will continue to have profound effects on methods of treatment as well as their cost.

Factors which influence the spread of biological threats:

- The construction of huge cities and metropolitan areas which lack the proper sanitary infrastructures or the means to deal with their own sanitation problems.

P.J. Stopa and M.A. Bartoszcze (eds.),
Rapid Methods for Analysis of Biological Materials in the Environment, 1–5.

• Changes in production technology regarding animal feed, and food products for human consumption.

• Products for consumption and prepared foods.

• The increased exposure of people to threats from the natural environment.

• Global deforestation resulting in drastic changes to the environment, which affect wild animals as well as insects.

• The closer proximity of wild animals to civilized areas.

• The dramatic increase of tourist travel and international trade.

• The migration of people on a worldwide basis.

• Human activities which promote new routes of infection (i.e., Mad Cow disease).

• The inability of the medical profession to detect new or recurring threats or to adequately respond to these threats.

• Inadequate funding for medical purposes in both developed and undeveloped countries.

• Climate, catastrophes, floods, earthquakes and other such as tornadoes, typhoons, hurricanes, etc.

It is interesting to note that considering the overall biological threat facing the world today. WHO has designated its theme for 1997 as Emerging Infectious Diseases Global Alert, Global Response.

When looking at the present situation concerning natural diseases, it is not difficult to imagine the scenario after the intentional introduction of man-made biological threats to the environment. This threat comes primarily from militarily conceived biological weapons and their possible use by terrorist groups.

In 1972, we erected a legal barrier to prevent the proliferation of these weapons in the Biological Weapons Convention. Although more than 100 nations signed this treaty, we cannot be sure that all signed participants will adhere to this agreement. Additionally, since many of the world's countries have not agreed to the convention, it offers very little consolation indeed.

As of 1995, seventeen (17) countries were suspected of conducting research, which could lead to the development of biological weapons. A completely different question arises when we consider the possibility that biological weapons will probably fall into the hands of terrorist organizations some time in the near future. The bombing attacks in the USA, in the World Trade Center and in Oklahoma City, and the Sarin attack in the Tokyo underground can only underscore the seriousness of this threat. Ominously, some terrorist organizations have apparently already taken steps in this direction. The Japanese sect had reportedly been planning to use the Ebola virus in the same fashion.

Why are biological weapons a focal point for military and terrorist organizations?

Biologicals are a focal point for military and terrorist organizations for the following reasons:

- Small amounts can effectively produce high casualties.

- Production of these types of threats compared to nuclear weapons is relatively inexpensive. Therefore, common opinion holds that biological weapons are suited for poorer nations.

- The mass production of these weapons can be achieved easily and very quickly.

- Facilities necessary for the production of these weapons are commonly found in companies, which produce vaccines as well as pharmaceutical products.

- Delivery to a designated target can be easily accomplished through a variety of methods, including missiles, airplanes, ammunition and even in uncomplicated devices such as atomizers, etc.

- Immediate detection and identification is extremely difficult due to the delayed appearance of symptoms in the victims of an unexpected attack.

- Individuals and/or terrorist organizations have easy access to dangerous pathogens. A case in point is the recent incident in the United States in which a technician attempted to obtain *Yersinia pestis*.

- Now through genetic engineering it is possible for militaries to create or obtain ideal biological weapons from the standpoint of lethal effectiveness as military weapons.

Regarding several of the points mentioned above, it has been estimated that the cost of operations for inflicting high casualties per square kilometer is $5,600 for nerve gas, but only about one dollar for biological agents.

Biological threats may be easily and rapidly spread in a variety of ways. During tests conducted in the New York City underground, biological material was detected at different points through the underground within minutes of its release into the subway system. Based on these observations it is easy to imagine the seriousness of an attack on a subway system used by approximately one million people every day. Not only aerosols proved to be dangerously effective, but also equally effective have been pathogens, which settle onto different surfaces into the environment. Gruinard Island on the coast of Great Britain is a case in point. Anthrax spores were able to survive for up to forty years after the British government undertook experiments there.

Biological can also be very effective from a psychological point of view. Evidence from the Gulf War suggests that the sight of dead animals in the region created a panic among soldiers there. A similar reaction could be observed in the civilian population. Still, this point cannot be overemphasized, as this panic would have long-lasting consequences. An infection to livestock with, for example, Foot and Mouth Disease would in a short time destroy the export industry of any given country. We have already seen the repercussions suffered by Great Britain in the wake of the Mad Cow scare. The following clean up would probably only compound the situation. As reported by the journal Prehospital and Disaster Medicine, an encephalitis scare in a flooded region of California made necessary a pesticide program, which in itself contaminated beef and poultry in 12 counties. In this case, the remedy forced the government to destroy approximately four million kilograms of contaminated food products. In this respect, the introduction of a biological weapon which could very easily contaminate air, water and soil would have disastrous effects for not only the short term, but for the long term future of a nation as well.

The most probable targets for terrorist attacks include water-supply systems in large cities, food production facilities and metros, as well as facilities for hosting gatherings, meetings, conventions, etc. These factors must be given serious consideration when one speaks of the complications we face in deterring these types of attacks.

Weapons for terrorist would not necessarily be limited to those falling into the category of low-technology/conventional weapons, but they could also feasibly produce specially-designed pathogens using techniques such as genetic engineering. Here we would note that well-financed terrorist groups may not only gain access to pathogens, but they can also launch their own production.

Biological weapons may become the weapon choice for right-wing political groups, nationalistic groups, religious sects or any extreme groups hoping to influence the politics of a given country either with strong-arm or scare tactics.

In order to effectively protect civilian populations and military personnel, we must establish new systems for early detection, identification and the destruction of such threats. However, these concepts will not be able to address this problem by themselves. One of the greatest roadblocks is the lack of understanding or the apathy of those in authority, both in the military and in the civilian governments. Adequate funding is of the utmost importance here.

In an article on this subject the American author Zilinskas presented a theoretical global warning detection program base don field teams, laboratory networks and a headquarters center for data analysis. The primary goal of this system would be the early detection of new biological threats as well as the creation of methods to deal with these threats. Another goal would be to conduct research on the subject. This type of system would be beneficial not only to the military, but also to the civilian population. Of course a very important link in the system would be its cooperation with the World Health Organization.

What can we do to protect ourselves from biological threats?

- A very important factor here is the work of our intelligence services.

- It is very important to enlarge the scope of international cooperation.

- We must intensify international pressure on countries suspected of presently conducting research on biological weapons.

- We must develop an effective method to control proliferation. Access to collections of biologically dangerous material must be restricted. It is interesting to note that countries have thus far been unable to begin biological weapons programs without outside help. By limiting access to these materials we can help to ensure that they types of programs will not be initiated.

- We must improve individual and collective methods of protection against biological threats and seek new methods of destroying biological threats.

- We must conduct exercises and educate military personnel and civilian populations in order to protect them from biological threats.

- A tight lid must be kept on information regarding biological defense programs and information, which could be dangerous if in the wrong hands. This type of information must be kept out of the public domain and off of the Internet. Attempting to control the Internet would be at the least, controversial, and would probably prove fruitless. Our job and the job of our government are to keep this type of information from every becoming available in the first place.

- There is a need for improvement of methods of monitoring detection and identification of different kinds of pathogens.

The basis of any warning system, and the basis of any preventive measures, lies in the field of detection and diagnostics. Therefore, the NATO Advanced Research Workshop on Biological Detection in the Environment is a very important event because it is a new approach in the improvement of current diagnostic techniques.

TECHNOLOGIES FOR MONITORING THE BIOLOGICAL AND TOXIN WEAPONS CONVENTION: AN EMERGING CONSENSUS?

MALCOLM R. DANDO
University of Bradford, Dept. of Peace Studies, West Yorkshire BD7 1DP, UK

1. Introduction

As official concerns over the possible proliferation of biological weapons have increased, they have been paralleled by efforts to strengthen the Biological and Toxin Weapons Convention (BTWC). These efforts began with the introduction of a system of Confidence Building Measures (CBMs) at the second five-yearly Review Conference in 1986 and the subsequent elaboration of these measures at the Third Review Conference in 1991. At the Third Review, in addition, an Ad Hoc Group of government experts (VEREX) was mandated to identify and examine, from a scientific and technical standpoint, potential measures that could determine:

- Whether a State Party is developing, producing, stockpiling, acquiring or retaining microbial or other biological agents or toxins, of types and in quantities that have no justification for prophylactic, protective or other peaceful purposes; and

- Whether a State Party is developing, producing, stockpiling, acquiring or retaining weapons, equipment or means of delivery designed to use such agents or toxins for hostile purposes or in armed conflict.

The final VEREX report was considered at a Special Conference of States Parties to the BTWC in 1994 and a further Ad Hoc Group (AHG) was then established to consider appropriate measures (including possible verification measures) and draft proposals to strengthen the Convention, to be included, as appropriate, in a legally-binding instrument. This AHG held a procedural meeting and two substantive meetings in 1995 and two substantive meetings in 1996. Its work was strongly endorsed at the Fourth Review Conference in late 1996 and it was encouraged to move to a negotiating format. This is due to happen at the second substantive meeting of 1997 in July (1).

Within these developments during the 1990s there has been consideration by officials and professional organizations/non-governmental groups of the place of sampling and analysis in strengthening the BTWC. There have also been strong expressions of concern from important involved parties about any such role. The aim of this paper is to make an assessment, through a consideration of some of these debates, of the role

P.J. Stopa and M.A. Bartoszcze (eds.),
Rapid Methods for Analysis of Biological Materials in the Environment, 7–41.

sampling and analysis might have in a strengthened BTWC - should that come about in the next few years. This assessment might then form a useful part of the context for our discussions of technologies in the rest of the ARW.

2. Compliance Assurance Measures/Future Verification Protocol and Proposals On Sampling and Analysis

The Confidence-Building Measures agreed at the 1991 Third Review Conference are set out in Table 1. As of mid-1996 eleven States Parties had made the required CBM declarations every year, and 75 had made at least one annual declaration. This disappointing level of response was compounded by the variable quality of the individual returns.

It is hardly surprising, therefore, that the examination of an alternative approach in the VEREX process was extremely thorough. For example, in an early working paper entitled The Biological Weapons Convention: A possible verification regime', Australia suggested that consideration of how such a regime might be structured would require study of how biological weapons (BW) would be produced (2). The paper stated:

> '...Such a study would embrace factors such as: the agents that would be the most likely candidates for BW; the stages required for BW production: what is involved in each stage; and the means by which work at each stage might be determined to be either legitimate or prohibited.

It continued:

> BW production can be considered as occurring in four stages: research, development, production and storage, and weaponization. Some of these stages obviously pose a greater risk to the Convention than others: research into suitable BW agents, for example, is a lower risk operation than weaponization. Consequently any credible verification regime should differentiate between high and low risk operations and to apply more stringent measures to higher risk operations.'

The paper went on to consider what it was necessary to verify in low-risk and high-risk operations, and to make an overall assessment of the requirements.

Another early paper, from Sweden, was entitled 'Outline of a systematic approach on technical verification measures and their applications for the BTWC' (3, 4). This contribution began by stressing the difficulty of the task confronting VEREX:

> ...Many of the activities needed to develop and produce biological agents for warfare use can be carried out covertly and are of a dual use character. This means that no single technical method will be enough to distinguish between permitted and prohibited activities to the BTWC...'

The paper suggested a three-stage approach: identification of targeted activities; identification of the geographical locations of the targeted activities; and identification of the technical verification measures which could be used to monitor the targeted activities. Significantly for our purposes, sampling was prominent amongst the list of technical measures noted in this contribution (Table 2).

Neither was the potential impact of such measures on the industrial world neglected at this early stage. A paper produced for the second VEREX session in late 1992, from the Netherlands, was entitled 'Elements for "brainstorming" discussion with companies: Informal translation' (5). This set out a range of questions including some related to sampling (Table 3).

It will be recalled that the final VEREX report judged the twenty-one potential verification measures against six different criteria, the first three mainly representing the *effectiveness* of the individual measures and the second three mainly their *requirements and costs* (6). The VEREX summary conclusions on the principal measures involving sampling are set out in Tables 4 and 5. The potential utility, particularly of on-site sampling and identification, is clear from the conclusion that this:

> ...could provide key information to resolve certain ambiguities about compliance because of the possibility of identifying the nature of the agent...'

However, the potential difficulties over the protection of Commercial Proprietary Information (CPI) with such a level of intrusiveness are also obvious from the report. Before discussing such concerns, however, we need to outline briefly how discussion of the overall package of measures has developed in order to note another element of direct concern to us here.

Whilst there is clearly not yet any consensus on the overall shape of a set of Compliance Assurance Measures/Verification Protocol for the BTWC, it is possible to discern a commonality in the proposals being put forward by strong supporters of the objective of strengthening the Convention. The suggestions, for example, made by Brazil in its working paper for the 1994 Special Conference, Strengthening the BWC:

Elements for a possible verification system' (7), clearly have many of the same elements as the proposals made by South Africa at the 19% Review Conference (8). So although the South African proposals did not achieve consensus, and inclusion in the Final Document (1), we can use them to illustrate a widespread current view of what will be required. The key sections of this viewpoint are set out in Table 6. At the first meeting of the AHG following the Fourth Review Conference, in March of this year, the South Africans developed and expanded these ideas and presented the full 'table of contents' of a protocol (9). This South African proposal is reproduced in Table 7.

Briefly, what seems to be envisaged is a system of declarations of high-risk facilities backed up by relatively unintrusive visits and by challenge inspections of facilities and investigations of alleged use. This additional element is quite clear from the mandate given to the present AHG by the 1994 Special Conference. This stated that the regime (10):

> ..would include, *interalia,* potential verification measures, as well as agreed procedures and mechanisms for their efficient implementation and measures for the investigation of alleged use...' [my emphasis]

The technical questions raised by such investigations have been extensively discussed, for example in a Canadian document in the mid-1980s (11), but have most recently been reviewed in detail by the Federation of American Scientists (FAS) Working Group on Biological Weapons Verification.

The FAS Working Group's subgroup concerned with 'Investigation of Alleged Use or Release of Biological or Toxin Weapons Agents' issued a report with a number of recommendations and proposed wording for an Annex to a Legally Binding Protocol to cover this issue (12). The sections of this proposed annex concerned with sampling are set out in Table 8. As the report points out, there is already a mechanism in place within the United Nations, under General Assembly Resolutions 42/37C of 1987 and 43/74A of 1988, for carrying out such investigations. But this mechanism is *ad hoc* and unfounded and the Special Conference's mandate to the AHG is an implicit recognition of the need for a more formal arrangement.

3. Sampling and Analysis Techniques

As has been well recognized at successive Review Conferences of the BTWC, the current rapid developments in science and technology do pose potential threats to the future maintenance of the Convention (1). Yet as David Franz has argued, these developments also present potential benefits for strengthening of the Convention and reinforcement of the norm of non-use of biological weapons (13):

> 'Genetic engineering and other new technologies, feared by some as the catalyst for the next generation of biological warfare agents, support nonproliferation by facilitating the development of better and faster means of diagnosis and identification, vastly improved, cheaper and less reactogenic vaccines, and allow all of these passive countermeasures to be developed and produced with minimal containment space and bioreactorlfermenter capacity...'

We are concerned here therefore with only part of a range of potentially significant positive developments.

In 1993 Richard Titball and Graham Pearson of the UK Chemical and Biological Defence Establishment reviewed the technologies for the identification of biological warfare agents and the application of such technologies in BTWC verification measures (14). Not surprisingly, given the ongoing VEREX discussions at that time, these authors argued first that:

> '...it is becoming clear that key elements in an effective verification regime would comprise focused declarations and selected on-site inspections...'

and then that:

> '...The effectiveness of on-site inspections will be dependent on the availability of techniques for the identification of BW agents...'

They also noted that since commercial sites were likely to be inspected, there were particular attractions in specific identification techniques for BW agents as this would not comprise CPI.

As in other careful discussions (12), these authors noted the wide range of sampling situations that might be encountered, and the necessity of systematic procedures for dealing with either on-site or off-site subsequent analysis. They then reviewed four identification systems as set out in Table 9. Conventional methods involving the culture of microorganisms and their later identification were, of course, well established and, although slow, were highly accurate. On the other hand, newer techniques based on antibodies, gene probes and receptors were seen to:

> '...have some advantages over traditional methods, particularly if political and safety requirements necessitate the inactivation of samples for analysis before the samples are transported from the site of collection...'

Moreover:

> '...These techniques are also well-suited to discriminating between illegitimate and legitimate uses (for example, the production of vaccines), as well as for identifying genetically modified agents.'

It was argued that antibodies reacting to the surface features of agents were available against many of the potential BW agents, but their suitability for inactivated agents had to be determined on a case-by-case basis. However, antibodies could be configured in sensitive but simple ELISA test formats:

> '...A simple ELISA can detect a substance when iO6~lO7 entities are present...whilst more complex 'amplified' ELISA systems can detect a substance when only 1O~ entities are present....botulinum toxins at low levels (20-3Opg) have been identified in foodstuffs.'

Gene probes, rather than reacting to surface features, react directly with the nucleic acid of the infectious or toxin-producing agent.

Given this fact, gene probe analyses could be simplified and greatly amplified through the use of the polymerase chain reaction (PCR) to increase the amount of DNA available in the sample. The authors noted that:

> '...Such PCR tests have already been developed for the identification of *Bacillus anthracis*. . .*Brucella* species...the Staphylococcal enterotoxin B gene...and many other microorganisms...'

The technique was clearly seen as being particularly useful in a verification regime for dealing with, for example, soil samples and genetically modified organisms. Extensions of the technique in sequence matching of genes were seen as a powerful approach. Use of receptors involved in the initial binding of classes of pathogens or toxins to the host tissues was, however, thought to be in too early a stage of development for consideration at that time.

In short, this review of available techniques concluded that:

> 'The use of gene probes in the way described above should not compromise commercial confidentiality. Thus, samples for testing could be rendered nonviable before transport from the site of inspection, and the gene probes would be targeted towards DNA sequences of BW agent significance or sequences of non-commercial significance...

In the future, it was suggested that:

> 'It is in the area of 'on-site' tests that the greatest developments are likely to take place over the next decade. Automated biosensors will be developed that will incorporate gene probes or receptors, and the reliability and fidelity of these devices will be improved...'

Thus the technologies required to underpin the verification regime were thought to be already available, and likely to improve rapidly.

Two years later, in 1995, another group from this UK facility made a specific assessment of the utility of gene probes in verification of the BTWC (15). This study concluded:

> ..Gene probes to speciate and assess the potential virulence of the principal pathogens of relevance to the BTWC in clean samples with a high degree of sensitivity and specificity are available now in specialized laboratories...'

and:

> ...assuming that the necessary research is carried out, these should be available in portable, automated assays within five years. Environmental samples and samples requiring strain differentiation tests will probably have to be performed off-site at accredited reference laboratories until further research is undertaken to develop on-site formats.'

Clearly, therefore, the appropriate technologies were seen to be developing rapidly by these officials.

In the UK in 1994, the Royal Society issued a major report on the problem of biological warfare and strengthening the Convention. This report contained a thorough review of 'Scientific aspects of verification' and paid particular attention to methods for on- and near-site identification of specific SW agents. The Royal Society authors considered that:

> '...inspections inside high-risk areas are the kernel of verification. Not only can they identify specific biological agents...but also provide indications of whether they are being used in circumstances that cannot be justified for legitimate purposes...

In order to simplify the problem the report suggested concentrating attention on the 'classical' agents which had been weaponized in the past (Table 10), as these would almost certainly be included in any proliferant's program.

A wide range of physico-chemical and biological methods of identification were examined and it was concluded that:

> .the simplest, cheapest and most effective method of providing unequivocal identification of all 'classical' agents at high levels of sensitivity in the immediate future (5-10 years) is *laboratory use of ELJSA and PCR.* The laboratory requirements are minimal and all are portable....Inspectors could be trained to use ELISA and PCR with portable equipment.'

These conclusions were therefore in line with, if even a little more optimistic than, those of the official UK studies cited here.

The Federation of American Scientists' 1995 report on 'Potential for new approaches to microorganism identification' (17) accepted that the scenario just discussed - 'verification of the presence or absence of a particular microorganism or one of a small group of microorganisms' - was likely to be a requirement of the BTWC verification regime. However, this report suggested that a much more difficult scenario, identification of an unknown microorganism from 'a "universe" of microorganisms', was also likely to be encountered. The report therefore looked forward to new approaches that might be developed. For example:

> '...Procedures for utilizing fluorescent-labeled DNA probes on microscope slides are in routine use. One interesting new approach utilizes a postage-stamp-sized, two-dimensional array (a 'chip') of DNA probes specific for different microorganisms....At present, 15,000 DNA probes can be affixed to a single chip, and Affymetrix expects to have production capabilities for 150,000 on a chip in the near future...

Clearly, such developments could have major implications in the future.

4. Industrial Reactions

Given current estimates of the extent of proliferation of offensive biological weapons programs and the clear Iraqi and Soviet/Russian examples of recent non-compliance (18), it has to be assumed that there may be States Parties which have no interest in adding any form of effective verification measures to the BTWC. Setting such potential opposition aside, the very difficult process of ratifying the Chemical Weapons Convention (CWC) in the US Senate suggests that the technically more complex issue of the strengthening of the BTWC will need considerable support within the United States. Yet whilst the chemical industry has strongly supported the CWC, the biotechnology industry has raised considerable concerns about the strengthening of the BTWC, and has expressed particular antagonism to sampling and analysis.

This is, in part, probably related to the previously long-held official US view that the BTWC could not be verified. As Dr. Edward Lacy, then Acting Assistant Director, Bureau of Verification and Implementation, US Arms Control and Disarmament Agency, told the Fall meeting of the Biological and Biotechnology Section of the Pharmaceutical Manufacturers Association in 1992(19):

> 'Effective verification measures, singly or in combination, should:
>
> • Provide confidence that the States Parties are in compliance with treaty provisions:
>
> • Deter violations of treaty provisions by significantly increasing the risk of detection and thereby raising the cost of cheating;
>
> • Enable the States Parties...to detect a violation in a timely fashion before it poses a significant military risk...
>
> Given this understanding of verification, our own analyses indicate that the BWC cannot be made more effective by adding verification measures known to us...

Dr. Lacy explained that this was the reason the US opposed the proposals for specific verification regimes at the 1991 Review Conference, but he noted that the US also opposed measures that might adversely affect US industry. In this regard he specifically discussed sampling.

Overall, Dr. Lacy's view of the VEREX process was that:

> '...the best way to prepare for any future decisions is to provide an experts report that reflects the technical and scientific realities that challenge any efforts to elaborate a viable and effective verification regime for the BWC...'

and he hoped that the US industry would help in this task.

Whilst the official US view has moderated (1), skepticism about the possibility of verifying the BTWC remains strong in influential US circles (20). Specifically in regard to sampling, Alan Zelicoff, a member of the US delegation at the AHG meetings on strengthening the Convention, has argued that a standard Bayesian analysis shows certain criteria (Table 11) that have to be met if sampling is to be of use (21):

> 'Among potential scenarios of concern to the BWC, at least two fulfill these criteria: investigation of alleged use and investigation of outbreaks of unusual disease. Other scenarios undoubtedly exist, such as the evaluation of suspicious weapons fill...'

But, in his opinion:

> 'Sampling is likely to be of much less use for routine visits or inspections unless some other clear suspicions of violations exist. Sampling at...pharmaceutical facilities...is likely to be confounded by the absence of specificity in tests applied to samples, as well as by the very low pre-test probability of violation at these sites.'

He concluded therefore that the high false positive rates, which would occur if sampling were used at pharmaceutical sites, would render the tests inconclusive as well as causing unnecessary public relations problems for industry. Further causes of concern in the US resulted from the perceived lack of success of the UN biological weapons inspections in Iraq (22) and of sampling and analysis in those inspections (23) and particularly from worries about protection of Commercial Proprietary Information during inspections of non-military facilities in the trilateral process (24,25).

Against that background, it is hardly surprising that representatives of the US biotechnology industry have stressed their concerns over the potential impact of the process of strengthening the BTWC on their dominant sector of the growing worldwide industry. As Alan Goldhammer noted in an early contribution *(25)*:

> 'The principal concern from industry's perspective is with potential loss of proprietary information....Inspectors could come on site and take a proprietary organism thus compromising a company's trade secret...'

A detailed briefing concentrating on the pharmaceutical part of the industry (26) explained the time and expense involved in producing a new drug:

> '...It takes an average of 10-12 years from the time the first laboratory studies are conducted until a drug is successfully marketed. It is estimated that the research and development cost to develop a single new drug (which includes the cost of 100,000 failures) is $350 million alone. This figure does not include the cost of the facilities and equipment to manufacture it.'

The value of information obviously increases as research and development proceed:

> '...Out of 100,000 compounds that are discovered/created, only about 10% or 10,000 show any promise. Of these, only about 10% or 1000 will actually go on to clinical testing to determine if they are safe, effective, and worthy of being developed further. Of these, only 1% or 10 will actively become marketed medicines. Of these 10 drugs that do make it to the market, only 3 recover their research and development costs and only two make a profit...'

Moreover, as the document points out, as research and development take some 10-12 years, and patents run out in about 17 years. there are only 5-7 years in which profits can be made before 'copies' are legally available. Thus loss of proprietary information could be extremely damaging to a company. However, whilst manufacturing plant, in particular, is subject to government inspection, the document argues that the agencies involved have a proven record of being interested narrowly in their legal concerns and of 'not sharing proprietary information with other competitors.

A senior UK industry executive summarized concerns similar to those expressed in the US over the use of information gained in inspections (Table 12), and over sampling in particular (Table 13), in a seminar presentation in 1996 (2'7). This indicates that industrial worries are not confined to the United States.

The end result of the US industry's consideration of these issues was that whilst the paper it produced for the Fourth Review Conference of the BTWC (28) stressed that the Pharmaceutical Research and Manufacturers of America (PhRMA) were opposed to biological weapons, and wished to play a key role in reducing the threat, they had strong views on what characteristics a protocol to the BTWC needed to have in order to be acceptable to them (Table 14). Clearly, they oppose any form of non-challenge inspections and, in regard to sampling, the paper stated that:

> 'Samples of production organisms are of special interest to PhRMA. The production organism at a commercial facility is usually the single most valuable piece of proprietary information on site...'

Therefore:

> ...PhRMA members will not permit the transport of production microorganisms off the inspected site but will endeavor to address any questions such a sample would have addressed.'

In short, whilst not being totally negative, PhR.MA saw BTWC verification inspections as being fundamentally different from CWC inspections and, indeed, appeared to be putting barriers in the way of an effective system (Table 14) possibly because it

> ...Does not believe that there is any characteristic or combination of characteristics which discriminate a legitimate industrial facility from a potential violator of the 1972 BWC.'

In addition to these arguments against sampling, the paper also pointed out that modern clean-in-place systems could remove 'virtually all traces of any development, manufacture or storage of these [biological] warfare agents. ..from a facility within a few hours.

Naturally, in view of such industry concerns, both official (29) and non-governmental organizations (30) have, during the 1990s, investigated means by which they might be accommodated. The question would appear to be one of whether any accommodation has been reached or is likely to be reached on this crucial issue in order that industrial support for strengthening the Convention is more forthcoming.

5. Comprise?

The whole issue of the utility of sampling and analysis for monitoring compliance with the BTWC was the subject of a wide-ranging workshop late in 1996 (31). The topics discussed at the workshop are set out in Table 15. Whilst noting the future promise of new approaches, the review of technologies for sampling and analysis came to broadly the same conclusions as previous analyses quoted here. It was argued that the three potential technologies available for compliance monitoring of the BTWC are classical bioassay techniques, immunological assay techniques and genetic analysis. Moreover, the risks of false positives and false negatives could be reduced by the use of two independent analytical techniques. The review of the UNSCOM experience with sampling and analysis was generally positive with examples presented of where it had been useful, but the need for careful political awareness of the implications of any sampling exercise was strongly stressed.

The crucial paper, however, was clearly the one presenting an industry viewpoint. This paper stressed that although it gave a personal assessment, this reflected widespread opinion within the US pharmaceutical industry. The kinds of information that might be lost were enumerated (Table 16) and the economic damage that could result from such loss was strongly emphasized. Indeed, according to the author

> 'Because of these compelling economic interests, sampling in any form is anathema to the pharmaceutical industry. I think most of us feel that no matter how innocuous an on-site inspection and sampling regime may appear, we are sure to lose significant CPI in the course of a challenge inspection...

Nevertheless, the author then added:

> '...Although US companies would prefer by far that there be no sampling whatsoever, we recognize this stance is unrealistic in a world in which some countries do develop and produce biological weapons in violation of international norms.

The question, therefore, is not whether there is to be a compromise, but where that compromise will be found.

The managed access procedures, developed for the Chemical Weapons Convention, the paper argued, could allow the inspected party to exercise the necessary control to protect its CPI. A range of examples was given of different levels of sensitivity and even in the most difficult example, testing by a company employee in the presence of inspectors to show that the sample was not of a proscribed nature was seen to be a possible solution. In general, however, the paper argued for the kind of safeguards set out by PIIRMA (Table 14) such as having only challenge inspections and a green light' procedure to initiate such inspections in the first place.

A paper from the Chemical and Biological Arms Control Institute discussed the various aspects of the 'Parameters and Procedures for Sampling and Analysis' which will have to be taken into account as the role of sampling and analysis is decided. The paper-suggested that the role will eventually not be seen as central:

> '...While a number of procedures can be developed to facilitate the conduct of sampling and analysis, their contribution will be circumscribed by technical constraints and political realities, including procedures to protect proprietary information...'

If non-challenge visits are eventually agreed as part of the new compliance measures it was argued that sampling and analysis would provide such small returns in relation to the potential loss of CPI that such procedures should be minimized. Even in regard to challenge inspections, given the climate of discussion in the AHG, it was judged that sampling and analysis will only be an exceptional measure and not something to be routinely and automatically employed.

On the other hand, it appears that evasion strategies based on modern microbiology would not be very successful at the present stage of development and that whilst cleaning up a facility could be done very quickly with modem methods the possibility of sampling and analysis would be a potential deterrent to a violator. An overall cost benefit analysis suggested that bioassay techniques would remain the mainstay for field investigations of outbreaks of unusual diseases but that:

> ...sampling and analysis during on-site inspections of biological facilities should rely on immunological and genetic techniques, which can reduce the risk of compromising CR by identifying dead microorganisms and denatured proteins...'

This final paper of the wide-ranging workshop concluded that restricting sampling and analysis to field investigations and challenge inspections would also limit the costs and focus attention where it was most required.

Similar thinking was also evident in official reports. A discussion paper produced by the UK for the AHG meeting in September 1996 began by arguing that (32):

> ...any system for investigating concerns about non-compliance with Article 1, or of BW use would in principle need some means of identifying the presence or absence of biological agents or toxins...'

and clearly, as sampling and analysis is the only measure that can meet this requirement:

> '...The UK therefore believes strongly that sampling and analysis needs to be available, both in investigations of alleged use, and in on-site inspections during the investigation of a concern about non-compliance with the Convention...'

However, the contribution also recognized the difficulties that could be posed, both in risks to legitimate national security or commercial proprietary information and from misleading results.

The UK's paper suggested that these problems could be addressed without destroying the usefulness of sampling and analysis. Three categories of action were outlined: guidance for inspectors on the use of sampling; technical measures; and managed access. It was argued first, in regard to guidance for inspectors that:

> '...The UK's conclusion from the experience of its practice challenge inspections was that, whilst sampling and analysis may be a vital tool in particular circumstances, these instances are likely to be comparatively rare, since most questions can be satisfactorily resolved by other means...

Thus the paper suggested that the principle should be established that samples were only to be taken where this was essential or of particular value (Table 17).

Technical means of addressing potential problems were seen to be of two types: methodological and scientific. Amongst the methodological suggestions designed to avoid misleading results and loss of sensitive information were:

> '...the use of standard sampling equipment and agreed and validated analysis techniques; use of duplicate samples and where possible analysis using more than one technique; the use of control samples; supervision of sample collection by the host side - or collection by the host side under the supervision of inspectors. Analysis laboratories could be designated by. a future BWC organization for any circumstances where off-site analysis was necessary.

Scientific means included use of:

> '...technologies, such as dip-sticks, allowing on-site analyses and avoiding the need to remove samples from the site; specific tests focusing, for example, on a specific agent, or even a specific strain or a particular gene of an organism, and thus not compromising other information.. .and analysis methods designed to avoid false positives...'

The UK stated that it was working on 'all of these technologies in connection with their possible use in the context of a future BWC compliance regime' and that the results of this research were very encouraging.

Finally, the paper argued that managed access, including the right to refuse a sample in any individual case, was needed to protect sensitive information. Moreover, there were many ways in which the host site could address the inspectors' concerns short of outright refusal of sampling:

> 'For example, the host could allow samples to be taken but not removed from the site...it could also insist on the use of specific tests which would not endanger legitimate...information. Other measures could include the host site itself taking and/or analyzing samples under the supervision of the inspectors; or offering the use of analysis facilities in the host state.'

In short, the crucial requirements of minimal sampling could be met with flexibility and imagination. Certainly also, advances in, for example, mass spectrometry, biosensors and dipstick technologies seem likely to ease problems in the future (16, 33).

6. Conclusion

On this evidence there does seem to be the possibility of an emerging corn promise position on sampling and analysis in monitoring the BTWC: whilst not central to any future verification system, it will be important in certain specific situations and under carefully controlled conditions. Moreover, developments in sampling technologies are likely to make sampling and analysis easier and more acceptable in the verification process.

That, of course, does not mean that a verification protocol will come about, or that differences between PhRMA and strong enthusiasts for the development of such a protocol will end. For example, making the present system of CBMs mandatory (Table 14) does not seem to go in any way far enough to produce the level of transparency required. Whilst declarations will have to be carefully crafted to avoid catching irrelevant information, it will surely be necessary to include production facilities which could be used as a cover for BW operations. Moreover, whilst a large-scale system of 'routine' non-challenge inspections appears unlikely, limited non-challenge information visits do seem a better alternative to having to rely on challenge inspections alone to clarify declarations. Crucially, the possibility of challenge inspection *at short notice* will have to be preserved if the verification system is to be of any real value and therefore, the 'red light' system used in the Chemical Weapons Convention (where a three-quarters vote is required to block a challenge) will surely have to be used rather than the 'green light' process proposed by PhRMA.

It has also to be remembered that in the background to all such discussions by those who favor strengthening the BTWC may be others who have no such objective in mind. Resolution of these current debates is thus only a stage on the way to a larger task. Meanwhile technology, not all of it benign (34), marches on.

Table 1: BTWC Confidence-Buildin2 Measures (CBMs) of 1991

1. Declaration form on 'Nothing to declare' or 'Nothing new to declare'.

2. Confidence-building measure 'A':

   - Part 1: Exchange of data on research centers and laboratories;
   - Part 2: Exchange of information on national biological defence research and development programs.

3. Confidence-building measure 'B':

   - Exchange of information on outbreaks of infectious diseases and similar occurrences caused by toxins.

4. Confidence-building measure 'C':

   - Encouragement of publication of results and promotion of use of knowledge.

5. Confidence-building measure 'D':

   - Active promotion of contacts.

6. Confidence-building measure 'F':

   - Declaration of legislation, regulations and other measures.

7. Confidence-building measure 'F:

   - Declaration of past activities in offensive and/or defensive biological research and development programs.

8. Confidence-building measure 'G':

   - Declaration of vaccine production facilities.

Table 2: Sampling suggestions in the list of technical verification measures in Sweden's paper of 1992 (4)

A. Off-site verification measures

- Sampling for agents, toxins or other compounds from, e.g.:
    - the air
    - sewage system
    - environment

B. On-site verification measures

- Sampling and analyses of, e.g.:
    - air, surfaces, sewage water, air filters and materials from different sites in a facility
    - culture collections
    - raw materials, products and effluents from a production line
- Automatic devices - that cannot be tampered with - for:
    - collection of samples on unregular intervals from, e.g., bioreactors

Table 3: Questions on samplin2 in The Netherlands paper of 1992 on discussions with companies (5)

Without sampling:

- What should be discussed: what should be looked at?
- What could be found out; what could be hidden in your company?

With sampling:

- Is it possible to take samples? From what ? (for instance, anywhere in freezers or air-filters)
- How to take samples?
- Is it possible to clean a plant in such a manner as to leave no traces? On what notice?

What would be your hardship?

In this context, what kind of government controls already exists in terms of declarations, inspections? To what extent do similar controls exist in other countries?

Table 4: Summary of VEREX conclusions for off-site samplin2 and identification (6)

Definition

To take samples of the area in the vicinity of a declared or undeclared facility without penetrating its boundary. (VEREX/9, Annex II, p. 83)

Criteria 1 - 3 ('Effectiveness')

The measure will usually provide information of rather poor quality, as the probability of obtaining a relevant sample is low. Using this measure alone can result in ambiguities, as e.g. the origin of any agent isolated may not be possible to clarify, and the risk of false positive as well as false negative tests may be very high. Different interpretations of the information are possible. Ability to differentiate between permitted and prohibited activities as well as resolving ambiguities is low. Could be of value in connection with open-air sites. (VEREX/9, Annex III, p.197 etc.)

Criteria 4 – 6 ('Requirements and Costs)

The costs will depend on the total number of inspections and subsequent number of samples. Small inspection teams will be required, but the chain of custody and laboratory analysis would be labor intensive. Safety problems for inspectors are generally low, except for open-air test sites. Assays for identification are not developed for some agents. Minimal impact on permitted activities and CPI.

Table 5: Summary of VEREX conclusions for on-site sampling and identification (6)

Definition

The act of taking samples on the inspected site, analyzing these samples either on the site using appropriate methods or to transfer these samples from the site for identification or further investigations in appropriate laboratories. (VEREX/9, Annex III, p.228)

Criteria 1 - 3 ('Effectiveness')

It could provide key information to resolve certain ambiguities about compliance because of the possibility of identifying the nature of an agent. Can provide information of significant quality and quantity, in particular because of the possibility of obtaining an independent confirmation of analytical results in the event that findings are disputed. A negative result does not necessarily rule out prohibited activities and may not resolve all cases of non-compliance ambiguities. The efficiency of this measure would be enhanced from a prior indication of the agents one is looking for. Ambiguous results would be reduced if more than one analytical technique and several samples from the same site were used. There is a need for an environmental profile of the site. Key issues are the chain of custody and the use of good sampling and identification practices (GSIP). (VEREX/9, Annex III, p.228 etc.)

Criteria 4 - 6 ('Requirements and Costs')

Currently available materials would allow many of the on-site presumptive tests to be performed. There is a need to establish infrastructure for training and deployment of inspectors. Creation and maintenance of a sophisticated field laboratory or an independent laboratory could be very costly. There is a risk of loss of CPI, but the use of equipment and methodology from the site could reduce the costs and protect confidentiality. The need to preserve intellectual, individual and commercial proprietary rights in the case of legitimate activities, means the obligation to use special technical and legal procedures for processing samples, particularly if there are grounds for removing samples from the site for subsequent analysis.

Table 6: The South African proposal for the basic elements of a legally binding verification protocol (8)

The Conference proposes that the following preliminary framework elaborating potential basic elements for a legally binding verification protocol be considered by the Ad Hoc Group:

- Compulsory annual Declarations preceded by a comprehensive initial Declaration;

- On-site measures including the investigations on non-compliance concerns;

- Voluntary confidence building measures;

- Measures to implement the provisions of Article X of the Convention;

- Definitions of objective terms and criteria where applicable.

Table 7: The South African proposal for the 'Table of Contents' of a verification protocol for the BTWC

Protocol on the Verification of the Convention on the Prohibition of the Development, Production and Stockpiling of Bacteriological (Biological) and Toxin Weapons and on their Destruction

Contents

Table 8: Sections on sampling for the proposed annex to a legally-binding BTWC protocol put forward by the FAS (12)

Sampling

25. The investigation team shall have the right to collect samples of types, and in quantities it considers necessary. If the investigation team deems it necessary, and if so requested by it, the host State Party shall assist in the collection of samples under the supervision of members of the investigation team. The host State Party shall also permit and cooperate in the collection of appropriate control samples from areas neighboring the site of the alleges use or release and from other areas as requested by the investigation team.

26. Samples of importance in the investigation of alleged use include biological toxins, infectious agents, delivery devices or remnants thereof, environmental samples (air, soil, vegetation, water, snow, arthropods, rodents and other wild animals as possible reservoirs and vectors, etc.), biomedical samples from human or animal sources (blood, urine, excreta, tissue, etc.), and tissue from affected and control plants. The collection of human samples will require informed consent, which the inspected State Party shall assist in obtaining.

Comment: the list of examples of sample types reflects the greater breadth of sampling likely to be required by a biological investigation. Note that human samples require informed consent; the inspected State Party should cooperate with the investigation team to obtain such consent, or to devise alternative means of obtaining information if cultural practices or religious beliefs prevent the team from obtaining samples.

27. The Director-General shall, upon nomination by States Parties, test and accredit reference laboratories to be certified to perform analysis of samples when called upon. A list of certified reference laboratories shall be communicated to States Parties, in writing, within ninety days after entry into force of the Protocol and after each change to the list.

Comment: this provision recognizes the need for a network of independent laboratories in different countries capable of performing the microbiological and toxicological analysis that will be required to confirm the field results of investigating teams, and to perform analytic tasks too sophisticated to be performed in the field.

28. If requested by the investigation team, the inspected State Party shall provide access to all biological samples, and to microbial cultures derived therefrom, taken at any time before the alleged use or release, to serve as reference or control samples.

Comment: this provision, not part of the CWC, is necessary to allow laboratory analysis to determine the particular strain of biological agent, or the toxic compound, has been naturally present in the area previously.

Table 9: Summary of the utility of identification systems in 1993

Identification System	Agent Type Application	Limitations
Antibodies	Classical infectious agents Toxin agents	May not work against inactivated agents
Gene probes	Classical infectious agents Genetically modified agents Unexpected BW agents	Can be used on inactivated materials
Receptors	Classical infectious agents Genetically modified agents Unexpected BW agents Toxin agents	May not work on inactivated materials
Sample Culture and	Classical infectious agents Unexpected agents	Not suitable for inactivated materials

Table 10: 'Classical'a2ents identified by the Royal Society as prime targets for identification technologies (16)

Bacteria

Bacillus anthracis
Yersinia (Pasteurella) pestis
Francisella tularensis
Brucella spp.
Vibrio cholerae

Rickettsia

Coxiella burnetti (Q fever)

Viruses

Venezuelan equine encephalitis virus **(VEE)**

Tick-borne (Russian spring/summer) encephalitis virus **(TBE)**

Toxins

Botulinum toxins (serogroups A to G)
Staphylococcus enterotoxin B
Ricin

Table 11: Where Bayesian analysis[1] suggests sampling is useful (21)

Scenarios in which:

- pre-test suspicions of violations approximate the magnitude of the false positive[2] rate of tests;

- tests affected by subtle protein or nucleotide substitutions are of marginal importance[3]; and

- spectrum bias[4] is not significant.

---

[1]A method of analysis in which the outcome of the test is assessed taking account of the pre-test probability of the outcome.
[2]An incorrect positive result.
[3]That is where minute variations in structure do not produce falsely negative results because of the very sensitivity of the test.
[4]Circumstantial variation in sensitivity and specificity: where specificity is the inverse of the false positive rate and sensitivity the inverse of the false negative rate.

Table 12: Industry concerns about abuse of information obtained during inspections (27)

Loss of intellectual property rights

e.g., patentable inventions

Company 'know-how'

e.g., production processes

Sensitive commercial information

e.g., scale of operations

Sampling

- on-site analysis
- off-site analysis

Disruption of research and/or production

Adverse PR triggered by an inspection

Table 13: Basis for concerns over sampling (27)

Microorganisms/cultured cells used in R & D or in commercial manufacture can be:

- valuable company assets
- results of many years R & D
- not necessarily protected by patents.

Cells can be replicated from small numbers.

Valuable information may be obtained from killed cells.

Table 14: Characteristics required by PhRMA in a protocol to the BTWC (28)

- Current confidence building measures may become mandatory but what is declared must not be expanded.

- On-site inspections must be limited to challenge inspections.

- Allegations leading to a challenge inspection must be reviewed using the 'green light' process before proceeding[1].

- Managed access must be employed during any on-site inspection.

- A non-governmental inspected facility has the final determination of what is proprietary.

---

[1]Three-quarters of members of the executive council of a future BTWC organization would have to vote in favor of the inspection to curtail frivolous claims/inspections.

Table 15: Topics covered in the 1996 Review of the utility of sampling and analysis (31)

- Technologies for biological sampling and analysis.

- Lessons from the UNSCOM experience with sampling and analysis.

- Industry views on sampling and analysis.

- Parameters and procedures for sampling and analysis.

- Evasion scenarios and countermeasures.

- Costs and benefits of sampling and analysis.

Table 16: CPI (Commercial Proprietary Information) that might be compromised by on-site sampling (31)

- The species and strain of the production microorganism

- The identity of the plasmid, or ring of DNA incorporating the recombinant gene for the product

- The relevant coding sequences in the plasmid DNA

- The precursor product (e.g., proinsulin)

- Specific steps in the manufacturing process

- Unannounced new products

Table 17: Circumstances where sampling is essential or of particular value (32)

- **Where** sampling is the only way to address a particular point of relevance to the non-compliance concern.

- **Where** there is other evidence acquired during the inspection, or otherwise available to the inspectors (e.g., from a declaration), which suggests that sampling could provide significant information which may help resolve a point of relevance to the non-compliance concern.

- Where sampling is suggested by the host side, e.g., as a quick way of confirming information provided.

7. References

1. Dando, M. R. and Pearson, G. S. (1997) The Fourth Review Conference of the Biological and Toxin Weapons Convention: Issues, outcomes and unfinished business. *Politics and the Life Sciences*, March, 1-22.

2. Australia (1992) *The Biological Weapons Convention: A Possible Verification Regime*. BWC/CONF.III/VEREX/WP.1O. United Nations, Geneva.

3. Sweden (1992) *Outline for a Systematic Approach on Technical Verification Measures and their Application for the BTWC*. BWC/CONF.III/VEREX/WP. 11. United Nations, Geneva.

4. Sweden (1992) *Revised Version of Appendices in Swedish Working Paper*. BWC/CONF.III/VEREX/WP.1 1/APPENDICES/Rev.1. United Nations, Geneva.

5. The Netherlands (1992) *Elements for 'Brainstorming' Discussion with Companies: Informal Translation*. BWC/CONF.III/VEREX/NONE.32. United Nations, Geneva.

6. Final Report (1993) *Ad Hoc Group of Governmental Experts to Identify and Examine Potential Verification Measures from a Scientific and Technical Standpoint*. (Attachment to Summary Report - Table). BWC/CONF.III/VEREX/8. United Nations, Geneva.

7. Brazil (1994) *Strengthening the BWC: Elements for a Possible Verification System*. BWC/SPCONF/WP.4. United Nations, Geneva.

8. South Africa (1996) *Consideration of the Work of the Ad Hoc Group: Proposed Wording for Inclusion in the Final Declaration*. BWC/CONF.IV/COW/WP.8. United Nations, Geneva.

9. South Africa (1997) *Protocol on the Verification of the Convention on the Prohibition of the Development. Production and Stockpiling of Bacteriological* 14 *(Biological) and Toxin Weapons and on their Destruction*. BWC/AD HOC *GROUP/WP*.133. United Nations, Geneva.

10. Final report (1994) *Special Conference of the States Parties to the Convention on the Prohibition of the Development. Production and Stockpiling of Bacteriological (Biological) and Toxin Weapons and on their Destruction*. BWC/SPCONF/1, September 19-30. United Nations. Geneva.

11. Canada (1985) *Handbook for the Investigation of the Use of Chemical or Biological Weapons*. November. Ottawa. Canada.

12. Working Group on Biological Weapons Verification (1996) *Report of the Subgroup on Investigation of Alleged Use or Release of Biological or Toxin Weapons Agents*. April. Federation of American Scientists. Washington.

13. Franz, D. R. and Smith. J. F. (1996) *Biotechnology: In support of the BWC*. Pugwash Study Group on the 'Implementation of the Chemical and Biological Weapons Conventions'. Pugwash Meeting No. 219, Geneva, Switzerland.

14. Titball, R. W. and Pearson. G. S. (1993) BWC verification measures: Technologies for the identification of biological warfare agents. *Politics and the Life Sciences*. August. 255-263.

15. Leslie, D., Phillips. T.. Miller. L. and Pearson, G. (1995) Gene probes for the verification of the Biological and Toxin Weapons Convention. Paper presented at a seminar on CBW verification in conjunction with the 5th International Symposium on Protection Against Chemical and Biological Warfare Agents. June. Stockholm.

16. Royal Society (1994) *Scientific Aspects of Control of Biological Weapons: Report of a Royal Society Study Group*. The Royal Society. London.

17. Working Group on Biological Weapons Verification (1995) Potential for new approaches to microorganism identification. Paper presented to the 4th Pugwash Study Group on the 'Implementation of the Chemical and Biological Weapons Conventions'. Pugwash Meeting No. 212. Geneva. Switzerland. 2-3 December.

18. Leitenberg, M. (19%) Biological weapons arms control. *Contemporary Security Policy,* 17(1), 1-79.

19. Lacey, E. I. (1992) *Address to the Fall Meeting of the Biological and Biotechnology Section of the Pharmaceutical Manufacturers Association, Baltimore, Maryland.* Disarmament Agency Washington DC, US Arms Control.

20. Bailey, K. C. (1995) Responding to the threat of biological weapons. *Security Dialogue,* 26 (4), 383-397.

21. Zelicoff, A. P. (1995) The Biological Weapons Convention: What is the role of sample collection in a legally binding regime? *Politics and the Life Sciences,* February, 79-84.

22. Zilinskas, R. A. (1995) UNSCOM and the UNSCOM experience in Iraq. *Politics and the Life Sciences.* August. 230-235.

23. Mohr, A. J. (1995) Biological sampling and analysis procedures for the United Nations Special Commission (UNSCOM) in Iraq. *Politics and the Life Sciences,* August, 240-243.

24. Lacey, F. J. (1994) Tackling the biological weapons threat: The next proliferation challenge. *The Washington Quarterly,* **1 7 (4),** 53-64.

25. Goldhammner, A. (1992) *Verification and Inspection Protocols under the Biological Weapons Convention and their Potential Impact on the Pharmaceutical Industry.* Industrial Biotechnology Association. Washington. DC.

26. Industrial Biotechnology Association and Pharmaceutical Manufacturers Association (undated) *Industrial Information Paper on Concerns Regarding Disclosure of Proprietary Information to the International Community.* Washington, DC.

27. Imrie, R. C. (1996) An industry view of BTWC verification. Paper presented to a NATO Advanced Research Workshop on 'The Technology of Biological Arms Control and Disarmament', Budapest. Hungary. 28-30 May.

28. Woollett, G. (1996) *Reducing the Threat of Biological Weapons - A PhRMA Perspective.* Statement to the Fourth Review Conference of the BTWC. PhRMA, Washington DC.

29. United Kingdom (1992) *Commercial Confidentiality Concerns Associated with Sampling and Analysis during On-Site Inspections under the BWC.* BWC/CONF.III/VFREX/NON.28. United Nations, Geneva.

30. Rosenberg, B. H. (1995) Sampling and analyses of proprietary microorganisms while protecting confidential proprietary information. Paper presented to the 4th Pugwash Study Group on the 'Implementation of the Chemical and Biological Weapons Conventions'. Pugwash Meeting No. 212, Geneva, Switzerland, 2-3 December.

31. Tucker, J. B. (ed.) (1996) *The Utility of Sampling and Analysis for Compliance Monitoring of the Biological Weapons Convention.* Monterey Institute of International Studies and Center for Global Security Research, Lawrence Livermore National Laboratory.

32. United Kingdom (1996) *Sampling and Analysis in Facility Inspections: Means of Addressing Confidentiality and Other Concerns.* BWC/AD HOC GROUP/WP. 109. United Nations, Geneva.

33. Donlon, M. A. (1996) Future technologies for detection. Paper presented to a NATO Advanced Research Workshop on 'The Technology of Biological Arms Control and Disarmament'. Budapest, Hungary, 28-30 May.

34. Dando, M. R. (1996) Article I: Scope. In G. S. Pearson and M. R. Dando, (eds). *Strengthening the Biological Weapons Convention: Key Points for the Fourth Review Conference.* QUNO, Geneva.

ROBOTICS MISSION TO MARS FROM EXOBIOLOGICAL POINT OF VIEW

ALEXANDER V. ZAKHAROV
Space Research Institute, Profsoyuznaya 84/32, Moscow 117810, Russia

Abstract

One of the most perspective celestial bodies to search ancient or existing life is Mars. Although the Viking mission results may indicate that Mars has no life today, the possibility exists that Mars may hold the best record of the evidence that led to the origin of life. There is direct geomorphological evidence that in the past Mars had large amounts of liquid water on its surface. Atmospheric models would suggest that this early period of hydrological activity was due to the presence of a thick atmosphere and the resulting warmer temperature. From a biological perspective the existence of liquid water, by itself motivates the question of the origin of life on Mars.

1. Introduction

I believe I am a strawman at this Workshop. My field is space exploration, in particular Mars exploration. However, there are several links between Mars exploration and the topic of this Workshop and I will mention them in my report.

About four billions years ago Mars seem to have had an Earth-like climate with rivers, lakes and perhaps even oceans. In exactly the same epoch that Mars was warm and wet, life arose on Earth. Is it plausible that on two nearby planets with very similar environments, life arose on one and not on the other? The search of morphological or chemical fossils of past life on Mars is one of the most exciting goals of planetary exploration. If found, it might indicate that life arise quickly on all planets in the Universe where the conditions are right. This is a fundamental scientific argument. There are practical applications of Mars exploration. For example, Mars has a planet wide ozone hole. Ultraviolet light from the Sun strikes its surface unimpeded. This is thought to be the reason that not even organic molecules were found by the American Viking landers on the surface of Mars. The study of Mars, therefore, helps us understand what the extreme consequences of ozone layer depletion on the Earth might be. So, the major exobiological goals in Mars exploration are:

- defining the nature of Martian environment,
- understanding the geochemistry of biogenic elements and organic compounds,
- determining whether biosphere presently exist on Mars or has existed in the past.

P.J. Stopa and M.A. Bartoszcze (eds.),
Rapid Methods for Analysis of Biological Materials in the Environment, 43–47.

We know methods and experiments for the first and the second goals. But we have not experience for detecting microorganisms existing or ancient on Mars. Review of methods and techniques of biological measurements is the first link between Mars exploration and the topic of this Workshop.

Mars is the main object for future planetary exploration within the Russian Space Program. The basic argument is favor of such a choice were the following:

- Mars, like Venus, is one of the terrestrial planets and is of great interest from the viewpoint of investigating the nature and evolution of the Solar system.

- Earlier missions to Mars showed that this planet in its earlier history was much more similar to the Earth than that is now. There is evidence that, at an earlier time, Mars had large bodies of water and rivers on its surface.

- Studies of the evolution of the atmosphere and climate of Mars will help us understand better the history of our own planet and allow us to forecast its future.

- With the presence of a previous hydrosphere on Mars, a biosphere could have also formed. Among the other planets of the Solar system, it is Mars where life very possibly exists or existed earlier.

Because of these reasons, Mars indeed is the first planet where we will send astronauts to visit. However, before such manned exploration, it is necessary to study this planet thoroughly using robotics missions.

2. Approach to Mars Missions

There are several phases of robotics missions to Mars to search for ancient life on this planet.

- remote sensing from spacecraft on the orbit of Mars with global coverage of the Martian atmosphere, surface, internal structure,

- in-situ measurements on choice landing sites of the Martian surface, elemental and mineral composition, volatile content, microscopic structure, physical properties,

- sampling return from Mars to Earth provides a possibility to detail studies of Martian samples in laboratories on the Earth, using different methods.

The scientific rational for obtaining samples of planetary bodies has been discussed in detail by many study groups as well as described in details in many publications (e.g., [1,2]).

Several recommendations have been endorsed by the exobiological community as a basis for implementing the present strategy to explore for evidence of past Martian life [3]. These include:

- identification of suitable landing sites by mapping the precise location of aqueous sedimentary deposits from the orbit,

- delivering capable rovers to within reachable distance of high priority targets, i.e., those that are most likely to have captured and preserved ancient biosignatures and/or prebiotic organic compounds in rocks, and

- conducting rover-based mineralogical and organic analyses to select the most appropriate rock samples for return to Earth.

The latter requirement is especially important because the amount of material that can be returned will be limited and we will need to prioritize samples in situ, to select those that are most likely to organic ancient biosignatures [4,5].

3. Current Program to Search for Ancient Life on Mars

Search for ancient life on Mars is one of the main goal of the current program of Mars exploration. All elements of this program: orbital missions, landers and sample return missions are different phases of this program. An active element of this program now is NASA's Mars Global Surveyor mission (Mars orbiter) launched in November 1996 and the Pathfinder mission (Sojourner mini-rover) launched in December 1996. The big Russian project to Mars, Mars-96, included an orbiter, two small stations on the Martian surface and two penetrators for subsurface study [6] was loosed during the launch in November 1996.

Orbital experiments should lead to identification of a number of sites characterized by promising lithologies, i.e., those potentially capable of preserving biosignatures if such were present during deposition of the lithologies. One of the key experiment of this program is high-resolution imagery of entire Martian surface with ultra-high-resolution imagery of selected areas. Conversion of the photographic images into three-dimensional map of the Martian surface requires precise topographic data. This will be provided by Laser Altimeter included in the Mars Global Surveyor payload. Mapping of global distribution of near-surface water from orbit requires neutron spectroscopy, which can be performed by gamma-ray-neutron spectrometer what is planned for inclusion in the Mars Surveyor mission in 2001. Location occurrence of appropriate lithologies on the Martian surface is best done using near- or mid-infrared spectroscopy. An example of such an instrument, the Thermal Emission Spectrometer, is included in the Mars Global Surveyor payload [5].

Ideally, landed spacecraft would be dispatched to several sites characterized by different lithologies to determine which would have the greatest potential for containing a biosignature. However, it is unrealistic to plan on multiple landers at a single launch opportunity, so, the current program includes one lander at a single promising site.

The first task of any rover is to find the most promising site at the terrain surrounding the lander and to analyze promising-looking rocks. Because of the large number of potential sampling targets, it will be necessary to perform some relatively rapid remote characterization of the accessible lithologies. This would be best done by a miniaturized IR spectral imager that would be capable of identifying the dominant mineralogy of any rock visible from the rover. It seems likely that elemental analysis and mineral identification can be combined within a single instrument, an X-ray fluorescence/X-ray diffraction spectrometer. These elemental and mineralogical data will suffice to identify any of targeted sedimentary lithologies that might be accessible to the rover. In addition, a close-up imaging system on the rover will exam the texture and fabric of the rock, which can yield information on the sedimentary environment in which the lithology was deposited, and even potentially reveal evidence for macroscopic biological activity in the form of stromatolites.

Finally, it will be desirable to conduct at least a limited series of isotopic and organic analyses of promising samples. A key isotopic measurement would be to measure the 13C/12C difference between reduced and oxidized carbon, which could be diagnostic of biological activity. Several different rock samples from the surface and from undersurface layers prioritize on the base of results of such exploration will be the first target for sample return [4].

One of the problem for a sample return mission is biological planetary protection, sterilization of spacecraft and instruments, what will be land on the Martian surface and sterilization of the return capsule on the Earth. The sterilization problem is the second link between Mars exploration and the topic of this Workshop.

In the conclusion, I would like to mention about another method of search of life on Mars. In August 1996 issue of the Science Magazine, an article was published about the search of past life on Mars using the meteorite, named Allan Hills 84001, which was classified as a Martian meteorite [7]. This meteorite was ejected from mars into space by an impact 16 million years ago and landed in Antarctica 13000 years ago. The authors of this paper used high-resolution scanning and transmission electron microscopy to study this meteorite. Fresh fracture surfaces of ALH8401 contain abundant polyciclic aromatic hydrocarbons (PAHs). These fresh fracture surfaces also display carbonate globules. The globules contain fine-grained, secondary phases of single-domain magnetite and Fe-sulfides. These globules are similar in texture and size to some terrestrial bacterially induced carbonate precipitates. The authors marked that although inorganic formation is possible, formation of the globules by biogenetic processes could explain many of the observed features, including PAHs. The PAHs, the carbonate globules, and their associated secondary mineral phases and texture could thus be fossil remains of the a past Martian biota.

I guess search of traces of extraterrestrial microorganisms on the Earth is the third link between planetary exploration and the topic of "Rapid Methods for Monitoring the Environmental for Biological Hazards" Workshop.

4. References

1. Drake, M., Boynton, W.. and Blanchard. D. (1987) The Case of Planetary Sample Return Missions: Origin of the Solar System, Eos. 68.
2. Gooding, J., Carr. M.. and McKay. Ch.. (1989) The Case for Planetary Sample Return Missions: History of Mars, Eos, 70. No. 31. 745-754.
3. NASA (1995) An exobiological strategry for Mars exploration, NASA Special Publication 530.
4. Farmer, J.D. (1997) Implementation a strategy for Mars Exploration, Proceedings of SPIE, Instruments, Methods, and Missions for the Investigation of Extraterrestrial Microorganisms, Ed. by R. Hoover, 3111, 200-212.
5. Kerridge, J.F. (1997) Search for evidence of ancient life on Mars, Proceedings of SPIE, Instruments, Methods, and Missions for the Investigation of Extraterrestrial Microorganisms, Ed. By R. Hoover, 3111, 51-60.
6. Zakharov, A.V. (1996) Mars-96 Project. Short description, Space Research Institute.
7. McKay, D.S., Gibson. E.K.. Thomas-Keprta. K.L., Vali, H., Romanek, C.S., Clemett, S.J., Chiller, X.D.F., Maechling, C.R., and Zare. R.N. (1996) Search for past life on Mars: Possible relict biogenic activity in Martian meteorite ALH84001. Science, 273. 924-930.

DORMANCY IN NON-SPORULATING BACTERIA: ITS SIGNIFICANCE FOR ENVIRONMENTAL MONITORING

[1]ARSENY S. KAPRELYANTS, [1]GALINA V. MUKAMOLOVA, [1]TATYANA V. VOTYAKOVA, [2]HAZEL M. DAVEY, AND [2]DOUGLAS B. KELL,
[1]*Bakh Institute of Biochemistry, Russian Academy of Sciences, Leninskii Prospekt 33, 117071 Moscow, Russia.*
[2]*Institute of Biological Sciences, University of Wales, Aberystwyth, Dyfed SY23 3DA, UK*

Abstract

In natural ecosystems, the total cell count obtained microscopically typically exceeds the viable count on non-selective media by orders of magnitude. The question therefore arises as to whether the "invisible", apparently nonculturable cells are dead, are killed by our isolation media, or are merely in a dormant state from which we might in principle be able to resuscitate them if only we knew how. In particular the suggested "viable-but-nonculturable" (VBNC) bacteria have been invoked to explain phenomena as divergent as the epidemiology of some infections and the persistence of genetically marked organisms in the environment (e.g. the failure to isolate Vibrio cholerae and Campylobacter jejuni from clearly implicated sources or reservoirs of infection could be accounted for on the basis of their being present in a VBNC or dormant state). Application of flow cytometry may be a useful tool to visualize bacteria without their growing and to discriminate between dead and dormant bacteria. We found that dormant Micrococcus luteus can be physically separated from dead cells by flow cytometry sorter after cell staining with rhodamine 123. Resuscitation of dormant M.luteus cells in liquid medium does not proceed in the absence of a culture supernatant from batch-grown cells. This suggests that viable cells can excrete a pheromone-like substance necessary for the resuscitation of dormant cells.

P.J. Stopa and M.A. Bartoszcze (eds.),
Rapid Methods for Analysis of Biological Materials in the Environment, 49–65.

1. Introduction

We have established that in the case of complete exhaustion of exogenous nutrients bacteria can go into an anabiotic (dormant) state, which helps the cells to survive for a long time without growth and multiplication. Dormancy may be defined as "*a reversible state of low metabolic activity, in which cells can persist for extended periods without division*" [1]; it often corresponds to a state in which cells are not "alive" in the sense of being able to form a colony when plated on a suitable solid medium, but one in which they are not "dead" in that when conditions are more favorable they can revert (by a procedure referred to as resuscitation) to a state of "aliveness" as so defined. The adoption by vegetative, nonsporulating bacteria of such dormant forms in natural environments is a subject of intense current interest (see e.g. [1-9]), and it has been suggested that such dormant bacteria may be represented by ultramicrobacteria [5] or the so-called viable-but-non-culturable (VBNC) forms [2] found in marine or river habitats (but cf. [1]). Although a characteristic property of dormant forms is their inability to produce colonies when plated *directly* onto solid agar media [1] the "VBNC" cells are usually taken to be revealed *via* by their metabolic activity [4] using the direct viable count technique [9].

It is increasingly evident that dormant (or nonculturable) forms are an important feature of hitherto-unrecognized disease states [10,11], and a gigantic biodiversity of both scientific and biotechnological interest [12], and their existence raises important questions concerning the famous Koch's postulates of microbial pathogenesis [13].

Foremost amongst the epidemiological mysteries are cholera and campylobacteriosis where the failure to isolate *Vibrio cholerae* and *Campylobacter jejuni* from clearly implicated sources or reservoirs of infection could be accounted for on the basis of their being present in a VBNC state. For both these organisms, environmental investigations have provided evidence for the presence of "nonculturable" cells in appropriate samples [14,15] while *in vitro* studies have demonstrated their capacity to form metabolically active cells which could not be grown immediately [16,17].

Thus there are many practical issues bearing on food and water safety, the distribution and influence of bacteria in the environment, the effects of antibiotics and the significance of declining colony-forming unit (cfu) counts which cannot be assessed until the authenticity of the putative dormant or VBNC state has been confirmed and how widely dormant cells are distributed. If some bacteria can differentiate into a "nonculturable" state, this undermines interpretation of all applications based on colony counting [4]. As discussed elsewhere [18,19], it is inevitable that the use of nucleic acid probe technology will reveal that many more diseases than are currently recognized do have a microbial aetiology [10,11], even when culture-based methods have not yet confirmed this [13].

The list of bacterial species which are claimed to be in dormant state, and where the apparent resuscitation of "nonculturable" cells has been reported is substantial (including such important Gram-negative pathogens as *Vibrio cholerae, Legionella pneumophila, Yersinia ruckeri, Salmonella enteritidis, Campylobacter jejuni, Helicobacter pylori* - Table 1), but the crucial question of whether such "dormant" forms could be converted to normal, viable cells remains open, despite the many experimental attempts that have been made to resolve it [1,20]. The central point of discussion in this area is now focused on the results of recovery or resuscitation experiments, almost all of which were done by cultivation of nonculturable cells on liquid media followed by plating onto agar plates. In the majority of studies, recovery has been difficult both to produce and to reproduce. Although some early experiments have purported to show the ability of "nonculturable" bacteria to grow on agar following resuscitation *in vitro* in appropriate liquid media, a limited number of operationally viable cells in the starved population could have been responsible for the growth which occurred. Indeed, in almost all published cases the populations of nonculturable bacteria diluted to an extent which might have been sufficient, statistically, to remove any viable cells, were not able to be resuscitated (Table 1) [21].

TABLE 1. A summary of some studies in which resuscitation of 'dormant' or a 'nonculturable' bacterium has been attempted.

Aeromonas salmonicida	starvation in sea water, 15°C	+	-	usage of rich medium (TSB) for resuscitation	[22]
Aeromonas salmonicida	starvation in sea water, 4°C	-	-	various media and conditions have been used for resuscitation	[20]
Aeromonas salmonicida	starvation in water, 10°C	-	+	various resuscitation media have been used	[23,24]
Campylobacter jejuni	starvation in physiological saline solution, 20°C	-	-	resuscitation in simulated stomach, ileal and colon environments (rich media)	[25]
Campylobacter jejuni	starvation in sterilized pond water, 4°C	+	-	resuscitation of some strains via passage in mice	[26]
Klebsiella pneumoniae	starved bacteria in phosphate buffer	+	-		[27]
Legionella pneumophila	starvation in pure water, 30°C	-	-	resuscitation in co-cultures with *T. pyriformis*	[28]
Micrococcus luteus	long storage in stationary phase, room temperature	+	+	resuscitation factor (supernate taken from active culture) required	[29]
Pasteurella piscicida	starvation in seawater, 6 and 20°C	+	-		[30]
Salmonella enteritidis	starvation in salt solutions, 21°C	-	+	usage of lactose broth (Difco) for resuscitation	[31]
Salmonella enteritidis	starvation in sterilized river water, 25°C	+	-	resuscitation by nutrient addition after 4 but not 21 days after culturability lost	[32]

TABLE 1. Cont'd

Pseudomonas fluorescens	starvation in soil, 24°C	+?	+	only several divisions of "VBNC" cells during resuscitation were found	[33]
Pseudomonas fluorescens	N-starvation in minimal medium, 25°C	+	-	usage of medium lacking a carbon source for resuscitation	[34]
Vibrio cholerae	starvation in autoclaved water	+	-	usage of passage through rabbit ileal loop	[35]
Escherichia coli *Vibrio cholerae*	starvation in autoclaved artificial sea water, 4°C	-	+	usage of nutrient- free medium for resuscitation	[36]
Vibrio, Aeromonas, Pseudomonas, Alcaligenes spp.	non-culturable ultramicrobac-teria from estuarine waters	+	-	resuscitation was found for a narrow range of nutrient concentrations	[37]
Vibrio parahaemolyticus	starvation in mineral medium under 3.5°C	-	+	usage of rich medium for resuscitation	[38]
Vibrio cholerae	starvation in salt solution , 15°C	+	++ ++	Conversion to the colony- forming cells was effected with a short heat shock	[39]
Vibrio cholerae	starvation in buffered saline, 4°C	+/-	+/-	Resuscitation in intestine after ingestion of non-pathogenic vaccine strains by volunteers. 2 sets of experiments were done; only one claimed resuscitation. Dilutions were probably not great enough to exclude presence of some viable cells	[40]
Vibrio vulnificus	starvation in defined media, 5°C	+	-	usage of natural estuarine environment for resuscitation	[41]

TABLE 1. Cont'd

Vibrio vulnificus	starvation in defined media, 5°C	+	+ |	*in vivo* resuscitation (injection in mice)	[42]
Vibrio vulnificus	starvation in defined media, 5°C	-	+	wide range of conditions for resuscitation *in vitro* were used	[43]
Yersinia ruckeri	starvation in sterile river water, 6 or 18°C	+	-	usage of rich medium for resuscitation	44]

" +" sign (third column) means that resuscitation was attempted and indeed claimed as judged by the appearance of increased numbers of culturable organisms, whilst the - sign means that no recovery or resuscitation was observed.

"+" sign (fourth column) means that the author(s) diluted the samples before performing resuscitation, in an attempt to remove genuinely viable cells present at the start of the resuscitation experiment, whilst a - sign means that they did not.

| The growth of cells on the medium used here may have underestimated the culturable fraction, since there is evidence that this system contains an injured fraction [43].

‡‡ Dilutions *per se* were not done but at one stage *no* viability was observed, although the resuscitation yielded 1000 colonies (again the total count of bacteria at the onset of resuscitation was not shown).

The difficulties in testing for the presence of dormant forms in environmental samples by conventional methods (due to their inability to grow on solid or liquid media) make the development of new approaches to detect bacteria in dormant states an important task for epidemiology and environmental monitoring.

2. Results

In recent work, we have found that cells of the nonsporulating, copiotrophic, Gram-positive coccus *Micrococcus luteus* starved for 3-7 months in spent growth medium following growth to stationary phase in batch culture can persist in a dormant state in cultures which overall exhibited a very low viability ($< 10^{-4}$) as estimated by plating on agar plates, while the total count remained close to its initial value. Using flow cytometry with appropriate probes and conditions we found that a high percentage of *M. luteus* cells in 3-month-old populations could be resuscitated to normal, colony-forming bacteria [8], under conditions which excluded any significant regrowth of initially viable cells. We confirmed this by using the Most Probable Number (MPN) method, when we resuscitated cells in media which, statistically, contained no "initially-viable" cells [29]. These and other data led to the conclusion that from a physiological point of view a significant number of cells in such starved *M. luteus* populations were not dead but were dormant and could be converted to normal, colony-forming bacteria.

However the assessment of the percentage of resuscitable cells in a population could, on the basis of these and related experiments, only be made *after* cell resuscitation where, under some conditions, an excellent correlation could be found between the *percentage* of dormant cells and the different extent to which rhodamine 123 could be accumulated in an uncoupler-sensitive fashion by different cohorts of cells. A particularly desirable goal would therefore be to establish which differences between the *individual* cells in a population *before* resuscitation can be correlated with whether such cells survive or not. To this end in the present study we used flow cytometry and cell sorting procedures, which allow one to analyze cell populations by flow cytometric procedures and then physically to separate subpopulations of cells with different properties for their further examination.

Figure 1A show the typical distribution of the fluorescence of non-starved *M. luteus* cells that had been stained with the membrane energization-sensitive cation Rh123 and studied in the flow cytometer. We discriminated the extent of staining as follows. Channel 136 was chosen as a threshold since almost no Rh123-stained, starved cells exhibited a fluorescence in a channel greater than this, and freshly harvested, viable cells in the presence of Rh123 and the absence of uncoupler exhibited a fluorescence between channels 80 and 136 (Figure 1A), which is fully uncoupler sensitive (not shown). Figure 1B shows a typical distribution of the fluorescence of *M. luteus* cells that had been starved for 5 months, stained with Rh123 and studied in the flow cytometer. A bimodal distribution in the extent of staining is evident. Region A (channel 0 to channel 80) represents cells which bind Rh123 nonspecifically: 98 % of fresh late logarithmic phase *M. luteus* cells stained with the same concentration of Rh123 followed by treatment with a suitable concentration of the uncoupler CCCP exhibited a fluorescence in this region. Although starved cells in region B (between channels 80 and 136) of Figure 1B had an elevated fluorescence relative to those in region A, its sensitivity to CCCP was also very low (only 2-5% of the cells in region B exhibiting a decrease in fluorescence after CCCP treatment).

FIGURE 1. Distribution of the fluorescence of non-starved (a) and starved (b) cells of *M. luteus* stained with Rh123 and assessed by flow cytometry. Cultures were grown in lactate minimal medium until late logarithmic phase, harvested, washed and resuspended in lactate minimal medium lacking lactate and stained with Rh123 (0.3 mM) (a), or starved for 5 months and diluted 20-fold in lactate minimal medium without lactate. The distribution of fluorescence was assessed by analytical flow cytometry using the Skatron Argus 100 instrument for the ability to accumulate Rh123.

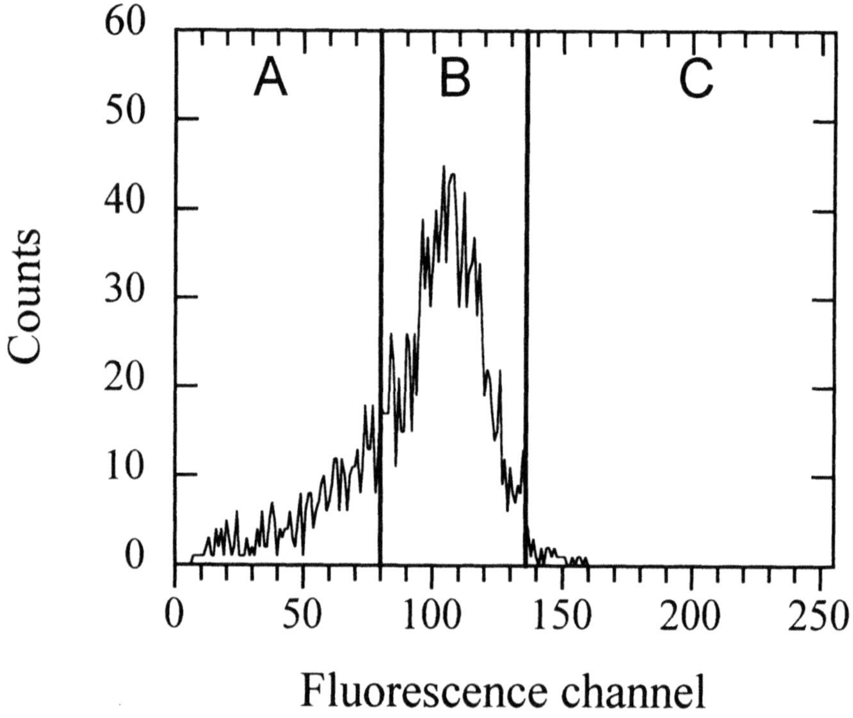

Figure 1A

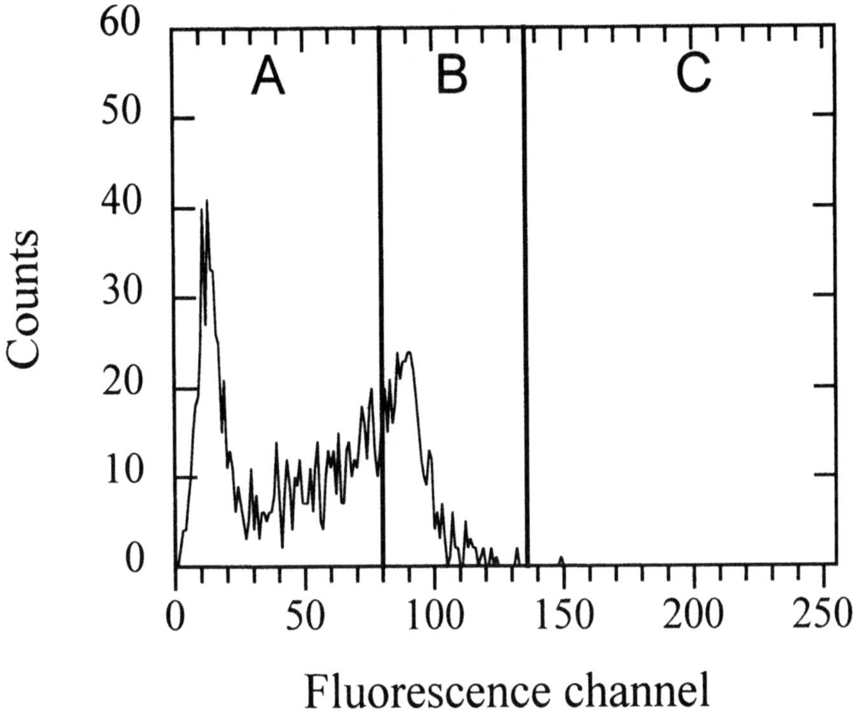

Figure 1B

We sorted cultures whose fluorescence was of the type displayed in Figure 1B and Figure 2 into two populations: (i) cells of which the rhodamine staining was sensitive or partially sensitive to CCCP (regions B+C of Figure 1) and cells whose rhodamine-dependent fluorescence was not sensitive to CCCP (region A of Figure 1). After sorting, cells were plated on nutrient agar for viable count determinations, while the total count of sorted samples was also examined. A control incubation of both starved and fresh cells in flow cytometer sheath fluid for 2 h had no influence on their viability (data not shown).

Table 1 shows the results of a typical experiment. Because of the dilution of samples during sorting (which varies from experiment to experiment) we express all viable count numbers as those normalized to the total number of cells in a particular region. As would be expected, the great majority of colony-forming cells originated from regions B+C, resulting in an increase of the viability of the population obtained from these regions. However, the number of colony forming cells in regions B+C and in the whole population after sorting (per 10^6 cells) was unexpectedly and significantly more (8-20 times in different experiments) than those before sorting. Similar results were obtained using plates with lactate minimal medium (not shown).

FIGURE 2. Distribution of *M. luteus* cells that had been starved for 5 months and stained with rhodamine 123 and subjected to sorting in a flow cytometric cell sorter. For experimental details see legend to Figure 1. The horizontal line shows the sorting region which was used to discriminate between regions A and B+C. Cells from the 2 regions were sorted differentially into separate test tubes, and examined by MPN assay, plating and assessed for their total counts. The scales are logarithmic and cover 4 decades.

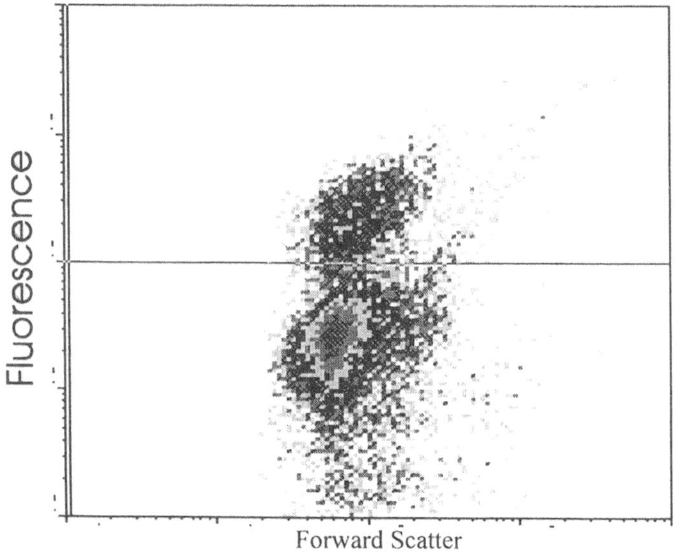

TABLE 2. Total count, viable count and the distribution in viability of cells taken from cultures of *M. luteus* that had been starved for 5 months, before and after cell flow cytometric analysis.

	sorting			
	before sorting	after sorting		
		region A	region B+C	regionA+B+C
total count	10^6	$8.\ 10^5$	2.10^5	10^6
viable count by plates	740	42	6200	6240
viability by plates	0.07%	0.005%	3.1%	0.62%
viability by MPN (no supernatant added)	0.065%	0.0044%	1.5%	0.3%
viability by MPN (in the presence of supernatant)	3%	0.004%	19.3%	3.9%

The distribution of the total count between regions A,B and C was estimated from the flow cytometric distribution pattern (as in Figure 7) normalized to a value for the whole population of 10^6 cells. Viable counts were calculated relative to the total number of cells in a particular region. Lab M medium was used for the count of viable bacteria on agar plates (second row). The viable count and hence viability by MPN assay was performed as described in Materials and Methods save that lactate minimal medium with 0.05% yeast extract for assay was used without (4th row) or with (5th row) supernatant taken from a fresh logarithmic-phase *M. luteus* culture. Coefficients of variation for the total and viable counts were 5.6% and 4.4% respectively.

We used the most probable number (MPN) method, which allows one to estimate viable cell numbers by their cultivation in liquid medium at high dilution [29]. The basic idea of such a measurement is that provided that the MPN medium supports resuscitation as well as growth, a dormant cell would score as viable, in contrast to its behavior as judged by direct plating. Under these conditions the growth of a limited number of viable cells during resuscitation is excluded. Appropriate resuscitation medium contained lactate minimal medium + yeast extract + supernatant from batch cultures of *M. luteus* grown in lactate minimal medium. The MPN assay was performed in a Bioscreen C optical growth analyzer (Labsystems, Finland). For the MPN assay supernatant taken from a batch culture of the organism grown in LMM (plus 0.5% lactate) to an OD of 2, slightly before the beginning of stationary phase, was added at a concentration of 0.1 ml of supernatant per well containing 0.1ml resuscitation medium with different concentration of yeast extract and starved cells. Under these conditions, we *did* observe cell growth in inocula taken from starved cells suspensions at very high dilutions (10^7- to 10^{10}-fold) after incubation of the test tubes for 72-120 h (Table3)

TABLE 3. Resuscitation of dormant *m. luteus* cells in liquid medium.

N culture	time of starvation	total count	viability by cfu	viability by MPN
1	2 months	$5.3\ 10^9$	$5\ 10^6$	$3.5\ 10^9$
2	4.5 months	10^{10}	$1.3\ 10^6$	$9.2\ 10^9$
3	6 months	$1.2\ 10^{10}$	$3.6\ 10^4$	$9.2\ 10^9$
4	9 months	$6.2\ 10^9$	$5.2\ 10^5$	$5.4\ 10^9$

The important condition for such an estimation was to add diluted supernatant from growing bacteria to the media in which the MPN assay was performed. We found that the effect of supernatant on the resuscitation of dormant cells was maximal when the supernatant had been taken from the late logarithmic phase of batch-grown bacteria (not shown). The active fraction from supernatant was purified by a combination of anion exchange media, and the final activity was eluted from a linear KCl gradient on a MonoQ column. This substance is a protein with a molecular mass of ca 17 kDa, which in concentrations of several ng/ml results in the resuscitation of starved cells and increases the viability of these culture at least 100-fold (Figure 3). This protein also stimulates the growth of viable *M. luteus* cells, and can therefore be considered as a growth factor or bacterial cytokine.

Table 2 demonstrates the resuscitation of cells under such conditions for the whole population before sorting (showing an increase in the viable count when MPN was done in the presence of supernatants by some 50-fold in these cultures that had been starved for 5 months). The same effect was observed when cells from sorted from regions B+C were subjected to resuscitation, while resuscitation of cells from region A was unsuccessful (Table 1).

For additional discrimination between different type of cells in starved population we monitored the state of the permeability barrier of the cells by staining with propidium iodide (PI, a DNA staining probe). We established that this probe does not penetrate through the cytoplasmic membrane of intact *M. luteus*, while octanol administration to the cell suspension (0.5%) resulted in 100% stained cells (Fig 4). Observation of different starved cultures of *M. luteus* revealed that in some cultures where the percentage of PI-positive cells is close to 100% the resuscitation of cells was not successful (even in the presence of supernatant). This can indicate the correlation between state of permeability barrier and ability of starved cells for recovery.

3. Discussion

In the present study starved cultures revealed two different type of cells, with low and elevated fluorescence. The fluorescence intensity of cells in population B was comparable with the fluorescence of (actively growing) *M. luteus* cells, although in the case of the starved cells the fluorescence was almost insensitive to uncoupler, in contrast to that of the actively-growing, viable cells. It is reasonable that the increased uncoupler-resistant fluorescence of some cells reflects the changes in surface (membrane) structure (see also 12) which could results in increased Rh123 adsorption or fluorescence quantum yield by the cells. Indeed, a characteristic property of starved bacteria (see 4) is an increase in their surface hydrophobicity.

FIGURE 3. Effect of purified 17-kDa protein on the resuscitation of dormant cells of *Micrococcus luteus*. *M. luteus* was starved and its viability assessed using an MPN assay as described [Kaprelyants et al 1994 FEMS]. Supernatants were mixed in the stated proportion with a lactate minimal medium containing 0.05 (w/v) yeast extract and used in the MPN assay.

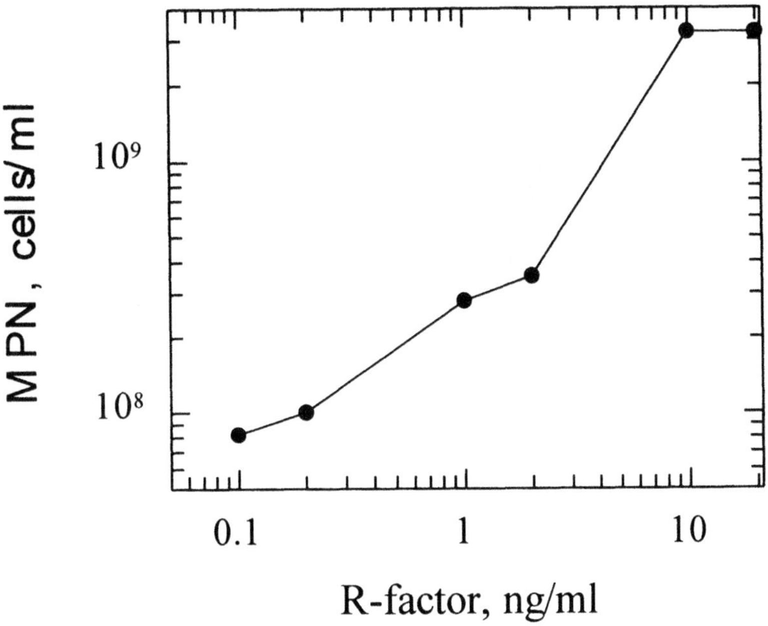

FIGURE 4. Distribution of the fluorescence of starved for 5 months cells of *M. luteus* stained with propidium iodide and assessed by flow cytometry. For experimental details see legend to Figure 1. Octanol was added to the cells before measuring to final concentration of 0.2%.

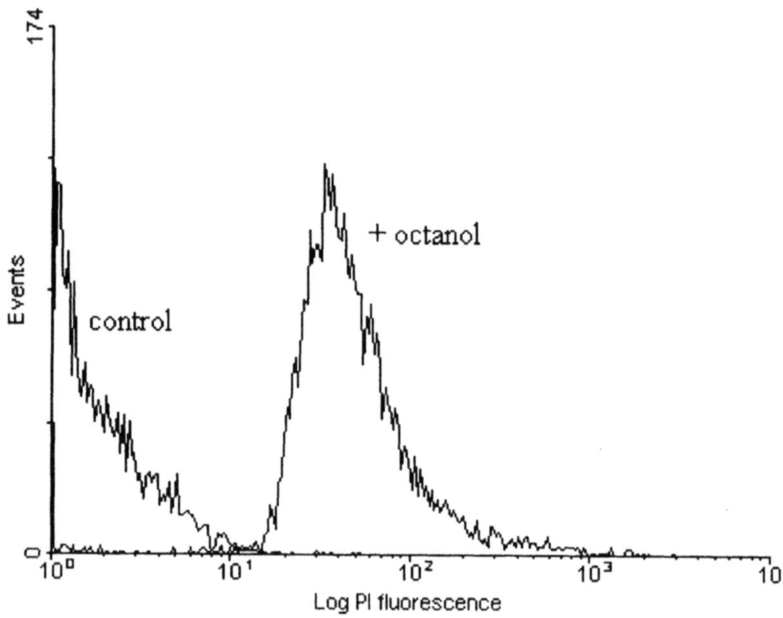

Cell sorting revealed that the viable cells in cultures starved for 5 months are concentrated mainly in regions B and C (Figure 1B). The viable count of the whole culture, as well as the population in regions B+C, increased after cell sorting. This result can be rationalized in the light of the existence of a "killer factor" which can be produced by starved cells during their resuscitation (13). The resuscitation of cells as judged by the MPN assay was successful for cells in regions B+C but not for cells in region A. This constitutes direct evidence that dormant cells are concentrated in regions B+C.

In conclusion, populations of the nonsporulating bacterium *M. luteus* that have been starved for an extended period are heterogeneous, and consist of viable, dormant and dead cells, which can influence each other during resuscitation, causing an underestimation of the number of potentially resuscitable cells in the culture. The active protein found in this study is to our knowledge the first purification of a factor which stimulates the resuscitation of bacteria after true dormancy. Since the factor could also stimulate the growth rate of viable cells, it is very likely that this protein is involved in the normal control of cell multiplication, although the molecular mechanism by which this factor facilitates resuscitation remains to be understood. We would also comment on the implications of the self-promoting mode of cell growth and recovery after dormancy for medicine and environmental microbiology, especially in the case of infections in which bacterial cells may persist for extended periods in latent or dormant states.

The two subpopulations of starved culture of *M .luteus* could be discriminated and sorted out from the latter on the basis of their ability to bind the weakly lipophilic cation dye Rh123 and to be stained by the membrane-impermeable propidium iodide. Flow cytometry is a powerful method for the analysis of heterogeneity in bacterial systems and promising tool for environmental monitoring.

4. Acknowledgments

The authors are indebted to the Chemicals and Pharmaceuticals Directorate of the U.K. BBSRC, the Royal Society under the terms of the Royal Society/ Russian Academy of Sciences Joint Project scheme, and the Russian Foundation for Basic Research (RFBR), for financial support of this work.

5. References

1. Kaprelyants, A. S., Gottschal, J. C., and Kell, D. B (1993) Dormancy in non-sporulating bacteria. *FEMS Microbiol. Rev.* **104**, 271-286.
2. Roszak, D. B and Colwell, R. R (1987) Survival strategies of bacteria in the natural environment. *Microbiol. Rev.* **51**, 365-379.
3. Matin A. (1992) Physiology, molecular biology and applications of the bacterial starvation response. *Journal of Applied Bacteriology Symposium Supplement* **73**, 49S-57S.
4. Barer, M. R., Gribbon, L. T, Harwood, C. R., and Nwoguh, C.E. (1993) The viable but non-culturable hypothesis and medical microbiology. *Rev. Med. Microbiol.* **4**, 183-91.
5. Morita, R. Y. (1990) The starvation-survival state of microorganisms in nature and its relationship to bioavailable energy. *Experientia* **46**, 813-817
6. Kaprelyants, A. S. and Kell, D. B (1992) Rapid assessment of bacterial viability and vitality using rhodamine 123 and flow cytometry. *J.Appl. Bacteriol.* **72,** 410-422.
7. Kell, D. B, Ryder, H. M, Kaprelyants, A.S., and Westerhoff, H. V. (1991) Quantifying heterogeneity: flow cytometry of bacterial cultures. *Antonie van Leeuwenhoek* **60**, 145-158.
8. Kaprelyants, A. S and Kell, D. B. (1993) Dormancy in stationary-phase cultures of *Micrococcus luteus*: flow cytometric analysis of starvation and resuscitation. *Appl. Env. Microbiol.*, **59**, 3187-3196.
9. Roszak, D. B. and Colwell, R. R. (1987a). Metabolic activity of bacterial cells enumerated by direct viable count. *Appl. Environ. Microbiol.* **53,** 2889-2983.
10. Domingue, G. J. (1995). Electron dense cytoplasmic particles and chronic infection - a bacterial pleomorphy hypothesis. *Endocytobiosis Cell Res.* 11, 19-40.
11. Domingue, G. J., Ghoniem, G. M., Bost, K. L., Fermin, C. and Human, L. G. (1995). Dormant microbes in interstitial cystitis. *Journal of Urology* **153,** 1321-1326
12. Amann, R. I., Ludwig, W. and Schleifer, K. H. (1995). Phylogenetic identification and *in situ* detection of individual microbial cells without cultivation. *Microbiological Reviews* **59,** 143-169.
13. Fredricks, D. N. and Relman, D. A. (1996). Sequence-Based Identification of Microbial Pathogens - a Reconsideration of Koch's Postulates. *Clinical Microbiology Reviews* **9,** 18.
14. Brayton, P. R., M. L. Tamplin, A. Huq and R. R. Colwell (1987) Enumeration of *Vibrio cholerae* O1 in Bangladesh waters by fluorescent-antibody direct viable count. *Appl Env Microbiol* 53, 2862-2865.
15. Pearson, A. D., M. Greenwood, T. D. Healing, D. Rollins, M. Shahamat, J. Donaldson and R. R. Colwell (1993) Colonization of broiler chickens by waterborne *Campylobacter jejuni*. *Appl Env Microbiol* **59**, 987-996.
16. Xu, H. S., Roberts, N., Singleton, F. L., Attwell, R. W., Grimes, D. J., and Colwell, R. R. (1982) Survival and Viability of Nonculturable *Escherichia Coli* and *Vibrio Cholerae* in the estuarine and marine environment. *Microbial Ecology* **8**, 313-323.

17. Rollins, D. M. and Colwell, R. R. (1986) Viable but nonculturable stage of *Campylobacter jejuni* and its role in survival in the natural aquatic environment. *Appl Env Microbiol* **52**, 531-538.
18. Kaprelyants, A. S. and Kell, D. B. (1996) Do bacteria need to communicate with each other for growth? *Trends Microbiol.*, **4**, 237-242.
19. Davey, H. M. and Kell, D. B. (1996) Flow cytometry and cell sorting of heterogeneous microbial populations: the importance of single cell analysis. *Microbiol Rev* **60**, 641-696.
20. Ferguson, Y., Glover, L. A., McGillivray, D. M., and Prosser, J. I. (1995) Survival and activity of Lux-Marked *Aeromonas salmonicida* in seawater. *Appl Env Microbiol* **61**, 3494-3498.
21. Kell, D. B., Kaprelyants, A. S., Weichart, D. H., Harwood, C. R., and Barer, M. R. (1997) On bacterial viability and culturability. *Microbiology*, in press.
22. Husevag, B. (1995) Starvation survival of the fish pathogen *Aeromonas salmonicida* in seawater. *FEMS Microbiol Ecology* **16**, 25-32.
23. Morgan, J., P., C. and R., P. (1991) Survival of *Aeromonas salmonicida* in lake water. *Appl Env Microbiol* **57**, 1777-1782.
24. Morgan, J. A.W., Clarke, K. J., Rhodes, G., and Pickup, R. W. (1992) Non-culturable *Aeromonas salmonicida* in lake water. *Microbial Releases* **1**, 71-78.
25. Beumer, R. R., Devries, J., and Rombouts, F.M. (1992) *Campylobacter jejuni* nonculturable coccoid cells. *Int J Food Microbiol* **15**, 153-163.
26. Jones, D. M., Sutcliffe, E. M., and Curry, A. (1991). Recovery of viable but non-culturable *Campylobacter jejuni. Journal of General Microbiology* **137**, 2477-2482.
27. Lappin-Scott, H. M., Cusack, F., Macleod, A., and Costerton, J. W. (1988) Starvation and nutrient resuscitation of *Klebsiella pneumoniae* isolated from oil well waters. *J Appl Bact* **64**, 541-549.
28. Yamamoto, H., Hashimoto, Y., and Ezaki, T. (1996) Study of nonculturable *Legionella pneumophila* cells during multiple nutrient starvation. *FEMS Microbiol Ecology* **20**, 149-154.
29. Kaprelyants, A. S., Mukamolova, G. V., and Kell, D. B. (1994) Estimation of dormant *Micrococcus luteus* cells by penicillin lysis and by resuscitation in cell-free spent medium at high dilution. *FEMS Microbiol Lett* **115**, 347-352.
30. Magarinos, B., Romalde, J., Barja, J., and Toranzo, A. E. (1994) Evidence of a dormant but infective state of the fish pathogen *Pasteurella piscicida* in seawater and sediment. *Appl Env Microbiol* **60**, 180-186.
31. Chmielewski, R. A. N. and Frank, J. F. (1995) Formation of viable but nonculturable *Salmonella* during starvation in chemically-defined solutions. *Lett Appl Microbiol* **20**, 380-384.
32. Roszak, D. B., Grimes, D. J., and Colwell, R. R. (1984). Viable but nonrecoverable stage of *Salmonella enteritidis* in aquatic systems. *Canadian Journal of Microbiology* **30**, 334-338.
33. Binnerup, S. J., Jensen, D. F., Thordal-Christensen, H. and Sorgensen, J. (1993) Detection of viable, but non-culturable *Pseudomonas fluorescens* DF57 in soil using a microcolony epifluorescence technique. *FEMS Microbiol Ecol.* **12**, 97-105.
34. Evdokimova, N. V., Dorofeev, A. G., and Panikov, N. S. (1994) Dynamics of survival and transition to dormant state of nitrogen- starving bacteria *Pseudomonas fluorescens. Microbiol (Russia)* **63**, 99-104.
35. Colwell, R. R., Brayton, B. R., Grimes, D. J., Roszak, D. B., Huq, S. A., and Palmer, L. M. (1985) Viable but non-culturable *Vibrio cholerae* and related pathogens in the environment: implications for release of genetically engineered microorganisms. *Bio/Technology* **3**, 817-820.
36. Ravel, J., Knight, I. T., Monahan, C. E., Hill, R. T., and Colwell, R. R. (1995) Temperature-Induced recovery of *Vibrio cholerae* from the viable but nonculturable state - growth or resuscitation. *Microbiol-UK* **141**, 377-383.
37. MacDonell, M. and Hood, M. (1982) Isolation and characterization of ultramicrobacteria from a gulf coast estuary. *Appl Env Microbiol* **43**, 566-571.
38. Jiang, X. P. and Chai, T. J. (1996) Survival of *Vibrio parahaemolyticus* at low temperatures under starvation conditions and subsequent resuscitation of viable, nonculturable cells. *Appl Env Microbiol* **62**, 1300-1305.
39. Wai, S. N., Moriya, T., Kondo, K., Misumi, H., and Amako, K. (1996) Resuscitation of *Vibrio cholerae* O1 strain tsi4 from a viable but nonculturable state by heat shock. *FEMS Microbiol Lett* **136**, 187-191.
40. Colwell, R. R., Brayton, P., Herrington, D., Tall, B., Huq, A., and Levine, M. M. (1996) Viable but nonculturable *Vibrio cholerae*-O1 revert to a cultivable state in the human intestine. *World J Microbiol Biotechnol* **12**, 28-31.
41. Oliver, J. D. and Bockian, R. (1995) *In vivo* resuscitation, and virulence towards mice, of viable but nonculturable cells of *Vibrio vulnificus. Appl Env Microbiol* **61**, 2620-2623.

42. Oliver, J. D., Hite, F., McDougald, D., andon, N. L., and Simpson, L. M. (1995) Entry into, and resuscitation from, the viable but nonculturable state by *Vibrio vulnificus* in an estuarine environment. *Appl Env Microbiol* **61**, 2624-2630.
43. Weichart, D. and Kjelleberg, S. (1996) Stress resistance and recovery potential of culturable and viable but nonculturable cells of *Vibrio vulnificus*. *Microbiol-UK* **142**, 845-853.
44. Romalde, J. L., Barja, J. L., Magarinos, B., and Toranzo, A. E. (1994) Starvation survival processes of the bacterial fish pathogen *Yers ia ruckeri*. *Syst Appl Microbiol* **17**, 161-168.

MONITORING THE BACTERIAL NEUSTON

VLADIMIR IVANOV
Ukrainian National University,
60 Vladimirskaya St.,Kiev 252017 ,Ukraine

Abstract

Bacterial neuston is a microbial cenosis of the biofilm on the air-water interphase. The concentration of hydrophobic and surface-active substances and bacterial cells within the film of the bacterial neuston may be some thousand times more than in bulk water. There are many cases when the biological hazards and water quality of natural aquatic systems is determined by the biodiversity of the bacterioneuston. Biodiversity of bacterioneuston includes the distribution of bacterial specia, genetic variants of one species, cells within the different phases of cell cycle, or different stages of the life cycle. Specific accumulation of specia, genetic variants, or cells of different age is determined by the hydrophobicity of their cell surface. For example, R- and S- genetic variants of bacteria differ significantly by the metabolic activity, pathogenic properties or resistance to unfavorable environmental factors. R-variants of *Bacillus licheniformis, B. thuringiensis,* and *Streptococcus lactis* are accumulated in the bacterioneuston. Cell surface hydrophobicity of R-variants was from 3 to 5 times more than one of S-variants. It was shown also that the bacterial cells from the beginning of the cell cycle are accumulated in the surface film of water. The hydrophobicity of cell surface is not stable character. The experimental data show that after 7 passages from the bacterial biofilm, the hydrophobicity of the cell surface of *Bacillus thuringiensis* and *Pseudomonas putida* is increased from 10-20% to 60-70%. Specificities of bacterioneuston's biodiversity can be applied in monitoring, control and bioremediation of aquatic systems. Accumulation of anabiotic, resistant or pathogenic variants in the bacterioneuston must be accounted in the monitoring of biohazards. This accumulation may also be used for the diminishing of risk from infections by the application the specifically selected antibiotic-producing bacteria with high hydrophobicity of cell surface into the natural aquatic system. Such biodisinfection effects were demonstrated by the application of some *Bacillus spp.* for the treatment of wastewaters containing pathogenic bacteria.

Key words: bacteria, neuston, biofilm, air-water interphase , biodiversity , monitoring.

Interaction between the bacterial cell and air-water interphase. Bacterial neuston is a microbial cenosis of the biofilm of the air-water interphase. The thickness of this biofilm is usually from 10 to 100 micrometers. It is a “biotop” where the hydrophobic and surface-active substances or particles are concentrated [6,15]. The surface of the

P.J. Stopa and M.A. Bartoszcze (eds.),
Rapid Methods for Analysis of Biological Materials in the Environment, 67–72.

microbial cell has hydrophobic sites, negatively charged, and positively charged sites, which determine the non-specific interaction between the bacterial cell and water-air interphase.

The cell structures that interact with the air-water interphase are the cell wall, capsule, fibrils, fimbriae, pili, and flagella. The physico-chemical parameters of the cell surface which determine this interaction are the hydrophobicity of cell surface, surface charge (zeta-potential), and surface free energy [1,15].

Hydrophobicity of the cell surface. The main physico-chemical parameter of the cell surface, which determines the interaction of a cell with the air-water interphase, is the hydrophobicity of cell surface. This parameter is correlated with the cell flotability, i.e., and the ability of the cells to adhere to the surface of the gas bubble and to move with the bubbles towards the gas - water interphase [10,11]. There are many methods to assess the hydrophobicity of the bacterial cell [1,2,4,7,15]. The simplest method employed, usually for the assessment of bacterial cell surface hydrophobicity is the vortexing of bacterial suspension with the hydrocarbon (hexadecane) and the counting of the number of cells removed by the hydrocarbon from the water [9]. Average hydrophobicity (H) of the cells is calculated by the following equation: $H = (No - N_f)/No$ (1), where No and N_f are the cell concentration in the water suspension before and after vortexing.

The rate of flotation of the cells (F) which correlated with the hydrophobicity of the cells [10] is determined by the equation: $F = \Delta \ln N/\Delta t$ (2), where N is the cell concentration in the suspension and t is the time .

The diversity of the cell surface hydrophobicity among bacterial specia is great. This parameter is variated from 0 to 90%. The examples of bacterial cell surface hydrophobicity are shown in Table 1.

TABLE 1. Examples of bacterial cell surface hydrophobicity. The data of different authors were taken from [15].

Species	Hydrophobicity,%
Escherichia coli	0
Pseudomonas aeruginosa	3
Bacillus cereus	
Vegetative cells	4
Spores	64
Bacillus thuringiensis	
Vegetative cells	4
Spores	62
Alcaligenes paradoxus	from 19 to 54 (unstable)
Pseudomonas putida	from 11 to 72 (unstable)
Methylobacterium sp.	from 8 to 57 (unstable)
Acinetobacter calcoaceticus	91
Staphylococcus aureus	92

This parameter does not have a stable genetic character. After 7 passages from the bacterial biofilm the hydrophobicity of cell surface of *Bacillus thuringiensis* and *Pseudomonas putida* was increased from 10-20% to 60-70% [14].

Sampling of bacterial neuston. The adhesion of the cells on a microscope slide, microbiological loop, or hydrophobic membrane may be used for the sampling of bacterial cells from the neustonic biofilm [8,15]. The biofilm obtained on the microscopic slide usually has three-layered structure [17]: the upper layer is the bacterial one, the middle layer is the layer of hydrophobic substance(s) and the bottom layer is the most concentrated layer of bacterial cells.

Bacterial accumulation in the neuston. The concentration of hydrophobic substances and bacterial cells within the bacterial neuston may be some thousand times more than in the water bulk. The examples of cell accumulation in the bacterial neuston are shown in Table 2.

TABLE 2. Examples of bacterial accumulation in the neuston (the data of different researchers were taken from [15]).

Microorganisms	Ratio of cell number in the bacterioneuston and in the bulk of water	Site of sampling
Total bacterial count	100-1000	Delta of Volga river
Total number of heterotrophic bacteria	80	Lake water
Total bacterial number	80% of total bacterial number in the water	
Pseudomonas spp.	65-85 % from total number pseudomonads in the sea water	Sea water
Pathogenic strains of Escherichia coli, Klebsiella pneumoniae, Pseudomonas aeruginosa , Salmonella sp., Staphylococcus aureus, Legionella pneumophila	10 –1000	Fresh and sea water
Mycobacterium spp.	100% of total number of mycobacterial cells	Liquid medium
Caulobacter spp.	10- 50	Swamp water

Biodiversity of bacterial neuston .There are many cases when the biological hazards and water quality of natural aquatic systems is determined by the biodiversity of bacterial neuston. Biodiversity includes the distribution of bacterial specia; genetic variants of one species; cells within the different phases of cell cycle, or different stages of the life cycle (Table 3).

TABLE 3. Biospecificity of bacterial neuston (the data of different researchers were taken from [15]).

Specificity	*Example*
Accumulation of R-forms	Salmonella typhimurium Bacillus licheniformis Streptococcus lactis
Accumulation of spores	Bacillus spp.
Accumulation of cells with fimbria	Salmonella typhimurium Acinetobacter calcoaceticus
Accumulation of cells with pili	Escherichia coli
Accumulation of pigmented variants	Oil-degrading bacteria
Accumulation of cells within B-period of cell cycle	Bacillus thuringiensis Bacillus megaterium Bacillus licheniformis
Accumulation of pathogenic variants	Mycobacterium tuberculosis Escherichia coli Pseudomonas aeruginosa Staphylococcus aureus Legionella pneumophila Salmonella sp. Streptococcus faecalis
Accumulation of starving cells	Pseudomonas sp. Vibrio sp.

Specific accumulation of specia, genetic variants, cells of different age is determined by the hydrophobicity of cell surface [2-5,12,13,16]. Cells of neuston are smaller ones than in the bulk of water [5,16]. They are the cells in B-period of DNA replication cycle. R- and S- genetic variants of bacteria differ significantly by the metabolic activity, pathogenic properties or resistance to unfavorable environmental factors. R-variants of *Bacillus licheniformis, B. thuringiensis,* and *Streptococcus lactis* are accumulated in the bacterioneuston [13]. Cell surface hydrophobicity of R-variants was from 3 to 5 times more than one of S-variants (Table 4).

TABLE 4. Hydrophobicity of R- and S- variants of bacterial specia [13].

Microorganism	*Genetic variant*	*Hydrophobicity,%*
Streptococcus lactis	R S	50 10
Bacillus licheniformis	R S	22 7
Bacillus thuringiensis	R S	26 7

The hydrophobicity of cell surface and UV-resistance of R- and S- variants are correlating parameters [13]. It means that the habitants of bacterial neuston have specific and correlated adaptations. These adaptations may be as following: hydrophobicity of cell surface; UV-resistance of cells; resistance to the heavy metals; resistance to the oxygen- induced damages; excretion of lipolytic enzymes; excretion of biosurfactants; synthesis of antibiotics; synthesis of protective pigments; ability to flocculate.

Applications in the monitoring of biohazards. Specificities of the bacterial neuston's biodiversity can be applied in monitoring, control and bioremediation of aquatic systems [17-20]. Accumulation of anabiotic, resistant or pathogenic variants in bacterial neuston may be used in the monitoring and diminishing of risk from pollution or biohazards of water. New way may be the application of selected antibiotic-producing bacteria with high hydrophobicity of cell surface into the natural aquatic system. The following calculations demonstrate the efficiency of such disinfection of bacterial neuston. If the concentration of antibiotic-producing bacteria to remove the biohazards from the water is 10^9 cells /l, the quantity of biomass needed for disinfection of 10,000 m^3 of water (water mirror is 1 ha) is 10^{16} cells or 10,000 g of dry biomass. The quantity of biomass needed to remove the biohazards from the bacterial neuston is only 10^{12} cells or 1 g of dry biomass.

The bacteria-disinfectants of bacterial neuston may be the antibiotic-producing strains of *Bacillus subtilis, Bacillus licheniformis, Pseudomonas fluorescens, Pseudomonas putida,* and some lactic-acid bacteria photosynthetic bacteria. The effect was demonstrated by the application of selected hydrophobic variants of *Bacillus sp.* (application dose was 10^6 cells/ml) for the treatment of wastewaters containing pathogenic bacteria (Table 5).

TABLE 5. Elimination of pathogenic bacteria in the waste water treatment facility by the application of hydrophobic strain of *Bacillus sp.* [17].

Detected microorganisms	CFU /ml in control	CFU/ml after 24 hours of experiment	CFU/ml after 48 hours of experiment
Staphylococcus sp.	$1.2x10^6$	$3.2x10^4$	< 10
Shigella sp.	$3.1x10^4$	$2.7x10^4$	<10

General Conclusions

The bacterial neuston must be monitored as the specific biotop of biological hazards. The biohazards of bacterial neuston may be monitored and treated specifically than the bulk of water.

References

1. Busscher, H. J., Handley, P. S., Rouxhet, P. G., Hesketh, L. M., and Mei, H. C. (1991) The relationship between structural and physicochemical surface properties of tufted Streptococcus sanguis strains, in: Nava Mozes, Pauline S. Handley, Henk J. Busscher, and Paul G. Rouxhet (eds), *Microbial cell surface analysis,* VCH Publishers, Inc., N.Y., pp.317-338.
2. Ivanov, V. N. and Ugodchikov, G. A. (1984) Cell cycle of microorganisms and heterogeneity of their populations, Naukova Dumka, Kiev (in Russian).
3. Ivanov, V. N., Pindrus, A. A., and Stabnikova, E. V. (1987) Fractionation of yeast cells of different species by flotation, *Applied Biochemistry and Microbiology (Moscow)* **23**, 706- 713 (in Russian).
4. Ivanov, V. N. (1990) *Exo- and endotrophy of a cell,* Naukova dumka, Kiev (in Russian).
5. Ivanov, V. N., Svechnikova, T. A., Stabnikova, E. V., and Gregirchak, N. N. (1995) Structure of cell cycle and age structure of bacterial populations, *Mikrobiologichesky Zhurnal (Kiev)* **57**, 3-11.
6. Marshall, K. C. (1991) The importance of studying microbial cell surfaces, in: Nava Mozes, Pauline S. Handley, Henk J. Busscher, and Paul G. Rouxhet (eds.), *Microbial cell surface analysis*, VCH Publishers, Inc., N.Y., pp. 3- 19.

7. Mei, H. C., Rosenberg, M., and Busscher, H. J. (1991) Assessment of microbial cell surface hydrophobicity, in: Nava Mozes, Pauline S. Handley, Henk J. Busscher, and Paul G. Rouxhet (eds.), *Microbial cell surface analysis*, VCH Publishers, Inc., N.Y., pp. 263-287.
8. Norkrans, B. (1980) Surface microlayers in aquatic environments, *Advances in Microbial Ecology* **4**, 51-85.
9. Rosenberg, M. (1984) Bacterial adherence to hydrocarbons: a useful technique for studying cell surface hydrophobicity, *FEMS Microbiol. Lett.* **22**, 289-295.
10. Stabnikova, E. V., Polyatevich, E. V., and Ivanov, V. N. (1989) Correlation between the hydrophobicity of surface of yeast cells and their flotability, *Mikrobiologichesky Zhurnal (Kiev)* **51**, 28-21 (in Russian).
11. Stabnikova, E. V., Polyatevich, E. V., Pindrus, A. A., and Ivanov, V. N. (1990) Flotability, hydrophobicity and viability of cells in yeast of different age, *Mikrobiologichesky Zhurnal (Kiev)* **52**, 46-50 (in Russian).
12. Stabnikova, E. V., Gregirchak, N. N., and Taranenko, T. O. (1991) Peculiarities of flotation of bacillar cells and spores, *Mikrobiologichesky Zhurnal (Kiev)* **53**, 38-42 (in Russian).
13. Stabnikova, E. V., Milko, E. S., and Gregirchak N. N. (1991) Flotative fractionation of cells of R- and S-variants of some bacteria, *Mikrobiologichesky Zhurnal (Kiev)* **53**, 52-54 (in Russian).
14. Stabnikova, E. V., Gregirchak, N. N., Taranenko, T. O., and Nudga, A.Yu. (1991) Selection of neuston forms of bacteria, *Mikrobiologichesky Zhurnal (Kiev)* **53**, 33-37 (in Russian).
15. Stabnikova, E. V., Gordienko, A. S., Ksenofontov, B. S., Poberezhniy, V. Y., Stavskaya T. Z., and Ivanov, V. N. (1991) *Interaction between the cells and water-air interface*, Naukova Dumka, Kiev (in Russian).
16. Stabnikova, E. V., Gregirchak, N. N., and Ivanov V. N. (1992) Age specificity of the bacilli cell interaction with water-air interface, *Mikrobiologia (Moscow)* **61**:110-114 (in Russian).
17. Stabnikova, E. V., Ivanov, V. N., Gregirchak, N. N., and Dulgerov, A. N. (1993) Utilization of neuston forms of bacilli for treatment and decontamination of surface waters, *Mikrobiologichesky Zhurnal (Kiev)* **55**, 88-94 (in Russian).
18. Stabnikova, E. V., Selezneva, M. V., and Dulgerov, A. N. (1994) Screening of microbial component for the soil bioremediation, *Mikrobiologichesky Zhurnal (Kiev)* **56**, 86-87 (in Russian).
19. Stabnikova, E. V., Selezneva, M. V., Ivanov, V. N., and Reva, O. I. (1995) Theoretical and experimental screening of microbial component of biopreparation using for bioremediation of soil contaminated with oil, *Applied Biochemistry and Microbiology (Moscow)* 31:534- 540 (in Russian).
20. Stabnikova, E. V., Selezneva, M. V., and Ivanov, V. N. (1996) Biopreparation "Lestan" for the soil bioremediation, *Applied Biochemistry and Microbiology (Moscow)* 32:219- 224 (in Russian).

THE PAST, PRESENT AND FUTURE OF LUMINOMETRIC METHODS IN BIOLOGICAL DETECTION

M. BARTOSZCZE AND A. BIELAWSKA
Military Institute of Hygiene and Epidemiology, 2 Lubelska Street, 24-100 Pulawy, Poland

Bioluminescent ATP is not a new method. It has been used for many years in breweries as well as in the pharmaceutical industry. However, widespread use of this method has previously been limited by time-consuming procedures, the size of equipment and instability of the reagents used. Only now are we going to see the real benefits of bioluminescent techniques.

Renewed interest in bioluminescence has come with new and improved procedures and modern equipment using new technology such as photo detectors and electronics devices. Additionally, progress has been made in bioluminescence because private as well as public organizations, such as the National Aeronautics and Space Agency, have shown great interest in its continued development. NASA especially, has found bioluminescence to be a tremendous help in its search for life on other planets.

Many public and private companies first become interested in this process in order to facilitate faster and more efficient hygienic control during production.

Currently, it is safe to say that bioluminescence is undergoing a renaissance of sorts. Its popularity is due to its high sensitivity and its speed in obtaining accurate results, and its ability to be used under difficult conditions in the field. However, the greatest factor contributing to its renewed popularity is its versatility.

The term "bioluminescence" comes from the Greek "bios" which means life, and from the Latin "lumen" which means light. Bioluminescence can be observed in a number of natural organisms including bacteria, mold, yeast, fungi, insects, etc.

The emission of light by these organisms is the result of an internal chemical reaction. The mechanics involved in this emission of light are based on the oxidation of luciferin in the presence of the enzyme luciferase. After the oxidation process the oxiluciferin returns to its normal state. At this moment ATP becomes AMP and during this transformation there is an emission of light. The light emitted is proportional to the quantity of ATP present in the biological material.

P.J. Stopa and M.A. Bartoszcze (eds.),
Rapid Methods for Analysis of Biological Materials in the Environment, 73–77.

The quantity of ATP in bacteria varies, but approximately the quantity is about 10 per gram. This means that within 1000 bacterial cells there is about 1 pg of ATP. The level of ATP in the bacterial cells fluctuates and depends on the kind of bacteria, the growth phase and environmental factors.

Kinetics of the Bioluminescent Reaction

The reaction process is extremely fast. From the moment that LL is added to a sample containing ATP and Mg, there is an intense emission of light, which reaches its peak within 3 to 5 seconds. A decrease in the intensity of the emission can be observed within 10 seconds. This drastic decrease in intensity is due to the reaction's final product, pyrophosphate, which serves to inhibit light production.

Many factors influence the bioluminescent reaction, such as temperature (optimal being 20-25 degrees Celsius) and pH (optimal 7.8). In such conditions, yellow-green light is emitted. The length of light emitted is about 562 nanometers.

Detergents, heavy metals and disinfectants influence the reaction process by inhibiting the emission of light. Care must be taken to account for these inhibitors during the interpretation of final results.

Although vegetative forms of bacterial ATP are easily detected, it is more difficult to detect ATP in bacterial spores. These difficulties are based on the different construction of spore walls, which resist the release of ATP from the spore cells. Additionally, there is a smaller quantity of ATP in spores.

In our experiments we used a preliminary heat shock on anthrax spores, after which we applied a special medium enriched with L-alanine and adenosine. During these experiments, we observed a dramatic emission of light five minutes after incubation. This procedure can be effectively used not only for spore detection, but also for the examination of the germination process in the spores.

Even though bioluminescence is a non-specific method, there are procedures combining bioluminescence and other techniques, which result in ideal methods of identifying specific bacterial agents.

In our experiments we used paramagnetic beads coated with salmonella antibodies. In a mixture of different bacteria, these antibodies captured only the salmonella bacteria, enabling the beads to effectively isolate this bacteria from others in the sample. After washing, the presence of salmonella on the surface of the beads could be measured using the bioluminescent method.

Thanks to the combined procedures the results of this process can be obtained within 20 minutes.

Also very promising are methods involving genetic engineering techniques using the luciferase gene. This gene as identified as that being responsible for bioluminescence and has also been successfully cloned. This L gene was inserted into the nucleic acid of the phage, which was specific for salmonella typhimurium. After infection, bacteria begin replication and its is during the actual transcription process that there is an intense emission of light.

Using phage P-22, which is specific for salmonella typhimurium, it was possible to detect 100 bacteria within 30 minutes. This work promises to be valuable in helping to prevent human infection through the contamination of food products.

Bioluminescent Immunoassays

To determine the source of biological contamination, we used different kinds of conjugates in combination with different enzymes such as horseradish peroxidase, alkaline phophatase and luciferase. Luciferase is a very attractive enzyme because of its unique characteristics. However, during the chemical staining of luciferase there is a decrease of enzyme activity which as a negative effect on the sensitivity of the detection method.

It is possible to avoid this type of difficulty using the protein hybrid, Protein A/luciferase. Protein A is known to bind the immunogloblin Fc fragment. The sensitivity of this method for IgG is about 10 mol.

Bioluminescence is of course a method for the detection of ATP. It is clear that using this method it is not possible to detect viruses because they do not contain ATP. However, it has been shown that it will be possible to detect viruses if we insert the Luciferase gene into the gene of a given virus. This was proved using pseudo-rabies virus as a model. Expression of the luciferase gene occurs during the replication process of the virus and can be measured by luminometers. Of course, after the addition of Luciferin and ATP. The emission of light is recordable on photosensitive film. A similar investigation was conducted using copies of cDNA luciferase inserted into the gene of the Vaccinnii virus. These techniques may be useful in the investigation of virus replication and pathogenesis of a viral infection.

Biosensors

Another perspective in the development of bioluminescence is in the area of biosensors. Enzymes, which can be used in bioluminescent reactions, are immobilized on the surface of activated polyamide membranes. The results enable the detection of ATP presence in examined samples and the sensitivity of this test is about 1 pg.

An additional improvement includes the automatization of the process, which gives us the possibility of continued measurement throughout an ongoing investigation. As a result of this research on ATP and NADH, we have seen the development of technology, which has led to high sensitivity and consistency in performance.

Additionally, new self-contained biosensors have been developed using two different enzymes at the same time.

In this type of investigation different membranes were used in the immobilization stage: collagen films and polyvinyl alcohol previously activated by glutaraldehyde and others.

Bioluminescent methods were used in a number of ways: in the medical services, food production and environmental monitoring. A typical example of its use in medicine is the adaptation of this method in urinalysis techniques for the detection of the amount of bacteria in urine samples (Bacteriuria).

This method was very effective when compared to classical methods of detecting microbiological growths. Using this method it is possible not only to detect aerobic bacteria, but also anaerobic bacteria in the samples. This method was also very useful for the detection of Mycobacterium tuberculosis growths, and it was possible to obtain the results within five days. The classical methods took thirty days. The growth of this type of bacteria is very slow.

Bioluminescent methods have also been used to detect the living form of Pneumocystis Carrinini. Very promising was work using this technique to detect bacteria within cerebra-spinal fluid. Thanks to this development, it is possible to differentiate between viral and bacterial forms of encephalitis, which is of course essential in treatment. Bioluminescent methods have also been used in hystopathological examinations. Additionally, very interesting work has been conducted using bioluminescent methods for protein blotting and hybridization methods.

Bioluminescent methods have also proved their usefulness in hospital operation wards where the rapid detection of microbiological contamination is essential to the safety of the patient. Using this method it is possible to examine equipment and even the air.

Bioluminescent methods may be useful also in the growth of bacteria and the dynamics of their growth in a culture. These methods have opened new possibilities in the estimation of a given antibiotic's potency as well as a bacteria's sensitivity to a drug. In the case of emerging or basically unknown diseases, time is of the utmost importance, and it is here that luminometric methods have their greatest benefit. These methods enable a doctor to quickly determine the effectiveness of an antibiotic on a given pathogen. The speed of an accurate diagnosis could be the difference between the life and death of a patient, or the life and death of thousands in the midst of an epidemic.

Other uses of these methods include food production, for example, in the examination of meat and dairy products. In our own experiments we showed that the method was very accurate in the assessment of milk contamination after the addition of helator reagents.

Many articles have also been published noting the use of bioluminescent methods the analysis of food products, which will play an important role in the implementation of the HCCP system.

To conclude, I can only emphasize that when considering the rapid progress of bioluminescent systems, as well as developments in genetic engineering and biosensor techniques, the future of bioluminescence seems to have no limits in its application.

ENVIRONMENTAL APPLICATIONS OF THE NRL FLOW IMMUNOSENSOR

ANNE W. KUSTERBECK[1], PAUL R. GAUGER[2], and PAUL T. CHARLES[1]
[1]Center for Bio/Molecular Science and Engineering, Biosensors and Biomaterials Laboratory, Code 6910, Naval Research Laboratory, 4555 Overlook Ave., S.W. Washington, DC 20375-5348
[2]Geo-Centers, Inc., Fort Washington, MD 20744

Abstract

An antibody-based biosensor developed at the Naval Research Laboratory (NRL) is being tested at several United States Environmental Protection Agency (USEPA) remediation sites to examine the utility of the technology for monitoring explosives levels in groundwater and soil. Using antibodies as the recognition element and fluorescently labeled explosive analogs as reporters of binding events, the biosensor can detect contamination at the sub-parts per billion level in under two minutes. A portable version of the device, known as the FAST 2000, has been engineered by Research International, Woodinville, WA, for convenient use in rapid on-site evaluation and characterization of contaminated areas. Alternatively, the device can be used to follow cleanup progress on a routine basis or to monitor air samples for the presence of explosives or other small-molecular-weight contaminants.

1. Introduction

Environmental security issues continue to be of major concern for military and civilian scientists working at United States or foreign installations. In many cases, a rapid, accurate means of monitoring areas for the presence of explosives or other small-molecular-weight chemicals is needed. We have developed a fieldable sensor that is lightweight, battery operated, and able to detect explosives at the sub-parts per billion [1,2]. The instrument has been successfully demonstrated in field trials at Umatilla Army Depot, Oregon and SUBASE Bangor, Washington, U.S. A. [3,4] and is now being tested at additional test sites in validation studies designed to obtain EPA certification of the sensor for field use. An overview of this work and applications of the technology to rapid monitoring of the environment are provided here.

P.J. Stopa and M.A. Bartoszcze (eds.),
***Rapid Methods for Analysis of Biological Materials in the Environment*, 79–85.**

2. Biosensor Description

In general, biosensors can be defined as small, portable, analytical devices that use the recognition of a biomolecule (antibody, enzyme, receptor) and some form of signal transduction (optical, electrical) to detect chemical or biological materials [5]. Biosensors provide the advantages of specific response and the sensitivity needed to measure concentrations of the target analyte within a complex sample.

Several limitations have previously prevented widespread use of biosensors in the environmental arena. These include the necessity that, unlike many analytical techniques, the samples must be tested in water, that the compounds to be detected may breakdown rapidly or may not be available for binding, and that the biological components of the system may not be stable enough for field conditions. However, recent advances in biochemistry, microfabrication and optical technology have allowed several biosensors to become practical for field use, with instruments being developed that are becoming small, portable, and rugged [6].

2.1. BIOSENSOR DESIGN AND OPERATIONS

In the NRL biosensor, the biological element is an antibody that is specific for a small molecule, such as an explosive. Similar to ELISA's, the antibodies are bound to a solid substrate (plastic beads, membrane capillaries) and allowed to bind fluorescently labeled antigens. When a sample is injected, a proportional number of fluorescent antigens are displaced by the sample and measured by a fluorimeter. Assays for explosives, pesticides and drugs of abuse have been developed for the NRL immunosensor. Figure 1 shows a typical dose response curve for a common military explosive, hexahydro-1,3,5-trinitro-1,3,5-triazine (RDX). Recent work has demonstrated that sensitivity levels can be achieved in the low picogram per milliliter (pg/mL) levels or part-per trillion (ppt)[3].

FIGURE 1. Dose response curve for RDX using the NRL biosensor. Increasing concentrations of RDX in 100 μl samples were injected and the integrated fluorescence areas were measured. Each point represents the average of three injections with a calculated standard error of the mean.

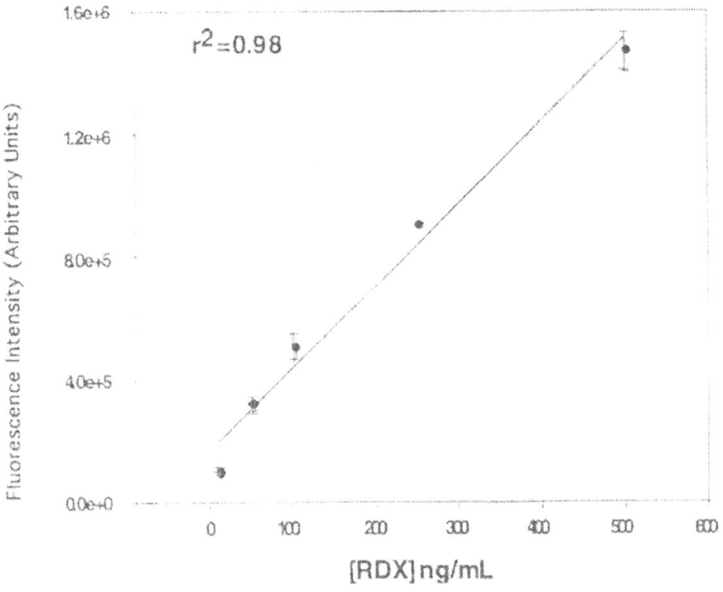

2.2. PORTABLE NRL BIOSENSOR

The field-portable version of the biosensor is the FAST 2000, engineered by Research International, Woodinville, WA. Approximately the size of a modem, the system is battery operated with electronics, fluidics, hardware and a fluorescence detector integrated with the laptop computer via a PC/MCIA card. Data analysis is done by the laptop computer (Figure 2).

To perform an assay, the disposable cartridge with the antibody/antigen complex is inserted into the instrument, buffer (10 mM sodium monophosphate, 2.5% ethanol and 0.01%Tween) is pumped at a set flow rate through the system. The baseline is monitored for approximately 15-30 minutes until the background fluorescence intensity is stabilized to a constant rate of change. At that point, 150 μL injections of standards and unknowns are made and the total increase in fluorescence is measured. Total assay times are less than 5 minutes per sample. A computer connected on-line with the instrument records and integrates the displacement of labeled-antigen as a positive change in background fluorescence. For quantitation of unknown samples, the integrated peak areas are compared to a range of known standards that are injected intermittently during the assay. Over 50 individual samples can be analyzed using a single cartridge, depending on the concentration of analyte in the sample.

FIGURE 2. Schematic of FAST 2000 Instrument engineered by Research International using NRL displacement assay technology.

3. Applications

Because antibodies exist for a wide variety of potential biological hazards, the system can be adapted to fit the specific detection need. To date, the NRL biosensor has been used in several detection scenarios. Specific applications have been demonstrated in on-site detection of explosives in groundwater and soil as well as air monitoring in areas suspected to contain explosives (cargo holds, land mine regions).

3.1. ENVIRONMENTAL MONITORING

The major research emphasis for the past few years has been to adapt the NRL biosensor for environmental monitoring. In one test of the sensor performance, soil samples containing explosives were first extracted with acetone. Dilutions of the acetone extract were made and biosensor determinations of RDX and TNT levels were made. As shown in Table 1, this studied demonstrated that the sensor performed well with environmental samples and exhibited negligible matrix effects. Results were comparable to analyses done using High Performance Liquid Chromatography (HPLC).

TABLE 1. Comparison of HPLC and NRL Biosensor Measurements of Explosives in Soil Samples from Umatilla Army Depot, Hermiston, OR.

Sample ID	HPLC TNT mg/g	HPLC RDX mg/g	NRL TNT mg/g	NRL RDX mg/g
G18-L6	15	4	0	0
G10-L3	21	10	39	0
G51-L1	25	24	0	19
HS-04	44	29	39	0
G18-L3	51	36	55	26
G21-L2	130	55	19	40
G16-L2	190	110	86	99
G18-L1	280	280	207	250
G17-L1	340	480	219	503
G55-X	1400	2800	896	1860

3.2. AIR SAMPLING AND DETECTION

When integrated with air samplers, the technology also has applications in the detection of small molecules in confined areas (for example, cargo holds). For example, the NRL biosensor performance is currently being tested utilizing the collection capabilities of a SCAEP (Space Charged Atomizing Electrostatic Precipitation) air collector from Team Technologies, Massachusetts, U.S.A. The SCAEP system collects gaseous and particulate impurities from the air through a glass tube that converges with a finely atomized liquid spray (SCAEP buffer). The charged liquid spray is electrostatically attracted to the walls of the glass tube where it precipitates, flows downward by gravity and is collected in a small glass vial. In tests with the collector and the NRL biosensor,

2,4,6-trinitrotoluene (TNT) was collected from explosives patches provided by the FAA. As shown in Figure 3, a detailed time course ranging from a quick wave (seconds) of the patch material over the air collection port up to 30 minutes was undertaken. Time intervals of 1, 2, 4, 5, 10, 20, and 30 minutes collection times were performed. Subsequent analyses of the samples were performed by the biosensor. Initial results revealed no significant response over background with the quick wave (seconds) of the patch material. However, the 1 min time interval showed a slight response over background. The most significant increase occurred with the 2 min interval. A constant increase in total TNT (μg) collected was also observed, with the exception of the 5 min time point.

FIGURE 3. Time course study using the SCAEP air collector and the NRL Biosensor to detect TNT.

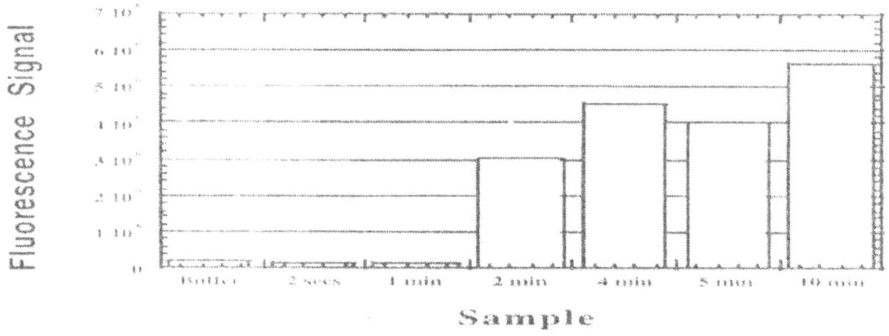

4. Conclusions

The NRL biosensor has been demonstrated successfully in field trials at U.S. military bases and is effective in determining concentration levels of explosives contamination in complex environmental water or soil samples. When integrated with air collector, explosives can be detected in samples collected from solid surfaces or small particulates present in the environment. The FAST 2000 instrument, which incorporates the NRL biosensor concept, provides a convenient way to analyze samples on-site within a few minutes. Future efforts will focus on expanding the NRL biosensor capabilities to multianalyte detection.

5. Acknowledgments

The authors would to thank Eddie Chang for a critical review of the manuscript. Funding for this works comes from the Office of Naval Research, the Environmental Security Technology Certification Program, and the Federal Aviation Administration. The views expressed here are those of the authors and do not represent those of the U.S. Navy, the Department of Defense or the U.S. Government.

6. References

1. Bart, John C., L. L. Judd, and A. W. Kusterbeck (1997) Environmental immunoassays for the explosive RDX using a fluorescent dye-labeled antigen and the continuous flow immunosensor. Sensors and ActuatorsB 38-39 (1997), 411-418.

2. Bart, John C., L. L. Judd, K. E. Hoffman, A. M. Wilkins, P. T. Charles, and A. W. Kusterbeck (1997) Application of a portable immunosensor to detect the explosives TNT and RDX in groundwater samples. Environmental Science & Technology, 31, 1505-1511.

3. Narang, U., P. R. Gauger, and F. S. Ligler (1997) Capillary-based displacement flow immunosensor. Anal. Chem., 67, (10), 1961-1964.

4. Narang, U., P. R. Gauger, and F. S. Ligler (1997) A displacement flow immunosensor for explosive detection using microcapillaries. Anal. Chem., 69, (14), 2779-2785.

5. Marco, M. –P., S. Gee, and B. D. Hammock (1995) Immunochemical techniques for environmental analysis, Trends in Anal. Chem., 14, (7), 341-350.

6. Rogers, K. R. and C. L. Gerlach (1997) Environmental Biosensors: A Status Report, Environmental Science & Technology, 30(11), 486A-491A.

CAPILLARY ELECTROPHORESIS/NUCLEIC ACID PROBE IDENTIFICATION OF BIOLOGICAL WARFARE AGENT SIMULANTS

CAMILLE A. BOULET, GEOFFREY HUNG, DOUGLAS E. BADER[1];
PETER DUCK, PAUL WISHART, AND ANGELA LAI-HOW[2]
[1]*Defence Research Establishment Suffield, PO Box 4000, Medicine Hat, Alberta, Canada T1A 8K6*
[2]*ID Biomedical Corporation, 8855 Northbrook Court, Burnaby, British Columbia, Canada V5J 5J1*

1. Introduction

The goal of this research is to develop rapid, robust assays and equipment that are amenable to field detection and identification of BW agent attacks. The system must detect BW agents at extremely low concentrations with no false alarms and operate under battlefield conditions [1]. Typically high volume air samplers are required to concentrate high volumes of air into liquid samples, often by a factor of 10^6, for analysis [2]. To provide the earliest possible warning and maximum protection to personnel, the field detection and identification equipment (and associated assay chemistry) must be highly specific, ultra-sensitive, and very reliable [3].

Capillary electrophoresis provides several advantages, which are attractive for field-portable instrumentation. It has a simple, easily automated design, has very low volume requirements, and high separation efficiencies. Capillary gel electrophoresis (CGE) was selected for initial method development because it has been used for determination of oligonucleotide purity, PCR product and forensic DNA analysis.

Cycling Probe Technology (CPT) is a nucleic acid identification technology based on target nucleic acid acting as a catalyst for the conversion of chimeric probe molecules to detectable products [4]. The single-stranded (ss) DNA target serves as the catalyst for the CPT reaction. In the presence of excess chimeric probe and RNase H, the RNA portion of the resulting probe-target complex is cleaved by RNase H at 65°C. The shorter cleaved fragments dissociate from the target thereby regenerating the target DNA for further cycling. With each cycle, the released fragments accumulate and can be detected. CPT has many advantages over other existing amplification methods such as polymerase chain reaction (PCR) which make it suitable for field identification of BW agents. It is fast (≤ 30 min), simple (one probe, one enzyme, one temperature), and is not prone to false positives due to sample carry-over since the target DNA is not amplified.

P.J. Stopa and M.A. Bartoszcze (eds.),
Rapid Methods for Analysis of Biological Materials in the Environment, 87–92.

GGE with laser-induced fluorescent detection (CGE-LIF) can provide the necessary specificity for the analysis of CPT products and identification of BW agents because of its ultra-low detection limits [5,6]. CE-based instruments are being developed to operate under field conditions [7]. This paper describes the fluorescent detection of the CPT reaction product and preliminary evaluation of the CPT reaction using CGE-LIF for BW agent identification.

2. Materials and Methods

2.1. DESIGN AND CONSTRUCTION OF DRES2A CYCLING PROBE

CPT probes and assay reactions were performed by ID Biomedical (Bumaby, B.C.). Candidate probes were selected from the sequence of the division initiation gene of *Bacillus subtilis, div*IB, and accession number M31800 [8]. The DRES2A probe was selected after testing candidate probes against non-specific target (e.g., *Escherichia coli* DNA) and specific target *(Bacillus globigii* DNA). The DRES2A probe had the following construction, 5' Fluorescein-$(DNA)_8$ $(RNA)_4$ $(DNA)_{16}$-Biotin 3', and was purified by HPLC and polyacrylamide gel electrophoresis. The 5'-fluorescein label was employed for the LIF detection in this work while the 3'-biotin label was required for other studies.

2.2. CYCLING PROBE REACTIONS

CPT probe reactions were performed with either 10 fmol or 100 fmol of DRES2A probe. The assay concentrations for the synthetic target DNA, the complement of the DRES2A probe sequence, are given in the figures. In addition, two control solutions containing only probe (CI) and probe plus 0.2 μg RNase H (C2) were prepared. Reaction and control solutions were heated to 65°C for 30 min and then promptly frozen to stop the reaction.

2.3. INSTRUMENTATION

CPT analyses were performed on a Beckman P/ACE System 5010, equipped with a P/ACE System Laser Module 488 (488 nm argon ion laser), and a P/ACE LIE Detector. Data was collected, stored, and analyzed using Beckman System Gold 7.0 software, Microsoft Excel 5.Oa, and Grapher 1.28 on a Dell Optiplex 466/L.

2.4. ELECTROPHORETIC ANALYSIS

Electrophoresis of the single stranded chimeric CPT probes was performed using 37 cm Beckman eCAP ssDNA 100 Gel capillaries containing a fixed polyacrylamide gel. Beckman Gel Buffer solution, consisting of 44% Tris, 56% borate and 7 M urea, at pH 8.4, was filtered through a 0.22 μm filter and sonicated before use. A 37 cm length of eCAP ssDNA 100 capillary was installed into a Beckman P/ACE System LIF Detection Capillary Cartridge. Analyses were performed in either constant voltage or constant current modes at 7.5 kV and 10.0 μA, respectively (30°C).

2.5. SAMPLE PREPARATION

The CPT reaction solutions were thawed and an aliquot of each was diluted to provide a working volume (10 μL) for effective pressure or electrokinetic injection. The 100 fmol probe reaction solutions were diluted in a 1:10 ratio with 1 μL of original reaction solution being mixed with 9 μl of deionized distilled water. The 10 fmol probe reaction solutions were diluted in a 1:5 ratio, with 2 μl of original reaction solution being mixed with 8 μl of deionized distilled water.

3. Results and Discussion

As the goal of this work is rapid identification, a modified CGE system was used to substantially reduce the analysis time. Initial experiments with a 30 cm L_d capillary column gave a migration time for the intact DRES2A probe of 18.5 min which was considered much too long for a practical field-identification method where near real-time identification is ideal. Since the L_d with cartridge-mounted capillary columns in the usual CE operating mode is 20 cm, much shorter capillaries were required to reduce the analysis time and provide faster detection. By changing the polarity and injecting the sample electrokinetically at the cathode (the outlet tray of the Beckman autosampler), a 7 cm L_d was possible and overall analysis times were substantially reduced.

A DRES2A CPT probe standard (1 fmol/μL) was used to establish reproducible CGE conditions. Initial experiments using constant voltage showed poor reproducibility of the 28 base DNA-RNA-DNA chimeric oligonucleotide's migration time. However, the migration times were found to be more reproducible using a constant current of 10.0 μA rather than a constant voltage of 7.5 kV. With a 7 cm eCAP 100 ssDNA capillary column, the migration time for the intact DRES2A probe was 6.1 min (Fig. 1).

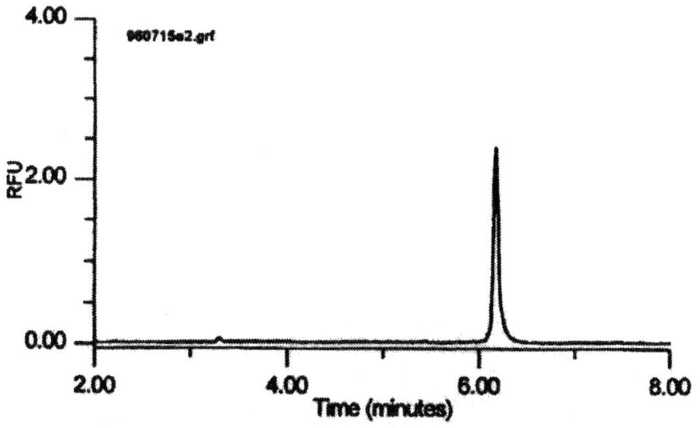

***Figure 1*. Electropherogram of DRES2A CPT probe standard (1 fmol/μL).**

The analysis of the 100 fmol probe reaction solution with synthetic target clearly demonstrated the appearance of a shorter, faster-migrating fluoresceinated product at 4.3 min (Fig. 2) which was considered to be the *5'* fluorescein-labelled fragment of the intact probe. The relative amounts of product to probe corresponded to the expected trend that increased amounts of synthetic target in the reaction solution increased the proportion of the intact probe that was detected as the product fragment. The electropherogram of the 100 fmol probe CPT reaction solution with 10^{-6} pmol of target displayed the migration time of the reaction product at 4.3 min, with a small amount of the unreacted intact probe evident at 6.1 min (Fig. 2A).

Decreasing the amount of probe from 100 fmol to 10 fmol in the reaction solution still produced a detectable amount of product for all target DNA concentrations tested (Fig. 3). As observed in the 100 fmol probe reactions, increasing the amount of synthetic target in the reaction solution increased the proportion of intact probe that was converted into cleaved product. All of the reactions essentially proceeded to completion with only trace amounts of intact probe evident.

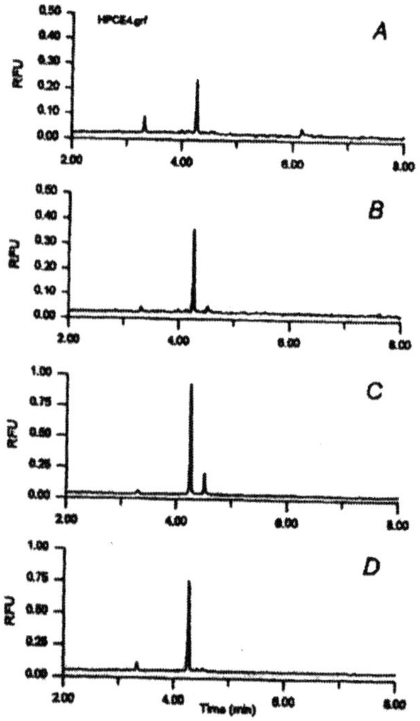

Figure 2. **Electropherogram of 100 fmol DRES2A probe, 0.2 µg RNase H CPT reaction solution with 10^{-6} (A), 10^{-5} (B), 10^{-4} (C) and 10^{-3} (D) pmol of synthetic DNA target.**

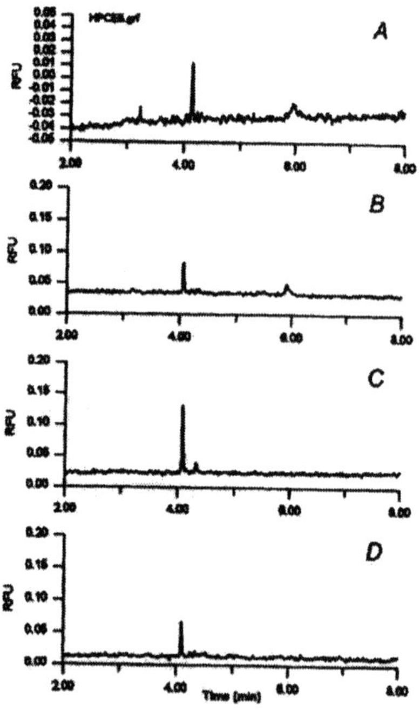

Figure 3. **Electropherogram of 10 fmol DRES2A probe, 0.2 µg RNase H CPT reaction solution with 10^{-6} (A), 10^{-5} (B), 10^{-4} (C) and 10^{-3} (D) pmol of synthetic DNA target.**

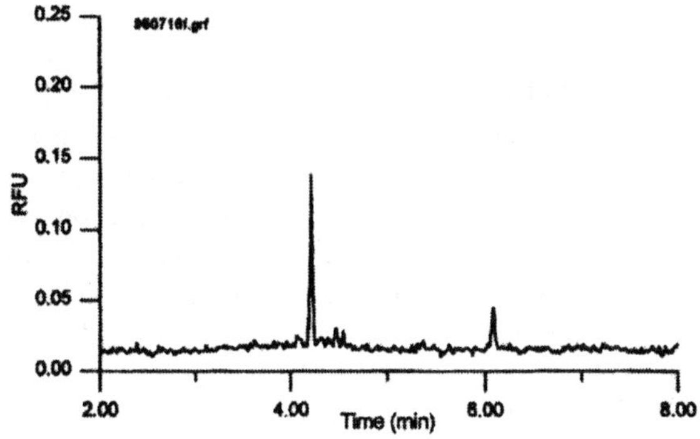

Figure 4. **Electropherogram of the CPT C2 control solution (100 fmol probe plus 0.2 µg RNase H).**

The presence of the CPT product fragment in the control solution C2 (probe + RNase H) suggests that the ribonuclease is reacting with the intact probe in the absence of the target sequence to produce probe fragments (Fig. 4). The cleaving of the probe in the presence of the ribonuclease but in the absence of the requisite target sequence may present problems in ultra-sensitive CPT reaction conditions with low amounts of extracted target DNA. Since proportionally more product was formed in the presence of target DNA than in the C2 control solution, the expected CPT reaction proceeded above the background production formation. In the absence of RNase H, the probe did not appear to undergo degradation in its ribonucleotide (Fig. 5).

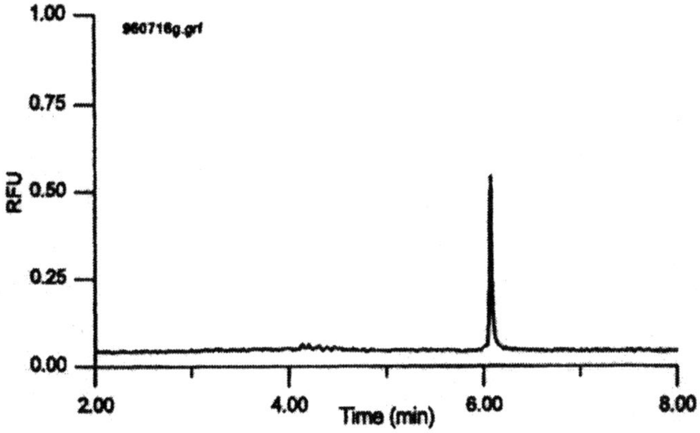

Figure 5. **Electropherogram of the CPT C1 control solution (100 fmol probe only).**

4. Conclusions

CGE-LIF provides a sensitive and effective method for the detection of CPT reaction products to detect BW agents. CGE can detect the CPT reaction products significantly faster than previously implemented methods of CPT product detection, such as light addressable potentiometric sensors (LAPS) or electrophoretic separation of radiolabelled oligonucleotides, which typically require much longer analysis times. CGE requires less sample preparation, avoids the use of radiolabelled oligonucleotides, and yields reproducible migration times that can be used for identification. This work demonstrates the detection of attomolar quantities of target DNA in under 7 min on a gel-filled 7cm capillary. The low detection limit capability of LIF allows for much earlier detection of CPT reaction products, more rapid field identification, and ultimately earlier warning of BW agent attacks. Continued development of CGE (as well as other CE methods including miniaturized instrumentation) and CPT can provide cheaper and more sensitive methods for BW agent detection.

5. References

1. For a review of the nature and threat of biological warfare see Cole. Leonard A. The Spectre of Biological Weapons, *Scientific American,* December 1996.

2. White, L. A., Hadley, D. J., Davids, D. E., and Naylor, R. *Appl Microbiol.* **29**, 335-339, *1975.*

3. Hewish, M. *Jane's International Defense Review* **3,** 30-48, 1996.

4. Duck, P., Alvarado-Urbina, G., Burdick, B., and Collier, B. *BioTechniques.* **9**, 142, 1990.

5. Xu, Yan. *Anal.Chem..,* **65** (12), 425R-433R, 1993.

6. Kuhr, W. G.; Monnig, C. A. *Anal.Chem.,* **64** (12), 389R-407R. 1992.

7. Boulet, C. A, Capillary Electrophoresis, *Proceedings of the NATO Advanced Research Workshop on Analytical Chemistry Associated with the Destruction of Chemical Weapons.* May 12-15, 1996, Brno, Czech Republic. M.J. Heyl, ed. in press, 1996.

8. Harry, E. J. and Wake, R. G.. *J. Bact.,* **171** (12), 6835. 1989.

RAPID METHOD FOR BACTERIAL COUNTING IN THE MILK BY USING BIOSENSOR BASED ON *E.coli* CELLS.

S.G. IGNATOV
State Research Center for Applied Microbiology,
Obolensk, Moscow Region, 142279 RUSSIA,

1. Introduction

The requirement for accurate analytical information is particularly marked in human health care, veterinary medicine, and environmental monitoring. Biosensors promise to provide analytically powerful and inexpensive alternatives to standard chemical and enzymatic methods used today.

Biosensors are analytical devices that respond selectively to analytes by use of a combination of a biological recognition system and a physical transducing element. The biological response is converted by the transducer to an electrical signal that is directly related to the concentration of the analyte of interest. A biosensor's recognition system may consist of one of many biological components including: enzymes, antigens, antibodies, nucleic acids, cells, cell organelles or receptor molecules. All of these biological components are capable of interacting very specifically with an analyte of interest and are what gives a biosensor its unique specificity.

All of these biological elements may be combined with various kinds of transducers, provided that the reaction of the biological element with the substrate can be monitored. The transducer types available and biological elements that have been combined with them to form biosensor are shown in Table 1.

Biosensors made from bacterial cells may have potential to provide more sophisticated sensing system containing multiple integrated molecular components, and have the advantage of allowing economical manufacturing of these complex system (1, 2). Bacterial biosensors have a number of advantages:

1) The great variety of bacteria makes them suitable for analyzing of a very wide range substrates.
2) Low cost.
3) The stability and activity for some enzymes is conferred by the natural medium of the cells and they can be immobilized by simple procedures.
4) Bacterial enzymes might be better shielded within microorganisms from interfering substances (such as heavy metals and others).

P.J. Stopa and M.A. Bartoszcze (eds.),
Rapid Methods for Analysis of Biological Materials in the Environment, 93–100.

5) Addition of exogenous cofactors is unnecessary in the case of whole organisms.

Determination of bacterial concentration in the milk is very important for food industry (3).

TABLE 1. Types of transducers and biological elements for biosensors

Biological element	Type of transduser	Examples	Referies
Enzymes Receptors Bacteria Plant and animal tissue Enzyme labeled antibody	*Amperometric*	Clark electrode	4, 5
Enzymes Receptors Bacteria Plant and animal tissue Enzyme labeled antibody	*Potentiometric*	Redox and ionoselective electrodes	5 – 7
Enzymes Bilayer lipids membranes (man made)	*Conductometric*	Pt and Au electrodes	8
Enzymes Receptors Antibodies	*Fluorescence*	Optodes	9, 10
Enzymes	*Luminiscence*	Optodes	10, 11
Receptors	*Evanescent wave*	Optodes	11
Antibodies Antigens Enzymes Nucleic acids Antibodies	*Surface plasmon resonance*	Gold or silver layer	12
Antibody Antigenes Enzymes Nucleic acids	*Acoustic*	Piezoelectronic devices	13
Enzymes Bacteria Cells	*Calorimetric*	Termistor	14

This paper describes the development of a biosensor based on bacterial cells to detect lactate concentration in the milk. This sensor can be applied for rapid bacterial counting in the milk using the correlation between lactate concentration and bacterial enumeration.

2. Materials and Methods

Escherichia coli cells were cultivated on a minimal medium M-9 (15). Lactate (Sigma, USA) at concentration 1% as C-source during cell growth was used to induce lactate oxidase complex in bacteria. Such cells possessed lactate oxidase activity and did not show glucose oxidase one.

3. Sensor Fabrication

To prepare lactate biochip, cells were put on cellulose disk ("Vladipor N1", Russia) at concentration 5 mg protein/disk. After cell adsorption, disk (diameter 8 mm) was carried on a top of the Clark-type pO_2 electrode. This lactate biochip fixed by nylon net with O-rubber ring (*Figure 1*).

FIGURE 1. The scheme of biosensor electrode. 1 - pO_2 electrode. 2 - nylon net, 3 - rubber O ring, 4 - biochip.

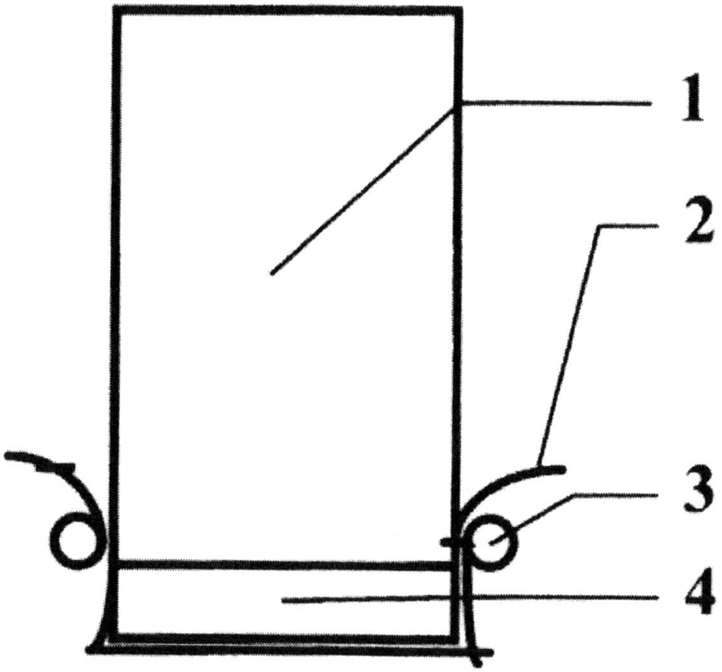

Lactate detection was made by using the measurement of O_2 consumption during aerobic lactate oxidation. The scheme of lactate biosensor device is shown on *Figure 2.*

FIGURE 2. The scheme of lactate biosensor device. 1 – the biosensor electrode. 2 – transducer. 3 – chamber. 4 – stirrer.

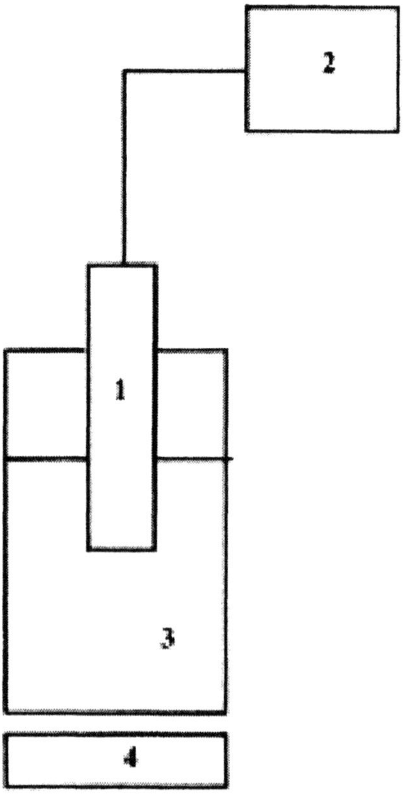

4. Results and Discussion

We studied the dependence of bacterial chip activity on pH from 6.0 to 9.0 and temperature in the range 20 - 50°C. Optimum pH was 7.4 for lactate detection. Temperature optimum was 37-40°C (*Figure 3*). These data were identical to results obtained by using enzyme lactate biochips (16).

FIGURE 3. Influence of pH (O) and temperature (●) on activity of lactate biosensor. % - activity in percent.

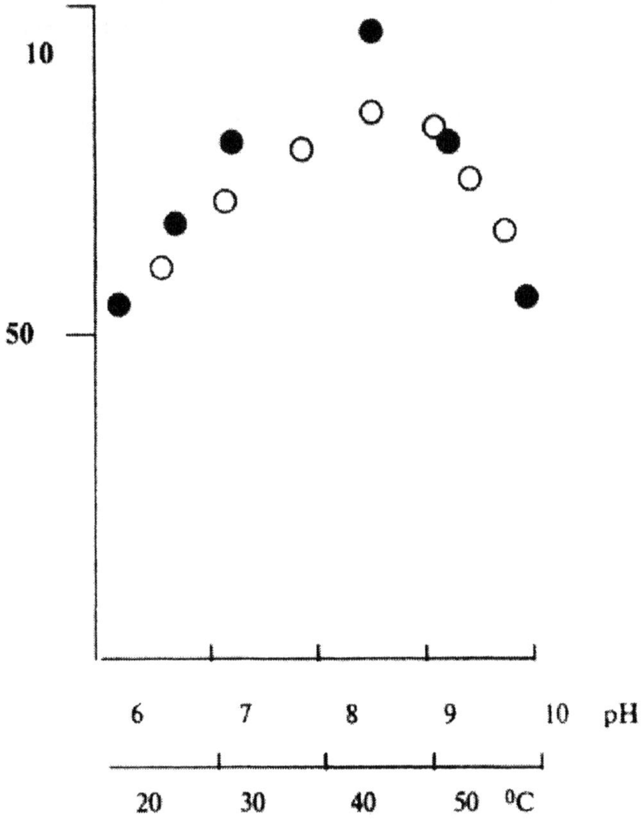

Linear range of lactate detection was obtained for the lactate concentration from 1 - 20 mM. Assays were carried out in 2 ml chamber during stirring. 30 mM phosphate buffer, pH 7.4 was used for calibration and analytical measurements.

The presence of sugars (glucose, galactose, fructose and mannose) at concentration 10 mM and aminoacids (lysine, serine, alanine and arginine) at concentration 10 mM did not interfere lactate detection in medium.

The fresh cow milk was used in all experiments for lactate detection and bacterial counting. To evaluate bacterial enumeration in the milk its samples (with different dilution) were inoculated on solid agar medium on Petry dishes. Bacterial counting was detected after 18 hours of cultivation. The correlation between lactate concentration and bacterial enumeration in milk using such biosensor was studied (*Figure 4*).

FIGURE 4. The correlation between lactate concentration and bacterial enumeration in the milk.

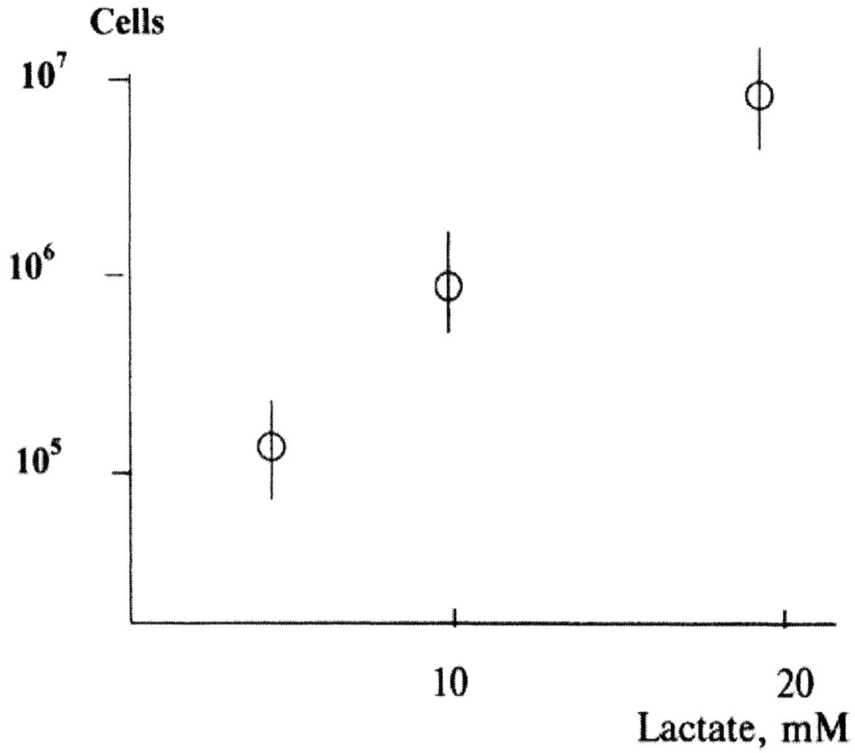

The raw milk was incubated at 37^{0}C. The milk samples were analyzed for lactate concentration and bacterial enumeration after 3 and 6 hours of incubation. In the fresh milk lactate concentration was 5 mM and bacterial concentration changed from $0,2x10^6$ - $0,4x10^6$ cells/ml. After 3 hours of incubation at 37^{0}C lactate concentration was 10 mM and bacterial concentration was $1,0x10^6$ - $2,0x10^6$ cells/ml. After 6 hours of incubation lactate concentration increased to 20 mM and bacterial concentration increased to 10^7 cells/ml.

It is known that critical bacterial concentration determining the milk quality is (1,0 - 2,0)x10^6 cells/ml. This bacterial concentration corresponds to 10 mM lactate concentration in the milk.

5. Conclusion

Method using biosensor based on bacterial cells for rapid bacterial counting in the milk has been developed.

6. Perspectives

Using of bacteria for sensor applications is disadvantaged by some problems in specificity, stability and multi-enzyme behavior, which result in decreased substance specificity.

Another very significant problem for bacterial and other types of biosensors is encountered during immobilization. Although immobilization can lead to the stabilization of biological elements (enzymes, cells, etc.) it can also lead to a loss of binding activity with analyte. Such activity losses can result from a large number of related factors including stearic hindrance at the binding site and changes of tertiary structure of proteins. From this point of view, it seems very attractive to create a new generation of biosensors where biochips will be based on cytoplasmic bacterial membranes. Cytoplasmic membranes (CPM) composed primarily of lipids arranged into two layers with embedded proteins. The use of cytoplasmic membranes (CPM) from bacteria which may provide a very convenient biosensing system. The composition and the enzyme activity of membrane-bound proteins can be specifically changed by varying bacterial growth media. Biosensors based on CPM should be more sensitive and stable than those based on whole cells. Moreover, the native enzyme organization in lipid bilayer should provide an advantage during immobilization on Clark, redox or optical electrodes.

Previous data for possibility to use bacterial CPM for biochip creation have been obtained.

7. Acknowledgment

This work was supported by NATO Collaborative Grant and ISTC #890.

8. References

1. Karube, I and Masaysu, S.W. (1990) Microbial biosensors. In: *Biosensors: a practical approach*. New York, 155-170.
2. Margenianu, D.G. (1990). Bioelectrochemical sensors with living cells. In: *Bioinstrumentations: research, development and application*, MA 1.
3. The milk and dairy products (methods).1979, Moscow, p.7 (in Russia).
4. Hendry, S.P. Higgins, I.J. and Bannister, J.V. (1990) Amperometric biosensors, *J. Biotechnol.* **15**, 229-238.
5. Kaufman, J.M. and Guilbaut G.G. (1992) Enzyme electrode biosensors: theory and applications, *Bioannal. Appl. Enzymes*, **36**, 36-113.
6. Kimura, J. and Kuriyama T. (1990) FET biosensors, *J.Biotechnol.*, **15**, 239-254.
7. Owicki, J.C. Bouse, L.J. Hafeman, D.G. Kirk, G.L. Olson, J.D. Wada, H.G. and Parce, J.W. (1994) The light-addressable potentiometric sensor. Principles and applications, *Ann. Rev. Biophys. Biomol. Struct.*, **23**, 87-113.
8. Janata, J. (1989) *Principles of chemical sensors*. Plenum Press, New York.
9. Arnold, M.A. (1990) Fibre-optic biosensors, *J. Biotechnol.* **15**, 219-228.
10. Patano, P. and Walt D. R. (1995) Analytical applications of optical imaging fibers, *Anal.Chem.*, **67**, 481A.
11. Blum, L.J. Gaunter, S.M. and Coulet, P.R. (1989) Design of luminescence photobiosensors, *J. Biolumin. Chemilumin.*, **4**, 543-550.
12. Fagestan, L.Q. and O'Shannessy, D.J. (1993) Surface plasmon resonance detection in affinity technologies, In: *Handbook of affinity chromotography*, **63**, 229-252.
13. Grata, G.W. Martin, S.J. and White R.M. (1993) Acoustic wave microsensors, *Annal Chem*, **65**, 987A-996A.
14. Danielsson, B. (1990) Calorimetric biosensors. *J. Biotechnol.*, **15**, 187-200.
15. Ignatov, S.G. Krasilnikov, V.A. Perelygin, V.V. Kaprelyants, A.S. and Ostrovsky, D.N. (1981) Study of functional and structural changes of E.coli membranes induced by low temperature freezing. *Biochimia*, , **46**, 1996-2003, (in Russia).
16. Casimiri, V. and Burstein C. (1996) Co-immobilized L-lactate oxidase and L-lactate dehydrogenase on a film mounted on oxygen electrode for highly sensitive L-lactate determination. *Biosen. & Bioelectr.*, **11**, 783-789.

TINY-TOF MASS SPECTROMETER FOR BIODETECTION

WAYNE A. BRYDEN, RICHARD C. BENSON AND
HARVEY W. KO
Applied Physics Laboratory
The Johns Hopkins University
Laurel, MD 20723

CATHERINE FENSELAU
Department of Chemistry and Biochemistry
University of Maryland Baltimore County
Baltimore Maryland 21228

ROBERT J. COTTER
Dept. of Pharmacology and Molecular Science
The Johns Hopkins University, School of Medicine
Baltimore MD 21205

Abstract

The mass spectrometer has long been known as the most powerful analytical tool in the laboratory for analysis of a broad spectrum of chemical and biological materials. The applicability of mass spectrometers to field detection problems has been quite limited given the large size, heavy weight, and prohibitive power requirements of the instrumentation. However, this situation is rapidly changing, as the need for field-portable detection systems becomes more critical. This paper reports on an effort to bring small, yet powerful, time-of-flight (TOF) mass spectrometer technology to bear on the chemical and biological detection problem. In order to achieve this goal, not only must small instrumentation be developed and fielded, but validated mass spectral signatures for all of the agents of interest must also be obtained. We will describe the development of a miniaturized time-of-flight mass spectrometer (*tinyTOF*) for field portable biodetection. Topics to be covered include: operating principles of the mass analyzer; miniaturization of the mass analyzer and other system components; mass spectral biosignatures; and the sample interface for airborne particulate samples.

P.J. Stopa and M.A. Bartoszcze (eds.),
Rapid Methods for Analysis of Biological Materials in the Environment, 101-110.

1. Introduction

At its most basic level, mass spectrometry is a technique for the determination of the masses of molecules and specific fragmentation products formed following vaporization and ionization. From detailed analysis of the mass distribution of the molecule and its fragments, molecular identification can be accomplished. These molecular measurements can be carried out at the attomole (10^{-18} mole) level of material using specialized laboratory-based instruments. The combination of specific molecular identification and extreme sensitivity makes mass spectrometry one of the most powerful analytical laboratory tools yet developed for detection and identification of chemical and biological substances.. While such capability has existed in the laboratory for many years, the development of a small, portable mass spectrometer for potential **field detection** of chemical and biological substances is described here.

2. Basic Principles of Time-of-Flight Mass Spectrometry

The time-of-flight (TOF) mass spectrometer is an excellent candidate for field portable operation due to a simple design that promises stability, ruggedness, low weight, and low power usage [1]. In a simple (linear) time-of-flight mass spectrometer (Figure 1),

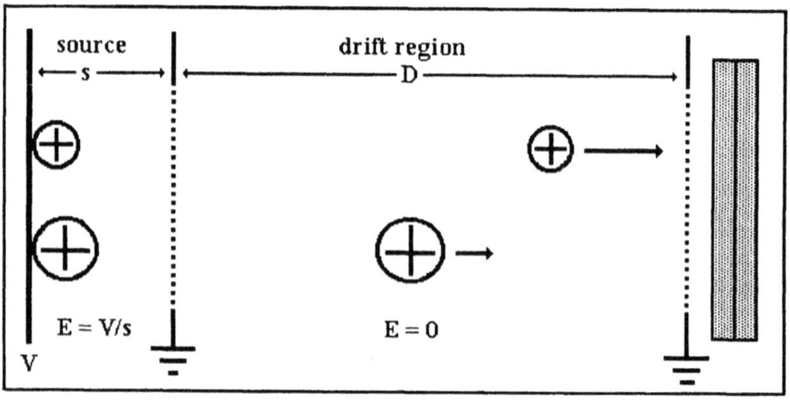

Figure 1. Diagram of a time-of-flight mass spectrometer.

ions are formed and accelerated in a short *source* region (s); and their flight times through a longer *drift* region (D) are measured to determine their masses:

$$t = \left(\frac{m}{2eV}\right)^{1/2} D \qquad (1)$$

where V = accelerating voltage. Because the ions have a distribution of kinetic energies $eV \pm U_0$, where U_0 is the initial energy distribution before acceleration, mass resolution is improved using a reflectron. For a *single-stage* reflectron instrument (Figure 2) the time of flight is:

$$t = \left(\frac{m}{2eV}\right)^{1/2} [L_1 + L_2 + 4d] \quad (2)$$

where L_1, L_2 and d are the regions before, after and inside the reflectron, respectively. The focusing action of a reflectron results from the fact that ions with higher kinetic energies penetrate the reflectron more deeply, so that these faster ions spend more time in the reflectron and arrive at the detector at the same time as the slower ions.

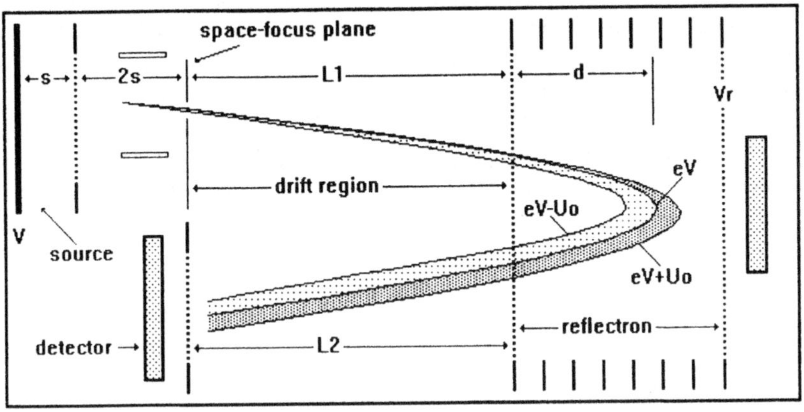

Figure 2. Diagram of a *single-stage* reflectron TOF mass spectrometer.

Single-stage reflectrons are generally constructed as a series of resistively-coupled lens elements x, whose voltages increase linearly ($V = ax$) to produce a constant retarding field. The voltage placed on the last lens element V_R is generally slightly larger than the accelerating voltage V, so that the average penetration depth d will be slightly shorter than the reflectron depth. Using this geometry, *first-order* kinetic energy focusing is achieved when:

$$L_1 + L_2 = 4d \quad (3)$$

In commercial instruments designed for biological research, V is typically 10-20 kV for proteins and oligonucleotides and 4-5 kV for peptides, while the total lengths of the mass analyzer range from 0.5 to 3 meters. *Dual-stage* reflectrons are also common, and use two retarding regions configured such that the ion kinetic energy is reduced by 2/3 in the first 10% of the reflectron. In this case, *second-order* kinetic energy focusing is achieved when the drift regions are 6-8 times the reflectron depth, thus permitting the construction of more compact reflectrons.

3. Tiny-TOF

Historically, time-of-flight mass spectrometers have employed long flight tubes (up to 8 m long), high accelerating voltages (up to 30 kV) and have required low background vacuum pressures (10^{-7} to 10^{-8} Torr). Long flight tubes resulted in longer flight times (that were more easily measured) and, along with high acceleration potentials, resulted in increased mass resolution. However, the long flight tubes, in addition to making the instrument quite large, required lower pressure levels to minimize the gas phase interactions that can degrade the mass spectrum. More recently, the development of matrix-assisted laser desorption/ionization (MALDI) has focused interest on the use of TOF mass spectrometers for the analysis of biological macromolecules, in which high accelerating voltages are utilized to improve ion transmission and detection efficiency, and long drift lengths preserve high time resolution for these faster moving ions.

However, if lower accelerating voltages are coupled with short drift lengths, very high speed electronics and fast pulsed lasers a compact (and portable) TOF instrument with excellent time (and mass) resolution can be produced. In our laboratories, we have developed the *tinyTOF* mass analyzer (Figure 3), using a coaxial design, which has a total length (drift + reflectron) of 3 inches, and accelerating voltages ranging from 100 to 1,000 volts. A mass spectrum of polyethylene glycol, characterizing typical performance of this small instrument, is shown in Figure 4. This instrument is capable of analyzing a broad range of chemical and biological materials with molecular weights up to 12,000 u (currently).

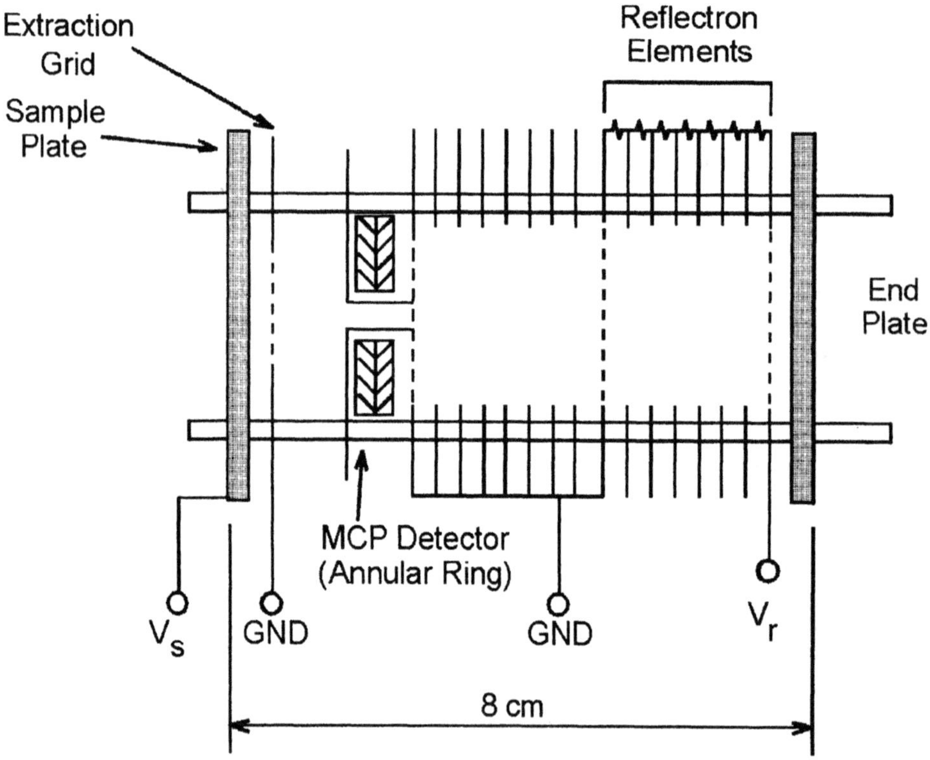

Figure 3. Diagram of the *tinyTOF* mass spectrometer.

While the *tinyTOF* performs very well in the laboratory, our ultimate goal is the development of a field portable *tinyTOF* mass spectrometer system. This development is proceeding by coupling a ruggedized version of the small mass analyzer with integrated control and acquisition electronics, miniaturized vacuum technology,

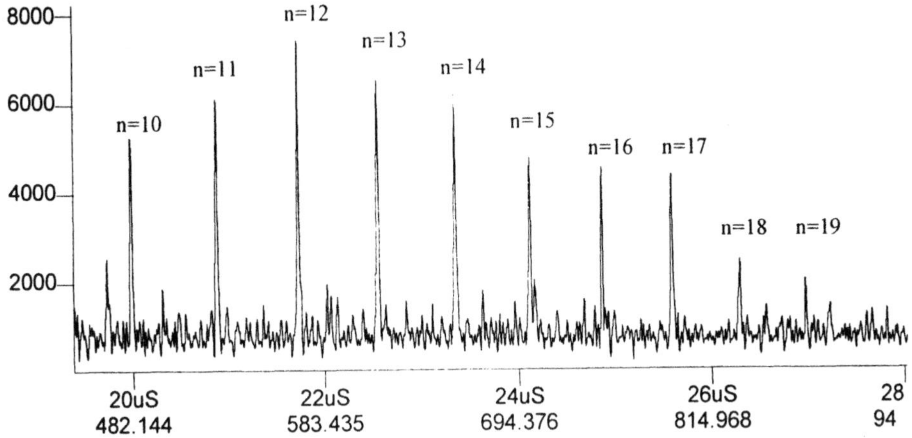

Figure 4. *TinyTOF* mass spectrum of polyethylene glycol.

advanced ionization sources and automated signal processing. Engineering estimates show that this instrument can be packaged in a volume roughly 30 X 10 X 10 cm^3 with a weight of less than 5 kg and with power requirements of less than 100 W. The *tinyTOF* system has potential applications for detecting and identifying chemical and biological substances for several applications including: clinical medicine, medical research, environmental monitoring, law-enforcement, and military troop protection. The focus of the current development is the detection of biological aerosols in a military environment.

4. Biodetection Using tinyTOF

We are currently developing a miniaturized time-of-flight mass spectrometer system for automated environmental bioagent detection. A schematic diagram of the conceptual system is shown in Figure 5. In operation, a trigger will initiate an aerosol sampler to sort, concentrate, and collect particulate onto a movable tape sample substrate. The sample is moved into the vacuum chamber of the mass spectrometer for analysis. Applying pulsed light from a laser onto the target creates sample ions. Depending on the peak power density of the laser, different ionization and molecular fragmentation conditions are obtained. For example, Hillenkamp, et. al. [2] introduced a high peak power laser microprobe that produced elemental analysis of biological samples. Using somewhat lower powers, Posthumus, Kistemaker, and Meuselar [3] introduced laser desorption techniques whereby nonvolatile and thermally labile substances could be desorbed into the gas phase with moderate fragmentation that allowed the determination of *molecular* identity. Finally, Karas, et. al. [4] and Tanaka, et. al. [5] introduced matrix-assisted laser desorption/ionization (MALDI). This is a much softer ionization technique, using much lower laser irradiances (typically 1 MW/cm^2). With this technique, biomolecules as large as 300,000 daltons can be ionized and desorbed intact into the gas phase for mass analysis. The basis for MALDI is the interaction of a pulsed laser beam with a laser-absorbing matrix material into which analyte molecules are dispersed. Pulsed laser energy is absorbed by the matrix and transferred to the analyte causing it to be ionized and desorbed into the gas phase. In the process, the analyte chemically interacts with fragment ions of the matrix, forming molecular adduct ions.

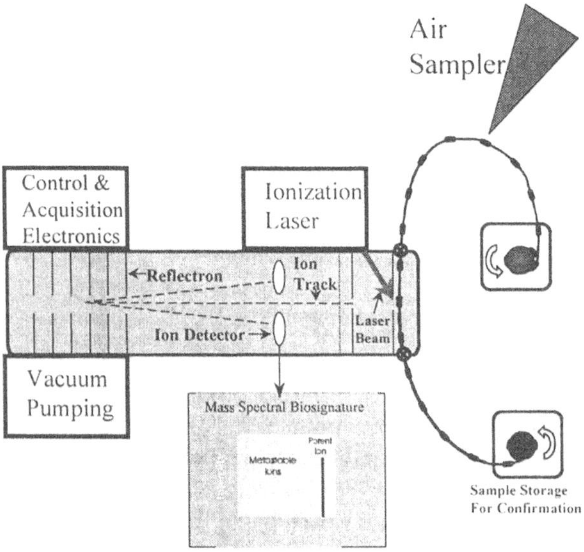

Figure 5. Schematic of *TinyTOF* biodection system.

The MALDI process generally involves wet chemical techniques. A solution of the matrix molecule is physically mixed with a solution containing the analyte. The resulting mixture is applied to a sample probe, allowed to dry, and introduced into the mass spectrometer for analysis. We are developing an alternative concept for MALDI processing that is much more amenable to automatic sample preparation. This approach embeds a sample collection and processing function into a cassette tape format. Physical and chemical pretreatment of a sample to prepare it for meaningful mass spectrometry will be incorporated into the tape system using a combination of approaches including: microencapsulation of solvents and other chemicals; the use of multiple layers of tape that will bring different functionality to the reaction zone; and nanoliter, ink-jet style, chemical applicators that are capable of applying focused spots of process chemicals with high volumetric accuracy. An added benefit to this approach is the potential of using the tape to maintain both an archival record of the analyses performed and the sample measured. If the samples are stored under the proper conditions, they can be reanalyzed at a later date with even more sophisticated analytical laboratory equipment. Such a system would have been an extremely useful tool in the ongoing investigation of Gulf War Illness and for other similar situations that are sure to arise in the future. The automation of the sample treatment methodology into a tape format will greatly reduce the logistics burden for fielding of the mass spectrometer system and provide a critical archiving function.

In the tinyTOF, the ions formed in the laser desorption process, are rapidly (~50 μ sec) analyzed using a small (<10 cm in length) reflectron time-of-flight mass analyzer. The ions are detected by a high gain ($\sim 10^7$) microchannel plate device and the signal is fed out to a high speed analog-to-digital converter. The digital data is rapidly acquired and stored in a small computer. The processing of the data will use intelligent algorithms that provide probabilities of detection for various agents of concern with a simple user interface. The sample is then moved out of the mass spectrometer, stabilized on the tape surface and stored for follow up analysis. The entire system is being miniaturized through the use of advanced lasers, vacuum pumps, electronics, and packaging techniques. This system will provide a field capability for bioaerosol clouds that is rapid, sensitive, provides broadband and specific information about the threat, has the capability to archive samples and data, and has a user interface designed for use by military personnel.

5. Mass Spectral Biomarker Development

While mass spectrometry is a prime candidate technology for rapid identification of microorganisms, the instrumentation can not be employed without the development of mass spectral signatures of the materials of interest. We are employing a chemotaxonomic approach to this problem that recognizes that all materials of biological origin are composed of common building blocks that include inorganic chemicals, small organic molecules, carbohydrates, amino acids, and nucleic acids [6]. These building blocks are linked together in precise fashion to form all manner of biological materials. These larger macromolecules are the prime signatures that can be used to identify materials of interest.

While conventional techniques for microbe identification have relied on morphological and metabolic characteristics, advances in biochemistry, molecular biology and chemical instrumentation have opened up new avenues of taxonomy based on the chemical makeup of the cells of the organism. Such an approach is commonly described as chemotaxonomy, which Priest and Austin [7] defined as the study of chemical variation in living organisms, and the use of these chemical characters for classification and identification. Chemical markers are being developed that are characteristic of broad classes of microorganisms as well as those that are specific for specific pathogens. These chemical markers can be extremely rapidly quantified using the techniques of modern mass spectrometry.

The approach we are taking is to (1) evaluate various biomarkers for the detection and characterization of microorganisms, (2) evaluate different techniques for mass spectrometric formation and analysis of biomarker ions from microorganisms that are compatible with field portable instrumentation, (3) optimize sample handling and preparation for the biomarkers and MS techniques with the greatest potential, (4) establish libraries that include both broad class and specific biomarkers, and (5) integrate the mass spectral knowledge base into intelligent software algorithms for automated machine processing.

The polar lipids present in bacterial cell (and several other) membranes are classically a useful chemotaxonomic biomarker. These biochemicals are also attractive (and useful) as mass spectral biomarkers. A straightforward approach to the mass spectra of these compounds is to record the molecular ions of the intact phospholipids, including both the polar head groups and the two fatty acid sidechains. Classical treaties [8-10] consider the polar head groups to be more closely under genetic control and the fatty acids to vary quantitatively, reflecting nutritional history and health. The structures and relative amounts of polar head groups and fatty acids in a mixture of phospholipid families may be separately deduced by recording and interpreting fragmentation of phospholipid ions. This high definition step is readily automated. Utilizing fragmentation in this way takes advantage of the high information content of mass spectra and provides information on both the chemotaxonomy and the production of environmental microorganisms. The structures and relative abundances of polar head groups vary between gram negative and positive bacteria, fungi, algae and our earlier studies indicated that semi-quantitative considerations based on mass spectra would allow broad classification of microorganisms [6].

The phospholipid biomarker libraries include both molecular ion fingerprints and polar head group analyses derived from fragment ions. We are determining to what extent polar head group ratios can be used to distinguish species and, in particular, target organisms. These quantitative analyses are based on constant neutral loss spectra measured using sophisticated laboratory instruments that are used to unambiguously determine the molecular identity of the ions and to test the reliability of signal processing algorithms that may be able to extract polar head group information from full-spectrum time-of-flight data.

While polar lipids are present in bacterial cells and enveloped viruses, they are absent in most viruses and all proteinaceous toxins. A more universal class of biomarker is based on peptides and proteins. The use of this type of biomarker is complicated by the large number (several thousand) of proteins present in a typical cell. However proteins are being identified that are characteristic of specific bacterial and viruses.

In addition to the obvious approach of determining patterns of protein molecular weights, peptide mixtures will be produced by selective cleavage. One candidate reagent is BNPS-skatole, which we have found cleaves rapidly and selectively at tryptophans [11]. Peptides produced in this way can be mapped more definitively than proteins, and with better sensitivity.

Finally, the genetic pattern (DNA and RNA) of bacteria and viruses is an absolutely characteristic biomarker. However, current mass spectrometer technology does not allow the measurement of long chain DNA and RNA. We are evaluating the chemical and enzymatic production of smaller fragments (oligonucleotides) for direct fingerprinting of viruses. Nucleotide mapping is more difficult for bacteria, where the genetic machinery constitutes a smaller fraction of the cellular mass and amplification is required [12,13]. Nucleic analysis using mass spectrometry is currently best accomplished by first enzymatically degrading the nucleic acid, using probes immobilized on a surface that capture specific sequences and then measuring the mass distribution on the surface. With properly designed probes, efficient generation of genetic information is provided by a mass spectrometer.

6. Conclusions

We have discussed the development of a field portable time-of-flight mass spectrometer detection system for biological materials in the environment. This system includes an aerosol sampling front end that concentrates airborne material onto a movable tape, a miniaturized laser desorption time-of-flight mass spectrometer, and an intelligent signal processing back end that is based on biological mass spectrometer signatures. When fully developed, this detection system will provide rapid and accurate information on the identity of bioaerosol clouds in a military environment.

7. References

1. Cotter, R.J. (1997) *Time-of-Flight Mass Spectrometry: Instrumentation and Applications in Biological Research*. American Chemical Society, Washington, DC.
2. Hillenkamp, F.; Kaufmann, R.; Nitsche, R.; Unsold, E. (1975) A High-Sensitivity Laser Microprobe Mass Analyzer. *Appl Phys.* **8**, 341.
3. Posthumus, M.A.; Kistemaker, P.G.; Meuselaar, H.L.C.; Ten Noever de Brauw, M.C. (1978) Laser Desorption-Mass Spectrometry Of Nonvolatile Bio-Organic Molecules, *Anal. Chem.* **50**, 985.
4. Karas, M.; Bachmann, D.; Bahr, U.; Hillenkamp, F (1987) Matrix-Assisted Ultraviolet Laser Desorption Of Non-Volatile Compounds, *Int. J. Mass Spectrom. Ion Processes* **78**, 53.
5. Tanaka, K.; Waki, H.; Ido, Y.; Akita, S.; Yoshida, Y.; Yoshida, T. (1988) *Rapid Commun. Mass Spectrom.* **2**, 151.
6. Fenselau, C. (1994) *Mass Spectrometry for the Characterization of Microorganisms*, (C. Fenselau, ed.) ACS Symposium Series 541, American Chemical Society: Washington, D.C., Chapter 1, pp. 1-7.
7. Priest F. and Austin B. (1993) *Modern Bacterial Taxonomy*, 2nd ed., Chapman & Hall, London.
8. Kates, M. (1964) In: *Advances in Lipid Research*, (Paoletti, R., ed.) Academic Press, New York, Vol. 2, pp 17-90.
9. Shaw, N. (1974) *Adv. Appl. Microbiol.* **63**, 108.
10. Lechevallier, M.P. (1977) *CRC Crit. Rev. Microbiol.* **5**, 109-210.
11. Vestling, M.M., Kelly, M.A. and Fenselau C. (1994) Optimization by Mass Spectrometry of Conditions for Cleavage of Proteins at Tryptophan Residues with BNPS-Skatole. *Rapid Commun. Mass Spectrom.* **8**, 786-790.
12. Muddiman, D.C., Wunschel, D.S., Liu, C., Pasa-Tolic, L., Fox, K.F., Fox, A., Anderson, G.A. and Smith, R.D. (1996) Rapid and Precise Characterization of PCR Products From Bacilli Using Electrospray Ionization FTICR Mass Spectrometry. *Anal Chem.* **68**, 3705-3712.

13. Wunschel, D.S., Fox, K.F., Fox, A., Bruce, J.E., Muddiman, D.C. and Smith, R.D. (1996) Analysis of Double-stranded Polymerase Chain Reaction Products from the Bacillus cereus Group by Electrospray Ionization Fourier Transform Ion Cyclotron Resonance Mass Spectrometry *Rapid Commun. Mass Spectrom.* **10**, 29-35.

INTELLIGENT SYSTEMS FOR THE CHARACTERIZATION OF MICROORGANISMS FROM HYPERSPECTRAL DATA

ROYSTON GOODACRE[1*], REBECCA BURTON[1], NAHEED KADERBHAI[1], ÉADAOIN M. TIMMINS[1], ANDREW WOODWARD[1], PAUL J. ROONEY[2] AND DOUGLAS B. KELL[1]
[1]*Institute of Biological Sciences, University of Wales, Aberystwyth, Ceredigion, SY23 3DA, Wales, U.K.* [2]*Bronglais General Hospital, Aberystwyth, Ceredigion, SY23 1ER, Wales, U.K.*

Summary

Three rapid spectroscopic approaches for whole-organism fingerprinting, *viz.* pyrolysis mass spectrometry (PyMS), Fourier transform infra-red spectroscopy (FT-IR) and dispersive Raman microscopy, were used to analyze a group of 59 clinical bacterial isolates associated with urinary tract infection.

Direct visual analysis of these spectra was not possible, highlighting the need to use multivariate methods to reduce the dimensionality of these hyperspectral data. First the unsupervised methods of discriminant function and hierarchical cluster analyses were employed to group these organisms based on their spectral fingerprints, but none produced wholly satisfactory groupings which were characteristic for each of the five bacterial types. In contrast, for PyMS and FT-IR, the artificial neural network-based approaches of multi-layer perceptrons and radial basis functions could be trained with representative spectra of the five bacterial groups so that isolates from clinical bacteriuria in an independent *unseen test set* could be correctly identified. ANNs trained with Raman spectra identified only 80% of the same test set.

These results demonstrate that modern analytical spectroscopies can provide rapid accurate microbial characterization techniques, but only when combined with intelligent systems.

1. Introduction

When a pathogen is isolated in a microbiology laboratory, the time taken for subsequent culture for identification and susceptibility testing may delay the administration of the most appropriate treatment.

For routine purposes the ideal method for microbial characterization would have minimum sample preparation, would analyze samples directly (i.e., would not require reagents), would be rapid, automated, and (at least relatively) inexpensive. With recent

P.J. Stopa and M.A. Bartoszcze (eds.),
Rapid Methods for Analysis of Biological Materials in the Environment, 111–136.

developments in analytical instrumentation, these requirements are being fulfilled by physico-chemical spectroscopic methods, often referred to as 'whole-organism fingerprinting' (Magee, 1993). The most common such methods are pyrolysis mass spectrometry (PyMS) (Goodacre & Kell, 1996), Fourier transform infrared spectroscopy (FT-IR) (Naumann *et al.*, 1991) and UV resonance Raman spectroscopy (Nelson *et al.*, 1992).

PyMS, FT-IR and dispersive Raman microscopy are physico-chemical methods which measure predominantly the bond strengths of molecules (PyMS) and the vibrations of bonds within functional groups (FT-IR and Raman) (Colthup *et al.*, 1990; Ferraro & Nakamoto, 1994; Griffiths & de Haseth, 1986; Meuzelaar *et al.*, 1982). Therefore they give quantitative information about the total biochemical composition of a sample. However, the interpretation of these multidimensional spectra has conventionally been by the application of "unsupervised" pattern recognition methods such as principal components (PCA), discriminant function (DFA) and hierarchical cluster (HCA) analyses. With "unsupervised learning" methods of this sort the relevant multivariate algorithms seek "clusters" in the data, thereby allowing the investigator to group objects together on the basis of their perceived closeness (Everitt, 1993); this process is often subjective because it relies upon the interpretation of complicated scatter plots and dendrograms. More recently, various related but much more powerful methods, most often referred to within the framework of chemometrics, have been applied to the "supervised" analysis of these hyperspectral data; arguably the most significant of these is the application of intelligent systems based on artificial neural networks (ANNs) (Bishop, 1995; Wasserman, 1989).

Urinary tract infection (UTI) or bacteriuria remains a major clinical problem and bacterial resistance to antibiotics is increasing. Indeed, many consultations in general practice are because of bacteriuria (Sleigh & Timbury, 1986). The number of culturable organisms which is regarded as significant clinical bacteriuria is $>10^5$ cells.ml^{-1}, and the bacteria typically associated with UTI are *Escherichia coli* (causative organism in 60- 90% of bacteriurias), *Staphylococcus saprophyticus* (30% of women with bacteriuria aged <25 years), *Proteus mirabilis* (10%), *Klebsiella* spp. (which are often multiply resistant to antibiotics), *Staphylococcus aureus* and *Pseudomonas aeruginosa* (especially after catheterization).

In this study a group of 59 bacteria commonly associated with bacteriuria were collected from the local hospital. All isolates were typed by conventional biochemical tests to belong to *E. coli*, *Prot. mirabilis*, *Klebsiella*, *Pseud. aeruginosa*, and *Enterococcus*. The aim of this study was to compare the phenotypic differentiation of these 59 bacterial isolates by PyMS, FT-IR and Raman spectroscopies, and to use ANNs to identify the bacteria from these hyperspectral measurements.

2. Materials and Methods

Strains and cultivation.
A group of 59 bacteria commonly associated with urinary tract infection, or bacteriuria, were collected from the local hospital. All isolates were typed by conventional biochemical tests to belong to *Escherichia coli* (17), *Proteus mirabilis* (10), *Klebsiella* spp. (4 *K. oxytoca* and 6 *K. pneumoniae*), *Pseudomonas aeruginosa* (10), and *Enterococcus* spp (12). Details are given in Table 1. All strains were cultivated axenically and aerobically on LabM Malthus blood agar base (37 $mg.ml^{-1}$) for 16h at 37°C. After sub-culturing three times to ensure pure cultures, biomass was carefully collected using sterile plastic loops and suspended in 1 ml aliquots of sterile physiological saline (0.9% NaCl). The samples were then analyzed by PyMS, FT-IR and dispersive Raman spectroscopies.

Pyrolysis mass spectrometry (PyMS).
Five microlitres of the bacterial samples (ca. 40 mg ml^{-1}) were evenly applied to clean iron-nickel foils which had been partially inserted into clean pyrolysis tubes. Samples were run in triplicate. Prior to pyrolysis the samples were oven-dried at 50°C for 30 min and the foils were then pushed into the tubes using a stainless steel depth gauge so as to lie 10 mm from the mouth of the tube. Viton O-rings were next placed approximately 1 mm from the mouth of each tube.

Pyrolysis mass spectrometry was then performed on a Horizon Instrument PyMS-200X (Horizon Instruments Ltd., Heathfield, U.K.). For full operational procedures see (Goodacre and Kell, 1996; Goodacre *et al.*, 1993; Goodacre *et al.*, 1994). Conditions used for each experiment involved heating the sample to 100°C for 5s followed by Curie-point pyrolysis at 530°C for 3s with a temperature rise time of 0.5s.

PyMS data may be displayed as quantitative pyrolysis mass spectra (e.g., as in Figure 1). The abscissa represents the 150 *m/z* ratios, while the ordinate contains information on ion count for any particular *m/z* value ranging from 51 to 200. To remove the influence of sample size *per se* data were normalized as a percentage of the total ion count. Total ion counts were typically in the range 1 - 3.10^6.

Diffuse reflectance-absorbance Fourier transform infrared (FT-IR) spectroscopy.
Ten microlitres of the bacterial samples were evenly applied onto a sand-blasted aluminum plate. Prior to analysis the samples were oven-dried at 50°C for 30 min. Samples were run in triplicate. The FT-IR instrument used was the Broker IFS28 FT-IR spectrometer (Broker Spectrospin Ltd., Banner Lane, Coventry, UK) equipped with an MCT (mercury-cadmium-telluride) detector cooled with liquid N_2. The aluminum plate was then loaded onto the motorized stage of a reflectance TLC accessory (Bouffard *et al.*, 1994; Goodacre *et al.*, 1996; Winson *et al.*, 1997).

The IBM-compatible personal computer used to control the IFS28, was also programmed (using OPUS version 2.1 software running under IBM O/S2 Warp provided by the manufacturers) to collect spectra over the wavenumber range 4000 cm^{-1} to 600 cm^{-1}. Spectra were acquired at a rate of 20 s^{-1}. The spectral resolution used was

4 cm^{-1}. To improve the signal-to-noise ratio, 256 spectra were co-added and averaged. Each sample was thus represented by a spectrum containing 882 points and spectra were displayed in terms of absorbance as calculated from the reflectance-absorbance spectra using the Opus software. Typical FT-IR spectra are shown in Figure 2.

ASCII data were exported from the Opus software used to control the FT-IR instrument and imported into Matlab version 4.2c. l (The MathWorks, Inc., 24 Prime Par Way, Natick, MA, USA), which runs under Microsoft Windows NT on an IBM-compatible personal computer. To minimize problems arising from baseline shifts the following procedure was implemented: (i) the spectra were first normalized so that the smallest absorbance was set to 0 and the highest to +1 for each spectrum, (ii) next these normalized spectra were detrended by subtracting a linearly increasing baseline from 4000 cm^{-1} to 600 cm^{-1}, (iii) finally the smoothed first derivative of these normalized and detrended spectra were calculated using the Savitzky-Golay algorithm (Savitzky & Golay, 1964) with 5-point smoothing.

Dispersive Raman spectroscopy.
Spectra were collected using the Renishaw dispersive Raman spectroscope (Ramascope) (Williams *et al.*, 1994) with a low power (30mW) near infra-red 780nm diode laser with the power at the sampling point typically at 3mW. The instrument was wavelength calibrated with a silicon wafer focused under the x50 objective and collected as a static spectrum centered at 520 cm^{-1} for 10 sec.

Samples were presented as 0.5ml bacterial suspensions in 2ml Supelco clear glass vials, covered with solid caps with aluminum liners. These glass vials were placed sequentially into the sample holder of a Renishaw Macropoint assembly. A 16mm focal length objective, fitted onto the objective system which fits into the standard microscope objective aperture and turns the beam through 90°, was then focused into the sample vial and the stage was locked. The spectrum was collected for 60 sec. The next sample was then placed into the sample holder and the spectral collection procedure was repeated.

The GRAMS WiRE software package running under Windows 95 was used for instrument control and data capture. Spectra were collected over the wavenumber range 200 cm^{-1} to 2300 cm^{-1}. The spectral resolution used was ~0.92 cm^{-1}. Each sample was thus represented by a spectrum containing 2283 points and spectra were displayed in terms of the intensity of Raman scattering (counts).

ASCII data were exported from the GRAMS WiRE software used to control the Raman instrument and imported into Matlab version 4.2c. l. To minimize problems arising from cosmic rays and noise due to short sampling times the following procedure was implemented: (i) any cosmic rays (which excite the CCD camera) were removed using a median filter with a window of 9 data points; (ii) these spectra were then smoothed using a fast Fourier transform denoising routine which briefly removes the high frequency bins (bins 1-110 were kept) from the Fourier-domain spectra, since these contain predominantly noise. These Fourier-domain spectra were then inversely transformed back to the Raman domain. Typical Stokes Raman spectra are shown in

Figure 3. Note that although the fluorescence is relatively low when cells are excited at 780 nm, the system can not discriminate photons arising by fluorescence from those scattered via the Raman effect.

Cluster analysis.

The initial stage involved the reduction of the dimensionality of the PyMS, FT-IR and Raman data by principal components analysis (PCA) (Causton, 1987; Jolliffe, 1986). PCA is a well known technique for reducing the dimensionality of multivariate data whilst preserving most of the variance, and Matlab was employed to perform PCA according to the NIPALS algorithm (Wold, 1966). Discriminant function analysis (DFA) then discriminated between groups on the basis of the retained principal components (PCs) and the *a priori* knowledge of which spectra were replicates (MacFie *et al.*, 1978; Windig *et al.*, 1983), and thus this process does not bias the analysis in any way. DFA was programmed according to Manly's principles (Manly, 1994). Finally, the Euclidean distance between *a priori* group centres in DFA space was used to construct a similarity measure, with the Gower similarity coefficient S_G (Gower, 1966), and these distance measures were then processed by an agglomerative clustering algorithm to construct a dendrogram (Manly, 1994).

Multilayer perceptrons.

All multilayer perceptrons (MLP) analyses (also known as back-propagation artificial neural networks (ANNs)) were carried out with a user-friendly, neural network simulation program, NeuFrame version 3,0,0,0 (Neural Computer Sciences, Lulworth Business Centre, Nutwood Way, Totton, Southampton, Hants), which runs under Microsoft Windows NT on an IBM-compatible personal computer. In-depth descriptions of the *modus operandi* of this type of MLP analysis are given elsewhere (Goodacre, et al., 1994; Goodacre *et al.*, 1994; Goodacre *et al.*, 1995).

The structure of the MLP used in this study to analyze the hyperspectral consisted of 3 layers, the first layer contains the spectra (made up of the 150 input nodes for PyMS, 882 for FT-IR, and 2283 for Raman), one "hidden" layer containing 8 (PyMS), 10 (FT-IR) or 12 (Raman) nodes, and five output nodes (encoded in binary fashion for the bacterial identities). These were binary encoded such that *E. coli* was coded as 10000, *Prot. mirabilis* as 01000, *Klebsiella* spp. as 00100, *Pseud. aeruginosa* as 00010, and *Enterococcus* spp. as 00001. Each of the input nodes were connected to the nodes of the hidden layer using abstract interconnections (connections or synapses) (see Figure 4 for a diagrammatic representation). Connections each have an associated real value, termed the weight (w_i), that scales the input (i_i) passing through them, this also includes the bias (ϑ), which also has a modifiable weight. Nodes sum the signals feeding to them (*Net*):

$$Net = i_1 w_1 + i_2 w_2 + i_3 w_3 + \ldots + i_i w_i + \ldots + i_n w_n = \sum_{i=1}^{n} i_i w_i + \vartheta$$

The sum of the scaled inputs and the node's bias, are then scaled to lie between 0 and +1 by an activation function to give the nodes output (*Out*); this scaling is typically achieved using a logistic "squashing" (or sigmoidal) function:

$$Out = \frac{1}{(1 + \exp^{-Net})}$$

These signals (*Out*) are then passed to the output node which sums them and in turn squashed by the above logistic sigmoidal activation function; the product of this node was then fed to the "outside world".

Before training commenced the values applied to the input and output nodes were normalized between 0.1 and 0.9. The scaling regime used for the input layer was to scale nodally, where the input nodes were scaled for *each* input node such that the lowest mass was set to 0.1 and the highest mass to 0.9. Finally, the connection weights were set to small random values (typically between -0.005 and +0.005).

The algorithm used to train the neural network was the standard back-propagation (BP) (Haykin, 1994; Rumelhart *et al.*, 1986; Wasserman, 1989; Werbos, 1994). For the training of the MLP each input (i.e. spectrum) is paired with a desired output (i.e., the identity of the bacteria); together these are called a training pair (or training pattern). A MLP is trained over a number of training pairs; this group is collectively called the training set, details of the training set are given in Table 2. The input is applied to the network, which is allowed to run until an output is produced at each output node. The differences between the actual and the desired output, taken over the entire training set are fed back through the network in the reverse direction to signal flow (hence back-propagation) modifying the weights as they go. This process is repeated until a suitable level of error is achieved. In the present work, a learning rate of 0.1 and a momentum of 0.9 were used.

Each epoch represented the connection weight updatings and a recalculation of the RMS error between the true and desired outputs over the entire training set (RMS error of formation; RMSEF). During training a plot of the error versus the number of epochs represents the "learning curve", and may be used to estimate the extent of training. All MLPs were trained until the RMSEF was 1.00.

Finally after training, all spectra collected from the bacterial isolates were used as the "unknown" inputs (test data); the network then calculated its estimate and for each sample the largest node in the output layer was taken as its identity.

Radial basis function neural networks.
All Radial basis function (RBF) analyses were also carried out with NeuFrame version 3,0,0,0 as detailed specifically by Saha and Keller (Saha & Keller, 1990).

RBFs are hybrid neural networks encompassing both unsupervised and supervised learning (Beale & Jackson, 1990; Bishop, 1995; Hush & Horne, 1993; Moody & Darken, 1989; Park & Sandberg, 1991; Saha and Keller, 1990; Wilkins *et al.*, 1994). RBFs are typically three-layer neural networks and in essence the sigmoidal squashing function is replaced by non-linear (often Gaussian or "Mexican hat") basis functions or kernels (Figure 5). The kernel is the function that determines the output of each node in the hidden layer when an input pattern is applied to it. This output is simply a function of the Euclidean distance from the kernel centre to the presented input pattern in the multi-dimensional space, and each node in the hidden layer only produces an output when the input applied is within its receptive field; if the input is beyond this receptive field the output is 0. This receptive field can be chosen and is radially symmetric around the kernel centre. Between them the receptive fields cover the entire region of the input space in which a multivariate input pattern may occur; a diagrammatic representation of this is shown in Figure 6, where a two dimensional input is mapped by seven radially-symmetric basis functions. This is a fundamentally different approach from the MLP, in which each hidden node represents a non-linear hyperplanar decision boundary bisecting the input space (Figure 4).

The outputs of the RBF nodes in the hidden layer are then fed forward via weighted connections to the nodes in the output layer in a similar fashion to the MLP, and each output node calculates a weighted sum of the outputs from the non-linear transfer from the kernels in the hidden layer. The only difference is that the output nodes of an RBF network are linear, whilst those of the MLP more typically employ a logistic (non-linear) squashing function.

The implementation of these RBF neural networks is exactly as described by Saha and Keller (Saha and Keller, 1990). Briefly the training proceeds in two stages:

Stage 1 involves unsupervised clustering of the input data, typically using the K-means clustering algorithm (Duda & Hart, 1973; Everitt, 1993; Hush and Horne, 1993) to divide the high- dimensional input data into clusters. Next, kernel centres are placed at the mean of each cluster of data points. The use of K-means is particularly useful because it positions the kernels relative to the density of the input data points. Next the receptive field is determined by the nearest neighbor heuristic where r_j (the radius of kernel j) is set to the Euclidean distance between w_j (the vector determining the centre for the jth RBF) and its nearest neighbor (k), and an overlap constant (*Overlap*) is used:

$$r_j = Overlap \times \min\left(\| w_j - w_k \|\right)$$

where || ... || denotes a vector norm, or Euclidean distance.

The overlap that gave best results was found to be 2, which means that the edge of the radius of one kernel is at the centre of its nearest neighbor; this optimum was also in agreement with the studies of Saha and Keller (Saha and Keller, 1990).

The output from nodes in the hidden layer is dependent on the shape of the basis function and the one used was that of the Mexican hat. Thus this value (R_j) for node j when given the ith input vector (i_i) can be calculated by:

$$R_j(i_i) = \exp^{-\left(\frac{\|w_i - i_i\|}{r_j}\right)^2}$$

Stage 2 involves supervised learning in a single layer MLP. The inputs are the output values for all n basis functions (R_1-R_n) for all the training input patterns to that layer (i_1-i_n), and the outputs are the bacterial identities binary encoded in 5 nodes as detailed above.

The output nodes are calibrated using simple linear regression. The optimum number of kernel functions was found by calculating the minimum error for the test set. Finally after training all spectra collected from the bacterial isolates were used as the "unknown" inputs (test data); the network then calculated its estimate and for each sample the winning node in the output layer was taken as its identity.

3. Results and Discussion

The raw spectra.

Typical normalized PyMS spectra for *Escherichia coli* isolate Ea and *Proteus mirabilis* isolate Pa are shown in Figure 1. These, and the spectra from all 59 bacteria, show an undulating, decaying feature with a periodicity of 14 atomic mass units; due to the loss of CH_2 units during pyrolysis (Meuzelaar, et al., 1982). The FT-IR diffuse reflectance-absorbance and Raman spectra of the same isolates are shown in Figures 2 and 3 respectively. These vibrational spectra and those from the other 59 bacteria all showed broad and complex contours; indeed for the Raman spectra it is very difficult to distinguish the Raman scattering from the background and/or any small levels of fluorescence by excitation using the 780 nm laser (although the contribution due to fluorescence should be reduced by the use of the near infra-red laser).

For all three spectral types there was very little qualitative difference between the spectra, although at least some complex quantitative differences between them were observed. Such spectra, uninterpretable by the naked eye, readily illustrate the need to employ multivariate statistical techniques for the analysis of PyMS, FT-IR and Raman data.

Unsupervised cluster analysis.

After collection of the three data types, each of the 59 strains, each represented by three replicate spectra, were coded to give 59 individual groups (see Table 1), and analyzed by DFA and HCA as detailed above: The resulting dendrogram from the analysis of the PyMS data is shown in Figure 7 where it can be seen that five clusters are recovered. Although the *Pseud. aeruginosa*, *Prot. mirabilis* and the enterococci strains form three well defined clusters, the *Klebsiella* spp. do not form one group and some of these cluster with the 17 *E. coli* (E) strains analyzed.

The analysis of the FT-IR data by DFA is depicted as a pseudo-3D ordination plot (Figure 8). In this figure it is again clear that the *Pseud. aeruginosa* (A), *Prot. mirabilis* (P) and the enterococci (C) strains form three distinct groups, however, the fourth cluster is a mixture of strains of the *E. coli* (E) and *Klebsiella* spp (O and K).

Finally, DFA was used to analyze the Raman spectra and the results are shown in Figure 9. Figure 9A shows the analysis of all the strains and the first discriminant function (DF1) indicates that the majority of the variation was between the *Pseud. aeruginosa* (A) strains and all the other isolates. This is likely to be due to small amount of fluorescence, since it is well known that *Pseud. aeruginosa* naturally fluoresce due to the production of pyocyanin (blue-green) and fluorescein (yellow) pigments (Sleigh and Timbury, 1986), and it is difficult to distinguish this electromagnetic radiation from Raman scattering as both are measured as a shift in wavelength from the 780 nm source laser. Therefore, these isolates were removed and the analysis rerun, the resultant DFA plot is shown in Figure 9B where it can be seen that the different isolates do not group together and only with *a priori* knowledge of the classes can any separation be inferred.

Supervised analysis using artificial neural networks.
Since none of the spectroscopies when analyzed by the various cluster analyses produced wholly satisfactory groupings which were characteristic for each of the five bacterial types, the next stage was to supervise the analysis using the artificial neural network-based approaches of multi-layer perceptrons (MLPs) and radial basis functions (RBFs).

As detailed above the first five organisms were used to train the MLPs and RBFs (see Table 2 for strain numbers). The input layers for the MLPs were either the full spectral data or the first few principal components; therefore for PyMS these were 150 m/z intensities or the first 10 PCs (which accounted for 97.16% of total variance), for FT-IR these were the absorbances at 882 wavenumbers or the first 20 PCs (which accounted for 96.88% of total variance), and for Raman were the counts at 2283 wavenumbers or the first 5 PCs (which accounted for 78.86% of total variance). Only the full spectra were used as inputs for the RBFs. The outputs were always the same for both MLPs, PC-MLPs or RBFs and were binary encoded such that *E. coli* was coded as 10000, *Prot. mirabilis* as 01000, *Klebsiella* spp. as 00100, *Pseud. aeruginosa* as 00010, and *Enterococcus* spp. as 00001.

After training each of the ANNs to a RMS error of 0.01 in the training set, each calibrated system was challenged with the training and test sets. For the PyMS data trained with a full spectral MLP the outputs for the training set are shown in Table 2 and the test set in Table 3. Using the criterion that the identity of an isolate from challenging a trained ANN is taken as the winning node (that is to say the largest value) in the output layer, this PyMS-MLP correctly identified all 25 bacteria in the training set and 33 of the 34 isolates in the *unknown* (unseen) test set. The incorrectly assigned isolate was *Klebsiella* sp. JX58 and was identified as an *E. coli*; this was not surprising since in the dendrogram (Figure 7) this *Klebsiella* isolate and *E. coli* were recovered together. Exactly the same result was seen for the PC-MLPs, by contrast the full spectral RBFs correctly identified all isolates in both the training and test sets. All three ANN-based methods correctly identified all isolates in the training and tests from their FT-IR data (data not shown).

Different results were seen for the ANN analyses of the Raman spectra. Whilst each method got 100% of the training set correct, each method only identified approximately 80% of the 34 isolates in the test sets correctly. MLPs correctly identified 25 (74%), RBFs 26 (76%), and PC-MLPs gave best prediction with 28 (82%). Generally the *E. coli, Pseud. aeruginosa* and *Enterococcus* spp. were always identified but approximately half of the *P. mirabilis* and *Klebsiella* isolates were incorrectly assigned.

4. Conclusions

Three rapid spectroscopic approaches for 'whole-organism fingerprinting' of PyMS, FT-IR and dispersive Raman microscopy were used to analyze a group of 59 clinical bacterial isolates associated with urinary tract infection.

Direct visual analysis of these spectra was not possible, highlighting the need to use multivariate methods to reduce the dimensionality of these hyperspectral data. Unsupervised learning methods of DFA and HCA were employed to group these organisms based on their spectral fingerprints, and although some groups were seen which were characteristic for each of the five bacterial types, wholly satisfactory clustering was not observed until *a priori* information was used in the interpretation of the complicated dendrograms (Figure 7) and ordination plots (Figure 8 and 9).

In contrast, for PyMS and FT-IR, the artificial neural network-based approaches of multi-layer perceptrons and radial basis functions could be trained with representative spectra of the five bacterial groups so that isolates from clinical bacteriuria in an independent *unseen test set* could be correctly identified. ANNs trained with Raman spectra identified only 80% of the same test set. It is likely that this was due to the sample presentation in that the concentration of cells in the slurries was low and future studies will therefore concentrate on analyzing the bacterial samples directly from colonies on petri dishes.

The ANNs for the very high dimensional Raman spectra (2283 wavenumbers) took a long time to train, and for the full-spectral MLPs this was 30 hours. However, we have previously used PCA as a method for reducing the inputs to ANNs (Goodacre *et al.*, 1997; Goodacre, et al., 1996; Timmins & Goodacre, 1997) and in the present study using principal components scores as inputs to MLPs reduced the training time to only 10 mins, with no degradation in the predictive ability of the PC-MLP. Finally the training time for the full spectral RBFs was very quick and only took 2 mins, with equivalent performance compared to the full spectral MLPs.

In conclusion, these results demonstrate that modern analytical spectroscopies, but only when combined with intelligent systems, can provide rapid accurate microbial characterization techniques.

5. Acknowledgments

R.G. and E.M.T. are indebted to the Wellcome Trust for financial support (grant number 042615/Z/94/Z). N.K., A.W. and D.B.K. thank the Chemicals and Pharmaceuticals Directorate of the UK BBSRC, Bruker Spectrospin, Glaxo Wellcome, Renishaw Transducer Systems and Zeneca Life Science Molecules for financial support.

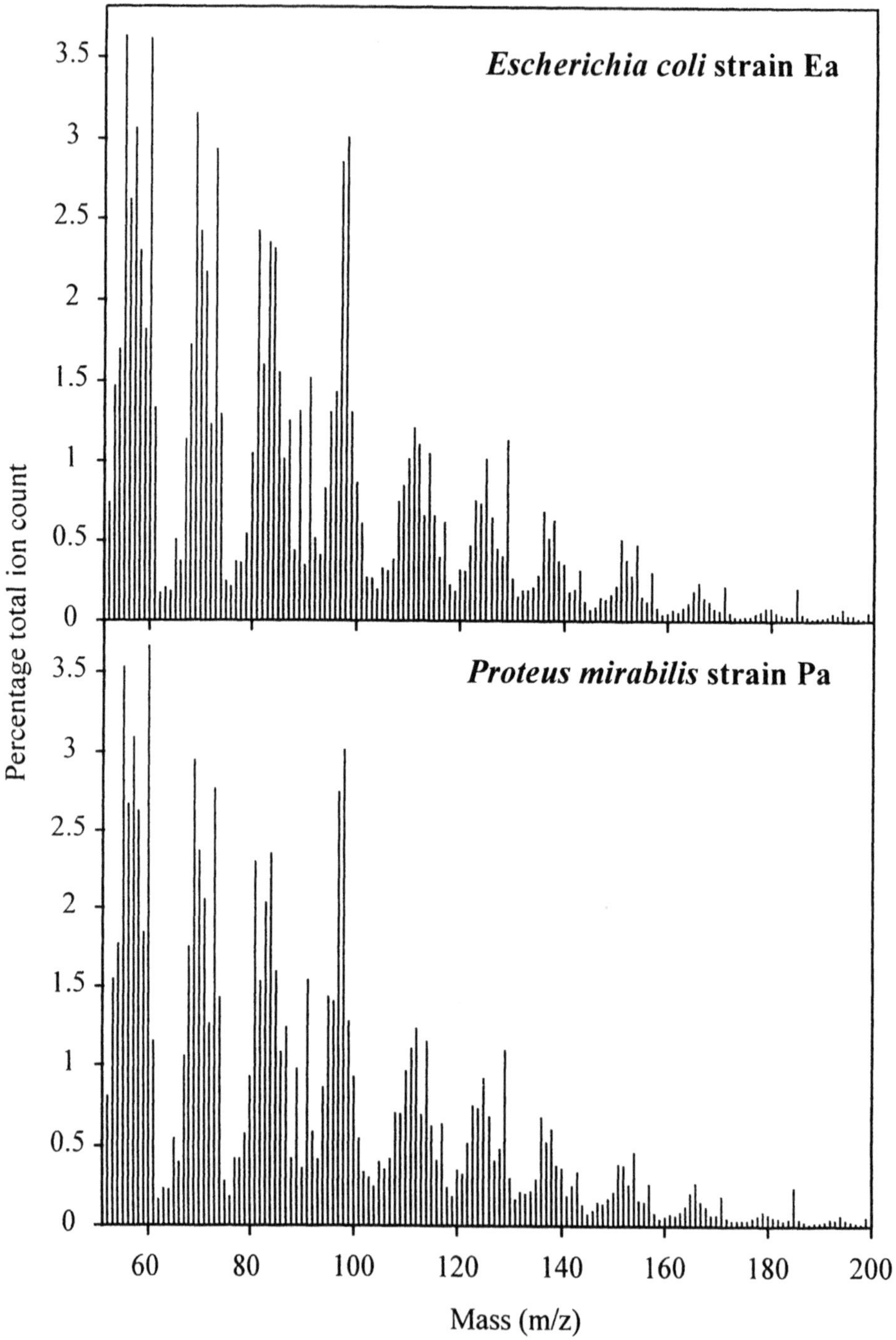

Fig. 1 Normalized pyrolysis mass spectra of *Escherichia coli* isolate Ea and *Proteus mirabilis* isolate Pa.

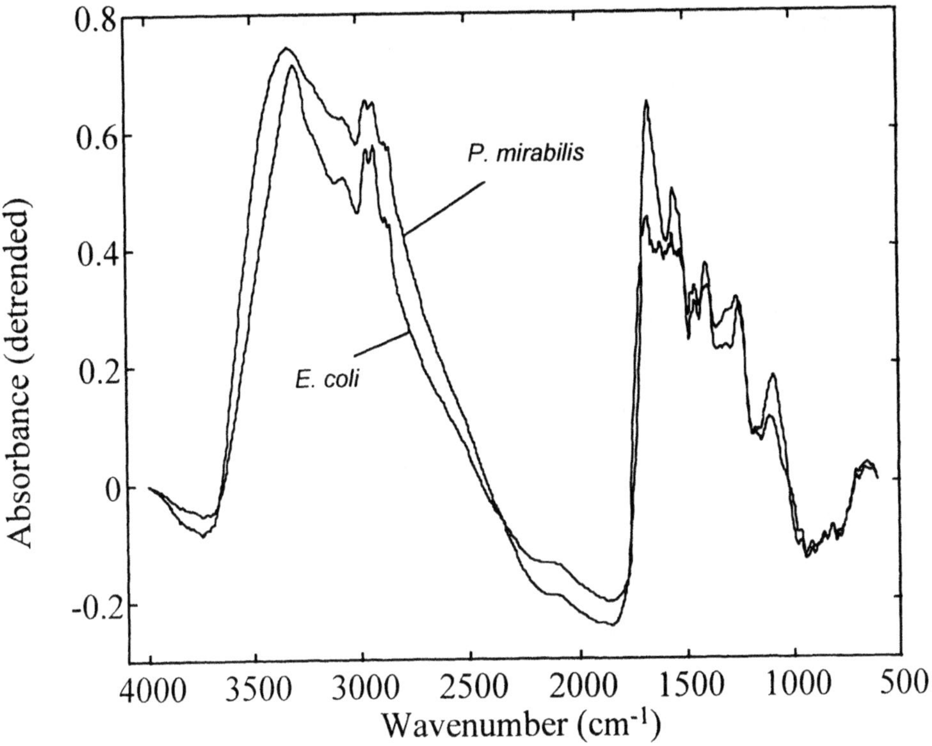

Fig. 2 FT-IR diffuse reflectance-absorbance spectra of *Escherichia coli* isolate Ea and *Proteus mirabilis* isolate Pa.

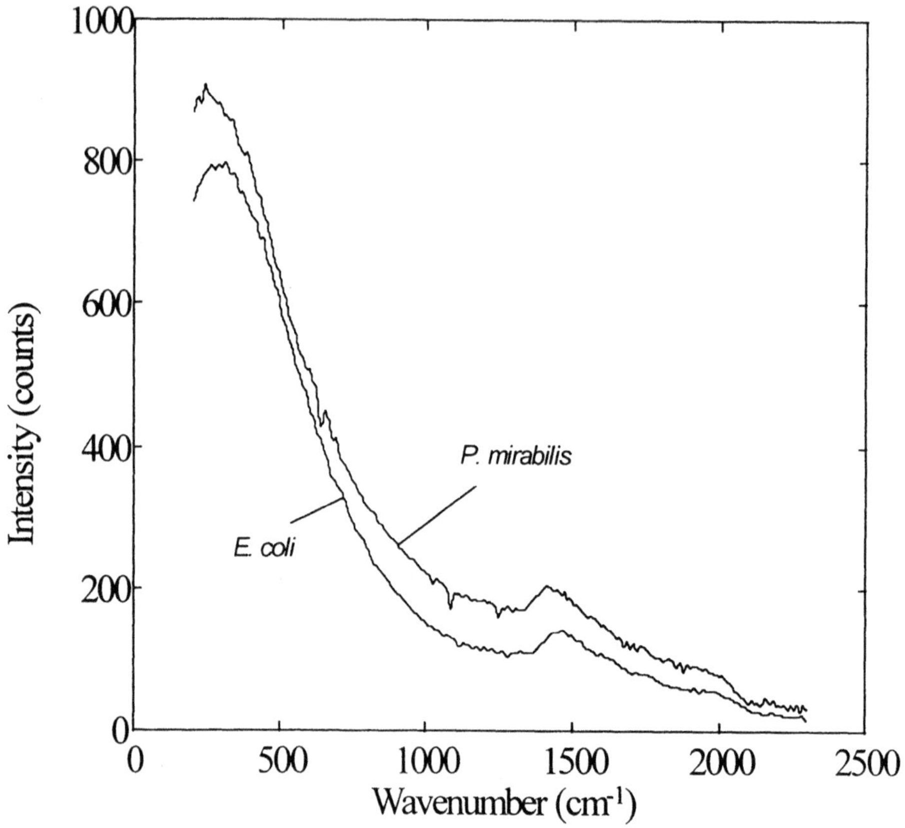

Fig. 3 Dispersive Raman spectra of *Escherichia coli* isolate Ea and *Proteus mirabilis* isolate Pa.

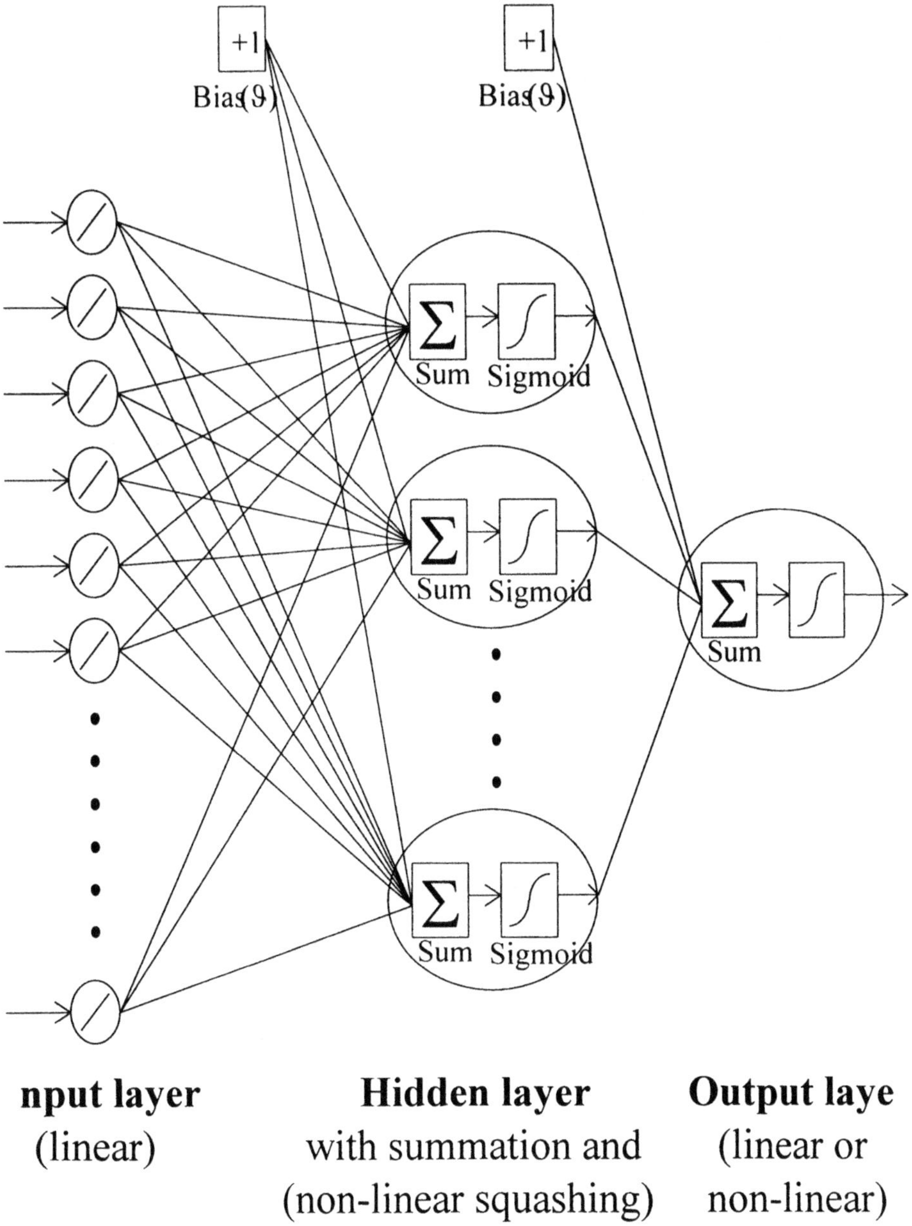

Fig. 4 A multilayer perceptron neural network consisting of an input layer connected to a single node in the output layer by 1 hidden layer. In the architecture shown, adjacent layers of the network are fully interconnected although other architectures are possible. Nodes in the hidden and output layers consist of processing elements which sum the input applied to it and scale the signal using a sigmoidal logistic squashing function.

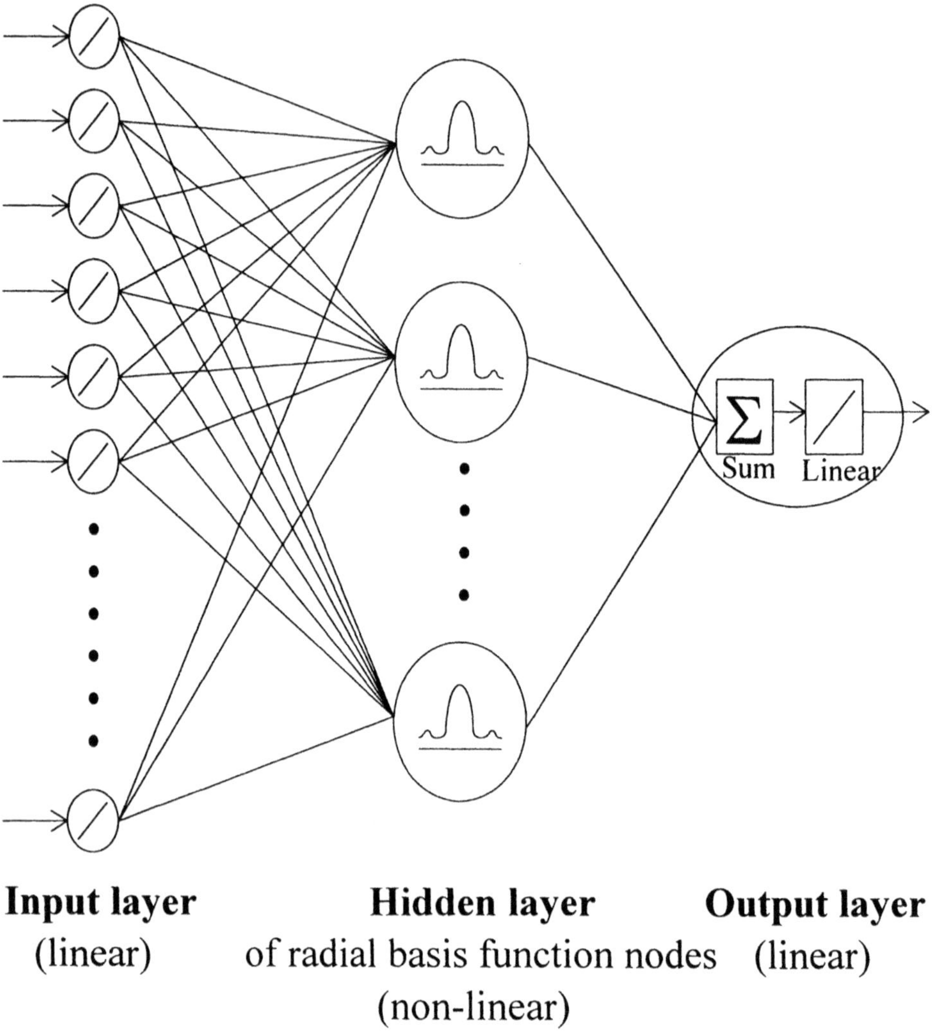

Fig. 5 Radial basis function neural net consisting of an input layer connected to a single node in the output layer by 1 hidden layer. The hidden layer consists of radially-symmetric gaussian functions, although others exist (e.g., Mexican hat and thin plate splines).

A

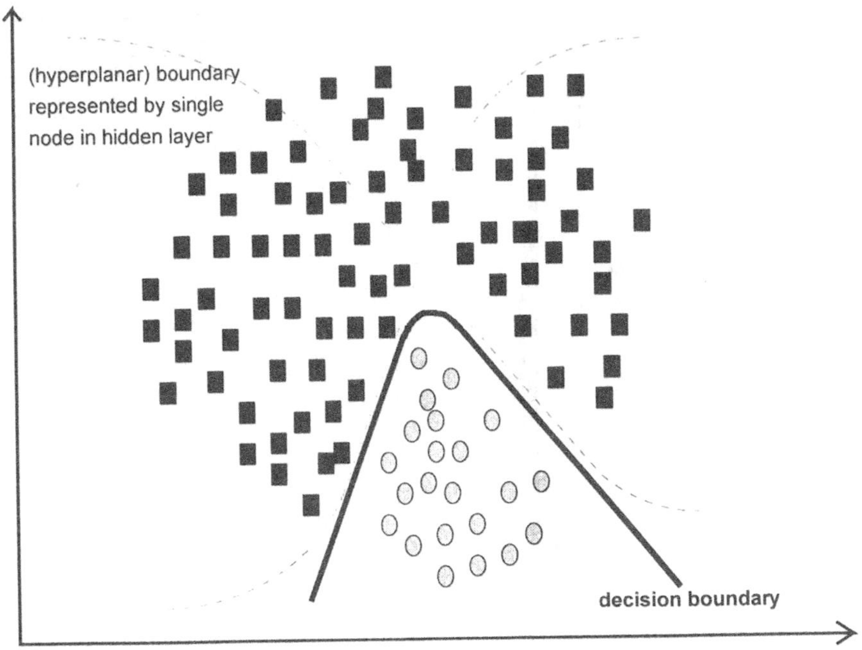

B

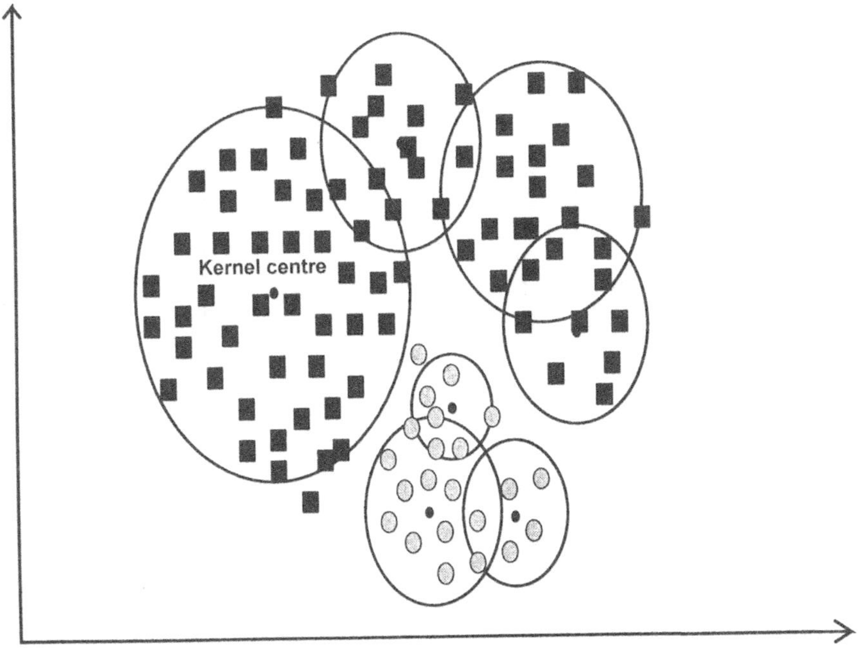

Fig. 6 (A) Typical decision boundary for a classification problem created between two data classes by a MLP with 2 nodes in the hidden layer, for 2 input nodes. Each hidden node represents a non-linear boundary and the nodes in the output layer interpolate this to form a decision boundary. (B) The same classification problem modeled by 7 radially-symmetric basis functions. The width of each kernel function (referred to as its receptive field) is determined by the local density distribution of training examples.

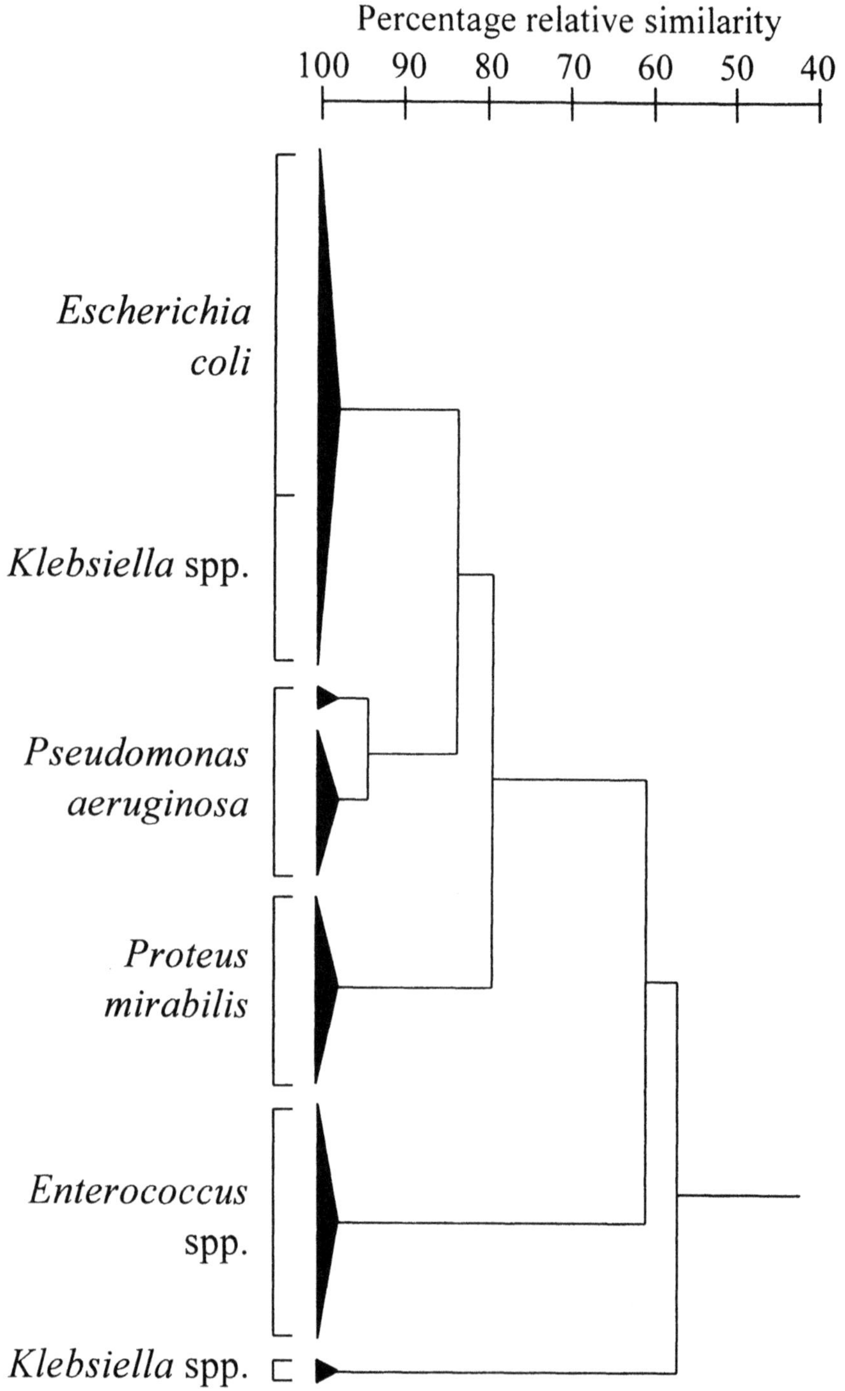

Fig. 7 Dendrogram based on PyMS data showing the relationship between the 59 bacterial isolates.

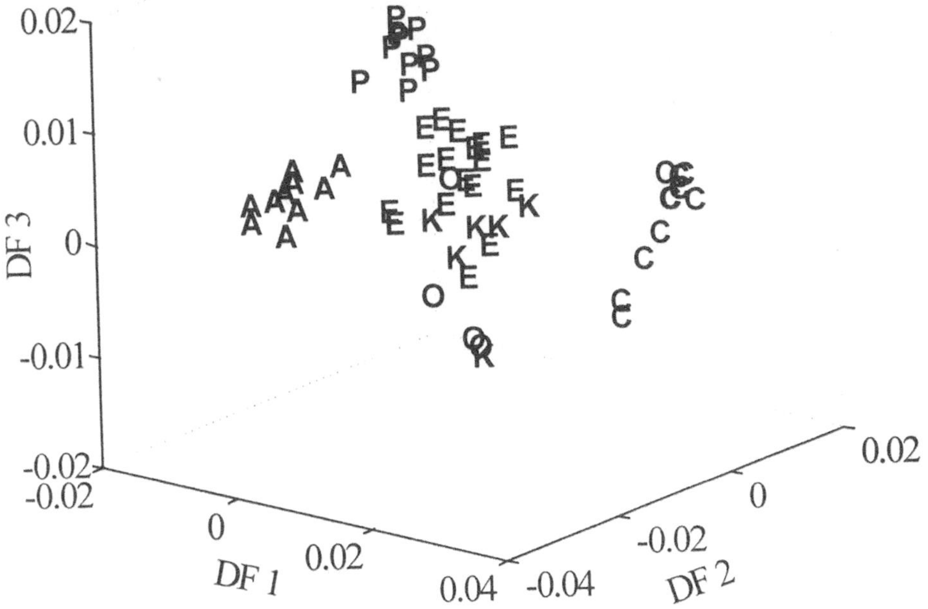

Fig. 8 Pseudo-3D discriminant analysis plot based on FT-IR data showing the relationship between the 59 bacterial isolates. The bacterial isolates are coded as follows; *E. coli* (E), *Prot. mirabilis* (P), *Klebsiella* spp (O and K), *Pseud. aeruginosa* (A), and *Enterococcus* spp. (C).

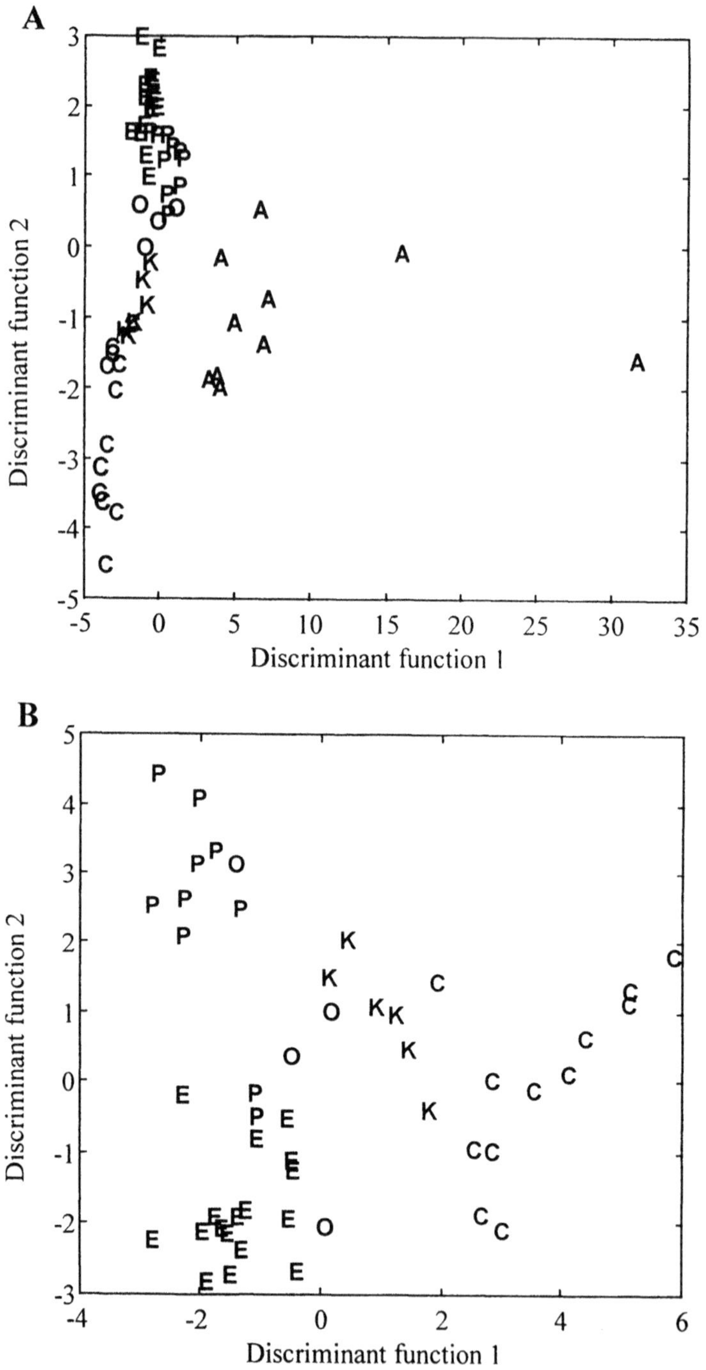

Fig. 9 Discriminant analysis biplots based on Raman data showing the relationship between the 59 bacterial isolates (A) and the same after removal of the *Pseudomonas aeruginosa* isolates (B). The bacterial isolates are coded as follows; *E. coli* (E), *Prot. mirabilis* (P), *Klebsiella* spp (O and K), *Pseud. aeruginosa* (A), and *Enterococcus* spp. (C).

TABLE 1. The 59 urinary isolates from clinical bacteriuria.

Organism	Bronglais Hospital identifier	DFA identifier
Eschericia coli	KX 22, KX 3, QX 2, RB 1, JX 47, JX 44, KX 35, QX 54, EX 8, AX 48, QX 34, QX 27, QX 57, QX 32, GX 3, QX 7, JX 39	Ea-Eq
Proteus mirabilis	EX 20, GX 13, JX 15, HX 23, KX 32, EX 25, HX 45, PX 11, HX 49, RX 44	Pa-Pj
Klebsiella oxytoca	WX 12, RX 15, HX 2, JX 58	Oa-Od
Klebsiella pneumoniae	VX 23, HX 45, DX 18, NX 1, RX 36, KX 44	Ka-Kf
Pseudomonas aeruginosa	VX 54, WX 21, KX 13, PX 8, GX 5, QX 53, HX 36, BX 38, DX 35, HX 24	Aa-Aj
Enterococcus spp.	TX 17, LX 44, CX 11, AX 44, QX 45, PX 6, JX 25, LX 9, VX 13, TX 28, WX 17, RX 51	Ca-Cl

TABLE 2. Identity of the bacteria used in the training set as judged by multilayer perceptron analysis of their PyMS data.

Strain	Bronglais Identifier	Estimates from MLP *E. coli*	*Prot. mirab.*	*Klebs.* spp.	*Pseud. aerug.*	*Enter.* spp.
E. coli	KX 22	**0.9**	0.0	0.1	0.0	-0.1
E. coli	KX 3	**1.0**	0.0	-0.1	0.0	0.0
E. coli	QX 2	**1.0**	0.0	-0.1	0.0	0.0
E. coli	RB 1	**0.6**	0.0	0.4	0.0	0.0
E. coli	JX 47	**1.0**	0.0	-0.1	0.0	0.0
Prot. mirab.	EX 20	0.0	**1.0**	0.1	0.0	0.0
Prot. mirab.	GX 13	0.0	**1.0**	-0.1	0.0	0.0
Prot. mirab.	JX 15	0.0	**0.9**	0.0	0.0	0.1
Prot. mirab.	HX 23	-0.1	**1.0**	0.0	0.1	0.0
Prot. mirab.	KX 32	0.0	**1.0**	0.0	-0.1	0.0
Klebs. spp.	WX 12	0.3	0.0	**0.7**	0.1	-0.1
Klebs. spp.	RX 15	0.0	-0.1	**1.0**	0.0	0.0
Klebs. spp.	VX 23	-0.1	0.1	**1.0**	0.0	0.0
Klebs. spp.	HX 45	0.0	0.0	**1.0**	-0.1	0.0
Klebs. spp.	DX 18	0.2	0.1	**0.7**	0.1	0.0
Pseud. aerug.	VX 54	0.2	-0.1	0.1	**0.8**	0.0
Pseud. aerug.	WX 21	0.0	-0.1	0.1	**1.0**	0.0
Pseud. aerug.	KX 13	-0.1	0.0	0.0	**1.1**	0.0
Pseud. aerug.	PX 8	0.0	0.1	-0.1	**1.0**	0.0
Pseud. aerug.	GX 5	0.0	0.0	0.0	**1.0**	0.0
Enter. spp.	TX 17	0.0	0.0	0.0	0.1	**1.0**
Enter. spp.	LX 44	0.0	0.0	0.1	0.0	**1.0**
Enter. spp.	CX 11	0.0	0.0	0.0	0.0	**1.0**
Enter. spp.	AX 44	0.0	0.0	0.0	0.0	**1.0**
Enter. spp.	QX 45	0.0	0.0	0.0	0.0	**1.0**

TABLE 3. Identity of the bacteria used in the test set as judged by multilayer perceptron analysis of their PyMS data.

Strain	Bronglais Identifier	Estimates from MLP *E. coli*	*Prot. mirab.*	*Klebs.* spp.	*Pseud. aerug.*	*Enter.* spp.
E. coli	JX 44	**0.7**	0.0	0.3	0.0	0.0
E. coli	KX 35	**1.0**	0.0	0.0	0.0	0.1
E. coli	QX 54	**1.1**	0.1	-0.1	0.1	-0.1
E. coli	EX 8	**1.1**	0.1	-0.1	0.0	-0.2
E. coli	AX 48	**1.0**	0.1	0.0	0.1	-0.1
E. coli	QX 34	**1.1**	0.0	-0.1	0.0	-0.1
E. coli	QX 27	**0.9**	0.1	0.0	0.1	-0.1
E. coli	QX 57	**1.1**	0.1	0.0	0.1	-0.2
E. coli	QX 32	**0.9**	0.0	0.2	0.0	0.0
E. coli	GX 3	**1.0**	0.0	-0.1	0.0	0.0
E. coli	QX 7	**1.0**	0.1	0.0	0.0	-0.1
E. coli	JX 39	**1.2**	0.1	-0.1	0.0	-0.2
Prot. mirab.	EX 25	-0.1	**1.0**	0.1	0.0	0.0
Prot. mirab.	HX 45	0.0	**1.1**	-0.1	0.0	0.1
Prot. mirab.	PX 11	-0.1	**1.1**	-0.1	0.0	0.0
Prot. mirab.	HX 49	0.1	**0.8**	0.2	0.1	-0.1
Prot. mirab.	RX 44	0.1	**1.0**	0.0	0.0	-0.1
Klebs. spp.	HX 2	0.2	-0.1	**0.9**	0.0	0.0
Klebs. spp.	JX 58	**0.7**	0.1	0.3	0.1	-0.2
Klebs. spp.	NX 1	-0.1	0.0	**0.9**	0.1	0.1
Klebs. spp.	RX 36	0.2	0.0	**0.7**	0.0	0.1
Klebs. spp.	KX 44	0.1	0.0	**0.9**	0.0	-0.1
Pseud. aerug.	QX 53	-0.2	0.0	0.3	**0.9**	0.0
Pseud. aerug.	HX 36	0.1	0.1	-0.3	**1.1**	0.0
Pseud. aerug.	BX 38	0.0	0.1	-0.1	**1.1**	0.0
Pseud. aerug.	DX 35	0.1	0.0	0.1	**0.8**	0.0
Pseud. aerug.	HX 24	0.0	0.0	0.0	**1.0**	0.0
Enter. spp.	PX 6	0.0	0.0	0.0	0.0	**1.0**
Enter. spp.	JX 25	0.0	0.0	-0.1	0.0	**1.1**
Enter. spp.	LX 9	0.0	0.0	0.0	0.0	**1.0**
Enter. spp.	VX 13	0.0	0.0	-0.1	0.0	**1.0**
Enter. spp.	TX 28	0.0	0.1	0.1	0.0	**0.9**
Enter. spp.	WX 17	0.1	0.0	-0.2	0.0	**1.1**
Enter. spp.	RX 51	0.0	0.0	0.0	0.0	**1.1**

6. Literature Cited

Beale, R. and Jackson, T. (1990). *Neural Computing: An Introduction*. Bristol: Adam Hilger.

Bishop, C. M. (1995). *Neural networks for pattern recognition*. Oxford: Clarendon Press.

Bouffard, S. P., Katon, J. E., Sommer, A. J. and Danielson, N. D. (1994). Development of microchannel thin layer chromatography with infrared microspectroscopic detection. *Analytical Chemistry* **66**, 1937-1940.

Causton, D. R. (1987). *A Biologist's Advanced Mathematics*. London: Allen and Unwin.

Colthup, N. B., Daly, L. H. and Wiberly, S. E. (1990). *Introduction to infrared and Raman spectroscopy*. New York: Academic Press.

Duda, R. O. and Hart, P. E. (1973). *Pattern Classification and Scene Analysis*. New York: Wiley.

Everitt, B. S. (1993). *Cluster Analysis*. London: Edward Arnold.

Ferraro, J. R. and Nakamoto, K. (1994). *Introductory Raman Spectroscopy*. London: Academic Press.

Goodacre, R., Hammond, D. and Kell, D. B. (1997). Quantitative analysis of the adulteration of orange juice with sucrose using pyrolysis mass spectrometry and chemometrics. *Journal of Analytical and Applied Pyrolysis* **40/41**, 135-158.

Goodacre, R. and Kell, D. B. (1996). Pyrolysis mass spectrometry and its applications in biotechnology. *Current Opinion in Biotechnology* **7**, 20-28.

Goodacre, R., Kell, D. B. and Bianchi, G. (1993). Rapid assessment of the adulteration of virgin olive oils by other seed oils using pyrolysis mass spectrometry and artificial neural networks. *Journal of the Science of Food and Agriculture* **63**, 297-307.

Goodacre, R., Neal, M. J. and Kell, D. B. (1994). Rapid and quantitative analysis of the pyrolysis mass spectra of complex binary and tertiary mixtures using multivariate calibration and artificial neural networks. *Analytical Chemistry* **66**, 1070-1085.

Goodacre, R., Neal, M. J., Kell, D. B., Greenham, L. W., Noble, W. C. and Harvey, R. G. (1994). Rapid identification using pyrolysis mass spectrometry and artificial neural networks of *Propionibacterium acnes* isolated from dogs. *Journal of Applied Bacteriology* **76**, 124-134.

Goodacre, R., Timmins, É. M., Rooney, P. J., Rowland, J. J. and Kell, D. B. (1996). Rapid identification of *Streptococcus* and *Enterococcus* species using diffuse reflectance-absorbance Fourier transform infrared spectroscopy and artificial neural networks. *FEMS Microbiology Letters* **140**, 233-239.

Goodacre, R., Trew, S., Wrigley-Jones, C., Saunders, G., Neal, M. J., Porter, N. and Kell, D. B. (1995). Rapid and quantitative analysis of metabolites in fermentor broths using pyrolysis mass spectrometry with supervised learning: application to the screening of *Penicillium chrysogenum* fermentations for the overproduction of penicillins. *Analytica Chimica Acta* **313**, 25-43.

Gower, J. C. (1966). Some Distance Properties of Latent Root and Vector Methods used in Multivariate Analysis. *Biometrika* **53**, 325-338.

Griffiths, P. R. and de Haseth, J. A. (1986). *Fourier transform infrared spectrometry*. New York: John Wiley.

Haykin, S. S. (1994). *Neural networks :a comprehensive foundation*. New York: Macmillan.

Hush, D. R. and Horne, B. G. (1993). Progress in supervised neural networks - what's new since Lippmann. *IEEE Signal Processing Magazine* **10**, 8-39.

Jolliffe, I. T. (1986). *Principal Component Analysis*. New York: Springer-Verlag.

MacFie, H. J. H., Gutteridge, C. S. and Norris, J. R. (1978). Use of canonical variates in differentiation of bacteria by pyrolysis gas-liquid chromatography. *Journal of General Microbiology* **104**, 67-74.

Magee, J. T. (1993). Whole-organism fingerprinting. In *Handbook of New Bacterial Systematics*, pp. 383-427. Edited by M. Goodfellow and A. G. O'Donnell. London: Academic Press.

Manly, B. F. J. (1994). *Multivariate Statistical Methods : A Primer*. London: Chapman & Hall.

Meuzelaar, H. L. C., Haverkamp, J. and Hileman, F. D. (1982). *Pyrolysis Mass Spectrometry of Recent and Fossil Biomaterials*. Amsterdam: Elsevier.

Moody, J. and Darken, C. J. (1989). Fast learning in networks of locally-tuned processing units. *Neural Computation* **1**, 281-294.

Naumann, D., Helm, D., Labischinski, H. and Giesbrecht, P. (1991). The characterization of microorganisms by Fourier-transform infrared spectroscopy (FT-IR). In *Modern techniques for rapid microbiological analysis*, pp. 43-96. Edited by W. H. Nelson. New York: VCH Publishers.

Nelson, W. H., Manoharan, R. and Sperry, J. F. (1992). UV resonance Raman studies of bacteria. *Applied Spectroscopy Reviews* **27**, 67-124.

Park, J. and Sandberg, I. W. (1991). Universal approximation using radial basis function networks. *Neural Computation* **3**, 246-257.

Rumelhart, D. E., McClelland, J. L. and The PDP Research Group (1986). *Parallel Distributed Processing, Experiments in the Microstructure of Cognition*. Cambridge, Mass.: MIT Press.

Saha, A. and Keller, J. D. (1990). Algorithms for better representation and faster learning in radial basis functions. In *Advances in Neural Information Processing Systems*, pp. 482-. Edited by D. Touretzky: Morgan Kaufmann Publishers.

Savitzky, A. and Golay, M. J. E. (1964). Smoothing and differentiation of data by simplified least squares procedures. *Analytical Chemistry* **36**, 1627-1633.

Sleigh, J. D. and Timbury, M. C. (1986). *Notes on Medical Bacteriology*. London: Churchill Livingstone.

Timmins, É. M. and Goodacre, R. (1997). Rapid quantitative analysis of binary mixtures of *Escherichia coli* strains using pyrolysis mass spectrometry with multivariate calibration and artificial neural networks. *Journal of Applied Microbiology* **83**, 208-218.

Wasserman, P. D. (1989). *Neural Computing: Theory and Practice*. New York: Van Nostrand Reinhold.

Werbos, P. J. (1994). *The roots of back-propagation: from ordered derivatives to neural networks and political forecasting*. Chichester: John Wiley.

Wilkins, M. F., Morris, C. W. and Boddy, L. (1994). A comparison of radial basis function and backpropagation neural networks for identification of marine phytoplankton from multivariate flow cytometry data. *Computer Applications In the Biosciences* **10**, 285-294.

Williams, K. P. J., Pitt, G. D., Batchelder, D. N. and Kip, B. J. (1994). Confocal Raman microspectroscopy using a stigmatic spectrograph and CCD detector. *Applied spectroscopy* **48**, 232-235.

Windig, W., Haverkamp, J. and Kistemaker, P. G. (1983). Interpretation of sets of pyrolysis mass spectra by discriminant analysis and graphical rotation. *Analytical Chemistry* **55**, 81-88.

Winson, M. K., Goodacre, R., Woodward, A. M., Timmins, É. M., Jones, A., Alsberg, B. K., Rowland, J. J. and Kell, D. B. (1997). Diffuse reflectance absorbance spectroscopy taking in chemometrics (DRASTIC). A hyperspectral FT-IR-based approach to rapid screening for metabolite overproduction. *Analytica Chimica Acta* , in press.

Wold, H. (1966). Estimation of principal components and related models by iterative least squares. In *Multivariate Analysis*, pp. 391-420. Edited by K. R. Krishnaiah. New York: Academic Press.

FIELD APPLICATIONS OF FLOW CYTOMETRY

PETER J. STOPA*[1], HENRIETTA KULAGA, PATRICIA ANDERSON, AND MICHAEL CAIN[2]
[1]Research and Technology Directorate, U.S. Army Edgewood Research, Development and Engineering Center, APG, MD 21010-5423 USA
[2]Science and Technology Corporation, P.O. Box 7390, Hampton, VA 23666-0390.
* Corresponding Author.

KEYWORDS:
Aerosol Detection, Flow Cytometry, Microbiology, Immunoassay, DNA dyes.

Abstract

Flow Cytometry (FCM), unlike conventional forms of spectroscopy, allows one to interrogate individual members of a sample population for size, shape, biological, and/or chemical properties. Over the past several years, we have investigated the use of flow cytometry to detect and identify microorganisms in aerosols. The effort to date has shown that changes in scatter patterns or DNA composition suggest that changes in the ambient air are evident. The use of macromolecular dyes and fluorescently tagged antibodies can be used to identify specific components in the aerosol. Moreover, it is possible to combine several of these modes into one step. The assays presented here take five minutes or less to perform.

Strategies for the use of flow cytometry to characterize biological aerosols will be discussed. Both laboratory and field studies are discussed which demonstrate the potential use of FCM as a means to detect and characterize microbiological materials in aerosols.

This paper is the opinion of the authors and does not reflect any official opinions of the Department of the Army, Department of Defense, or any other US Government Agency.

The use or naming of any products in this paper does not constitute an endorsement by the Department of the Army, Department of Defense, or any other US Government Agency.

P.J. Stopa and M.A. Bartoszcze (eds.),
Rapid Methods for Analysis of Biological Materials in the Environment, 137–158.

1. Introduction

During World War II, the US Army had a requirement to test gas masks and needed a capability to measure atmospheric particles. In addition, the measurement device also had to differentiate between dust and bacteria. Gucker and his colleagues devised a system whereby aerosols were introduced into a set of tubes, focused into a light beam, and refractive index changes of the particles were measured (1). This system is often referred to as the first modern flow cytometer. Unfortunately, the instrument lay virtually unknown until the late 1960's, when a resurgence of interest in single cell analysis of biological samples led to the development of the first fluorescence activated cell sorters.

Flow cytometry is a process in which individual cells are passed through a focus of exciting light in a linear stream. Most optical flow cytometers use a second sheath fluid flow to hydrodynamically focus the sample stream to a central core within the total fluid flow. The light source may be a laser or an arc lamp. The light is scattered and the distribution and intensity of scattered light is characteristic of the cell passing through the light focus (2). The scattering measurements at different angles provide different information. The size and refractive index of the biological cell primarily determine the amount of light scattered in the forward direction. Large angle or side scattered light provides an indication of particle shape or granularity and also depends on refractive index. When the size of the particles is similar in dimension to that of the light focus or beam waist in the direction of flow, the length or diameter of the particle can be estimated from the duration of the light pulse emitted by the particle. As the particle produces scattered radiation at the location of the laser beam waist, information about the shape of the particle and the location of fluorescing molecules on the particle is encoded on the light pulse (3).

Flow cytometers also perform a wide range of fluorescence measurements. Target cells can be labeled with a dye molecule with a high binding specificity for a particular compound or with a fluorochrome-conjugated antibody or ligand. On passing through the laser light focus, the cells emit a pulse of light at a different wavelength, which can be distinguished from scattered light. Using the same excitation source, the fluorescence may be split into different color components so that several different fluorophores can be measured simultaneously by separate detectors. Data can be automatically interpreted by sophisticated software packages without operator intervention (4).

Adaptation of FCM strategies to microbiological experimentation has become a major focus of current interest in the basic and applied science arenas. In the early 1990's, with the advent of tabletop instruments, oceanographers began to apply this technology to the study of marine microorganisms (5). It has also been shown that prokaryotic organisms can be identified in tissue and body fluids, as well as in environmental samples. This technology has been combined successfully with a number of microbiological and molecular strategies. For example, both bacterial viability and antibiotic sensitivity have been demonstrated on the FCM. At the other end of the spectrum, one integrated copy of HIV-1 can be detected within the genome of human cells. A plethora of microbiological FCM investigations can be found in the literature (6) and will not be detailed in this report.

The present report will discuss the detection of microbiological materials with dyes and identification of specific organisms with specific antibodies. It is of importance to note that all experimental procedures were developed with the following restrictions/conditions in mind: rapidity, reproducibility by non-laboratory personnel, the lack of auxiliary equipment, and multistep/multicomponent use.

2. Materials and Methods

2.1. PREPARATION OF BACTERIAL STOCKS

Bacteria used in this study were obtained from various sources. Isolates of environmental bacteria (*Nocardia globurela (OW23), Acinetobacter haemolyticus (B672), Shingomonas paucimobilis (SP), Pseudomonas aeruginosa* (CSB9), *and Pseudomonas aureofaciens (ML1 and ML2))* were obtained from Dr. Ilona Fry of the Environmental Sciences Team, ECBC. They were 24 hour cultures grown on nutrient agar (Difco Laboratories) and were harvested into phosphate-buffered saline (PBS) (0.01 M Phosphate, pH 7.4, with 0.15 M NaCl, Sigma Cat. # 1000-3) by flooding the plate with 10 ml of PBS, loosening the colonies with a swab, and pipeting off the liquid with a sterile pipet into a sterile centrifuge tube (50 ml, Falcon # 35-0297). Isolates were then washed twice with an additional 10 ml of sterile PBS. Cells were pelleted by slow speed centrifugation and the bacteria were resuspended into PBS. The concentration of the stock was then adjusted to 10^8 Colony Forming Units/ml (CFU/ml) by measuring the absorbance at 660 nm. A value of 0.8 AU corresponds to $1\text{x } 10^8$ CFU/ml.

The following bacteria were obtained in dry powder form from Sigma Chemical Co, St. Louis, Mo.- *Micrococcus luteus (ML)* (Cat # M-3770); *Escherichia coli (EC)* (Cat # EC-11303); and *Staphylococcus aureus (SA)* (Sigma S-2014). Powders were weighed and solutions of bacteria in PBS were prepared at a concentration of 1 mg/ml. Cell number was estimated by the absorbance method mentioned above.

*Bacillus globigii (BG) (*now known as *Bacillus subtilis var. niger)* and *Erwinia herbicola (EH)* were obtained from Dr. Bruce Harper of the Baker Lifesciences Center (BLC), US Army Dugway Proving Ground. BG was supplied as either a dry powder or as spore suspensions in water. EH was supplied as suspensions in either nutrient broth or saline. EH stocks were maintained in the laboratory by growth on nutrient agar at 30 C. Cell number was estimated by either plate dilution techniques, absorbance at 660 nm, or by turbidity (Bio-Log Instruments).

Additional cultures were obtained from the American Type Culture Collection (ATCC), Rockville, MD.

2.2. INTERFERING MATERIALS

Tables 1-3 summarize the pollen, fungal spores, and interfering materials that were evaluated in this study. It lists the common name; botanical name; commercial source; Catalogue #; and particle size.

Table 1. Pollens Used for Interference Studies

Common Name	Botanical Name	Source	Catalogue #	State	Size
Ragweed pollen	*Ambrosia sp.*	Duke Scientific Corp.	14187	Unknown	18-24 um
Ragweed pollen, desert	*Ambrosia dumosa*	Greer Lab	2EE272-4	Non-defatted	25-28 um
Sagebrush	*Artemisia tridentata*	Greer Lab	57GG61-8D	Defatted with Ether	25-30 um
Lycopodium spores	*Lycopodium sp.*	Duke Scientific	14190	Unknown	25-35 um
White Birch	*Betula populifola*	Greer Lab	57HH80-4	Non-Defatted	26-30 um
Black walnut	*Juglans sp.*	Duke Scientific	14188	Unknown	32-42 um
Rye, giant,wild	*Elymus condensatus*	Greer Lab	51AA264-6D	Defatted with Ether	45-50 um
Pine, yellow shortleaf	*Pinus echinata*	Greer Lab	24CC131-48D	Defatted with Ether	72-92 um

Table 2. Mold Spores Used for Interference Studies

Common Name	Botanical Name	Source	Strain Source/Number	Lot #	Size
Penicillium	*Penicillium brevicompactum*	Greer Lab, Inc.	ATCC#9056	MY86-1	2.5-3.5 um
Aspergillus	*Aspergillus fumigatus*	Greer Lab, Inc.	SN-1 #5167	MY3-42	2.5-5 um
Rhizopus	*Rhizopus stolonifer*	Greer Lab, Inc.	ATCC #24862	MY 23-21	5-10 um
Candida	*Candida albicans*	Greer Lab, Inc.	SN-1 #6946	MY15-72	5-8 um

Table 3. Inert Materials Used for interference Studies

Common Name	Source	Catalogue Number
Arizona Road Dust	Particle Technology Inc.	unknown
Bovine Serum Albumin (BSA)	Sigma Chemical Co.	A-4503
Casein	Hammarsten Biochemical Co.	44020
Nutrient Both	Difco Laboratories, Inc.	0003-01
Ovalbumin	Pharmacia	17-0442C
Tryptic Soy Broth	Difco Laboratories, Inc.	0370-17-3

2.3. FLOW CYTOMETER

Three flow cytometers were used in these studies. An Epics XL flow cytometer, manufactured by Coulter Corporation, Miami, FL; the FACSCalibur, and FACSCount instruments, manufactured by Becton-Dickinson, Inc., San Jose, CA. A recent book by Shapiro gives instructions on how to build your own instrument (7).

2.4. DYES

The Coriphosphine O (CPO) dye was obtained from Pfaltz and Bauer, Inc., Waterbury, CT (lot # 15E93). It was supplied as a 27.8 uM solution in sterile water. It was stored at 4 C in the dark.

YOYO-1 (cat # Y-3601) was obtained from Molecular Probes, Eugene, OR., and is supplied as a 1 mM solution in dimethyl sulfoxide (DMSO). It was diluted to a working strength of 10 uM in 70% isopropanol (IPOH).

2.5. ANTIBODY DEVELOPMENT AND SCREENING

Antibodies used were obtained from the Baker Lifesciences Center, US Army Dugway Proving Ground, Dugway, UT. Titers were determined in the Edgewood Laboratory by ELISA. A sample well was considered positive if its absorbance was greater than the mean plus 2 standard deviations of the pre-bleed (blank) samples. A serum, which showed a titer greater than 1:25,600 by this method, was considered to be suitable for immunoassay development.

2.6. FLOW CYTOMETER STUDIES

A Coulter Epics XL flow cytometer was used in the CPO and YOYO-1 studies. The instrument was allowed to stabilize as per its internal start-up procedures and was then calibrated using a set of calibration beads supplied by the manufacturer.

CPO dye was then diluted to a working dilution of 2.78 uM in sterile-filtered water (0.22 uM filter). 100 *ul* of sample is mixed with 500 ul of dye and placed into the instrument. The discriminator is set on green fluorescence (FL1) so that particles or cells that exhibit green fluorescence are measured. The low flow rate setting is used.

The "SAMPLE" protocol is used to obtain the data during the analysis. Peaks that appear in the "Cell" (Region A) or "Spore" (Region B) areas of the cytogram with a count greater than 400 in either histograms 3 or 4 are considered to be positive. The sample was evaluated for 5 minutes or until 10,000 events were recorded.

YOYO-1 dye was also used in these studies. 100 ul of sample was mixed with 100 ul of YOYO-1 dye in isopropyl alcohol (IPOH) and incubated for 3 minutes in the dark. 300 ul of sheath fluid was added to bring the volume up to 500 ul. Forward scatter (log FS) was used as the discriminator so as to analyze particles/cells in the sample that exhibited size and shape (side-scatter (SS)) parameters consistent with bacteria. The discriminator was usually set to channel 10 so that smaller particles were not visualized.

Dye studies were performed on the Becton-Dickinson instruments in a similar fashion.

2.7. ANTIBODY PURIFICATION, LABELING, AND IMMUNOASSAY DEVELOPMENT

Protein G was used to purify the antiserum. A commercially available kit (Pharmacia Fine Chemical, Piscataway , NJ) was used as per the manufacturer's instructions. The fractions are then pooled and passed over a Sephadex G-25 column for equilibration into PBS.

The pooled fractions were labeled using the N-hydroxysuccinimide ester of carboxyfluorescein, as is contained in the Immuno-Ligand Assay ™ (ILA) kit obtained from Molecular Devices Corporation, Sunnyvale, CA, and labeled as per the manufacturer's instructions.

The fluoresceinated antibody conjugates were then evaluated on the Coulter Epics XL flow cytometer against 10-fold dilutions of antigen. 100 ul of antigen and 100 ul of fluoresceinated antibody conjugate were mixed and incubated for 5 minutes. They were then diluted to 0.5 ml and read on the flow cytometer. It was determined that a concentration of 1-5 ng/ml of antibody at a F/P ratio of 4-6 was determined to be optimal.

2.8. AEROSOL STUDIES

The Coulter flow cytometer was used during Joint Field Trial II (held at Dugway Proving Ground, September of 1995). During this exercise, releases of biological simulant materials were made by a variety of techniques, including wet spraying and dry powder. Samples were also evaluated by the following reference methods: bacteria- by plate count methods; MS2 phage virus- by plaque assays on *E. coli*; and ovalbumin- by ELISA. The test support personnel of the Baker Lifesciences Center at Dugway Proving Ground, Dugway , UT, performed these assays.

3. RESULTS AND DISCUSSION

3.1. CPO DYE STUDIES

Nucleic acids, such as DNA and RNA, are common to all biological cells. The content and base-composition varies from species to species; therefore, one could potentially use nucleic acid dyes to differentiate among the various cell types present in a sample. In these studies we investigated the use of two different DNA dyes in an attempt to develop a rapid means to discriminate among different types of biological material in an aerosol sample.

It should be noted that Acridine Orange (AO), the parent compound of the CPO dye that was used in this study, is used to determine the presence of bacteria in blood and water. It is considered to be an impermeant dye and requires a fixative to enter live bacterial cells, particularly gram-negative types of bacteria (8,9). Because the objective was to use a specific protocol *verbatim* in these studies as a reference point, the use of a fixative with CPO was not investigated as part of this study. CPO is more permeant than most of these types of dyes, however, and may enter cells more readily than more lipophilic permeant dyes, such as the cyanines (Howard Shapiro; personal communication).

Figure 1a shows the pattern observed with unstained BG while Figure 1b shows the pattern observed with BG that is stained with CPO at the reported detection limit of 10^5 CFU/ml. In Histogram 1, the particles are split into “Cells” by Region A (lower box) or “Spores” (upper box). The components of theses boxes are then scrutinized separately in Histograms 3 and 4, as denoted by the letters next to the histogram numbers.

The data obtained with CPO showed that this dye stained both spore-forming and vegetative bacteria equally well. However, one might have expected the vegetative cell (region D) scatter to be to the right of the spore scatter (region C) in these histograms since vegetative cells are larger than spores and should have a corresponding larger scatter signature.

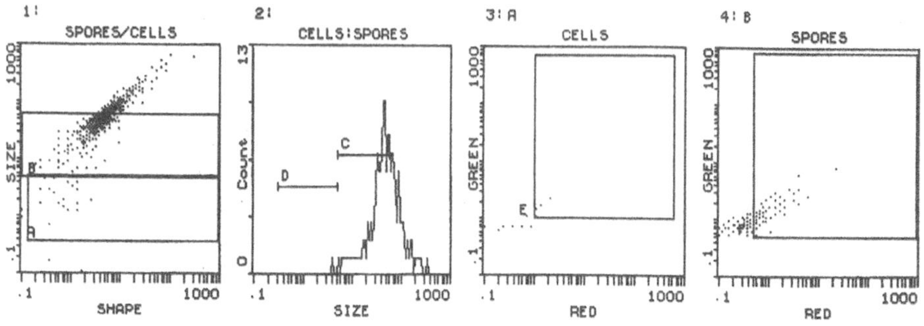

(a)

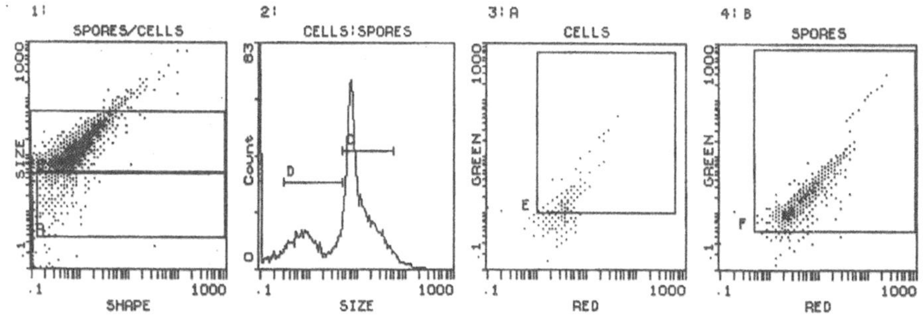

(b)

Figure 1. Example of an assay of the spores of *Bacillus subtilis* var. *niger* (BG) alone (a) and stained with CPO dye (b). Figure 1a shows that the scatter signal of bacteria does not lie totally within region B (Upper box of Histogram 1) and that the peak in region C is shifted over to the right. This data suggests that there is some intrinsic fluorescence for the BG sample. Figure 1b is an example of a sample that gives a positive result for BG (10^5 CFU/ml). Note that scatter lies totally within region B, the peak in region C shifts to the left, and the center of mass is within the box in region F.

3.2. YOYO-1 IODIDE DYE STUDIES

The data presented in this section describes the results obtained with YOYO-1 dye. Figures 2 and 3 show two parameter histograms of various bacteria and interferents. The data were obtained with bacteria or spores at a concentration of 10^6 CFU/ml or concentrations of 10 ug/ml of some of the environmental materials. The data show that upon the addition of the dye, there are clearly increases in the green fluorescence intensity (nucleic acid content) of the cells in the sample and that there are differences in the shape parameter (side-scatter).

Several trends are apparent. There seems to be a limited correlation between gram-positive and gram-negative bacteria. The gram-negative bacteria seem to cluster to the right of the histogram while the gram-positives cluster to the left. This may be due to the fact that the alcohol may have more of an effect on the gram-negative membrane, allowing more of the dye into the cell to react with the double stranded nucleic contained within. Alternatively, there may be a higher nucleic acid content in these particular species, but this needs to be investigated further.

The side scatter parameter also shows that the pattern for bacteria occurs in the second log-decade of signal intensity. There appears to be very little difference among the strains with respect to positional difference.

Figure 3 shows the results obtained with some of the interferences, such as mold spores. Intense fluorescence signals are observed, corresponding to the binding of the dye with the DNA within the cells of those materials that were biological in nature. Very little fluorescence is observed with materials that are either not particulate in nature or contain very little DNA or RNA. Although sagebrush pollen does have DNA, the nature of the outer coating of the pollen may make it impermeant to dyes, even after alcohol fixation.

The side scatter is of particular interest (Figure 3). In some cases, a full log decade increase in either side-scatter or fluorescence is observed as compared to that observed with bacteria (Figure 2). This suggests that one may be able to establish a set of criteria, based on side-scatter data, to differentiate between bacteria and mold spores. Therefore, internal complexity may be used to differentiate among them (10).

Another possibility is that the Arizona road dust that was used in these studies is not necessarily sterile. It may contain bacteria or mold spores that give rise to the relatively weak fluorescence signal that is observed.

This data also suggests that the dye intensity changes as a function of the content of nucleic acid in the organism. This point could be clarified further if one were to plot the intensity as a function of the number of base-pairs in the organism or total nucleic acid content. Since the goal of this study was to evaluate just qualitative differences, this was not performed.

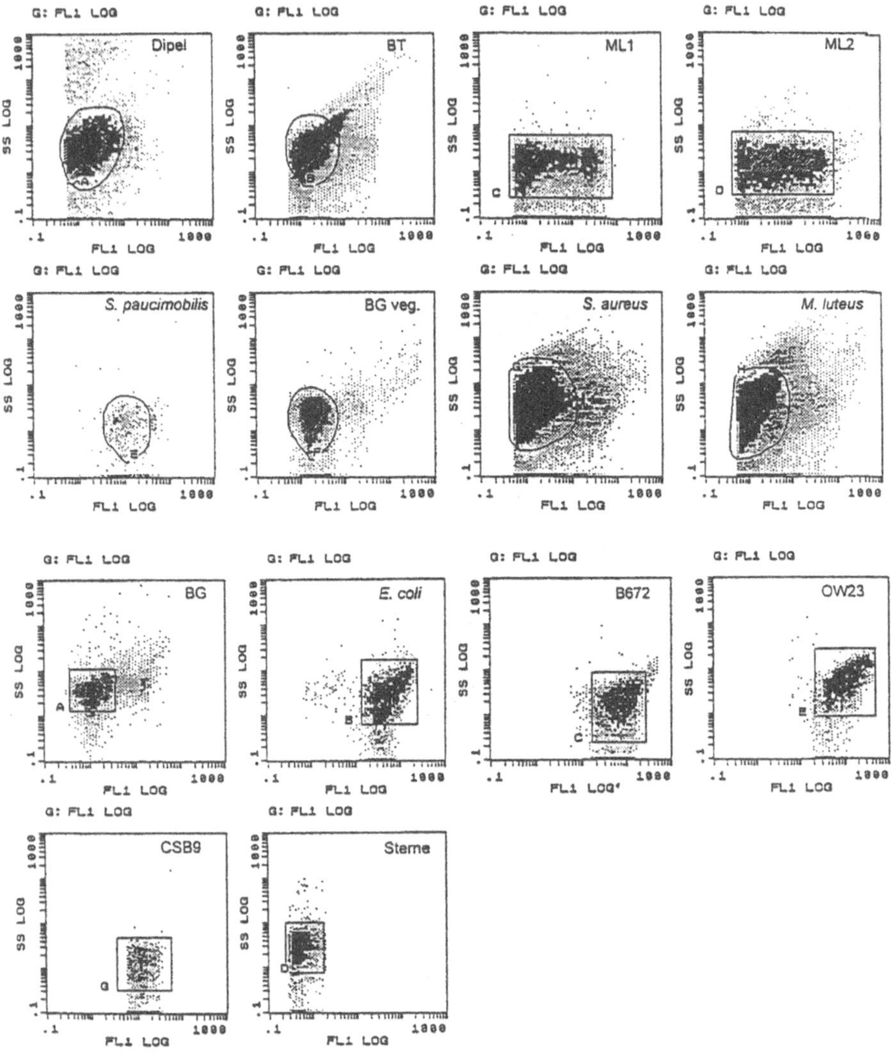

Figure 2. Dual parameter histograms of log side-scatter versus log green fluorescence of YOYO-1 Iodide stained bacterial spores and vegetative cells at 10^6 CFU/ml. Note that there are differences in the position and intensity of the data clusters.

Codes for the organisms are as follows: Dipel – Trade name of commercial preparation of *B. thuringensis* that consists of the cells, protein toxin, and additives; BT- *Bacillus thuringensis* vegetative cells; ML1 and ML2 – 2 different strains of *Pseudomonas aureofaciens*; BG- spores of *B. subtilis* var. *niger*; B672 –*Acinetobacter haenolyticus*; OW23 – *Nocardia globurela*; CSB9 – *Pseudomonas aeruginosa*; Sterne – spores of the Sterne strain of *B. anthracis*

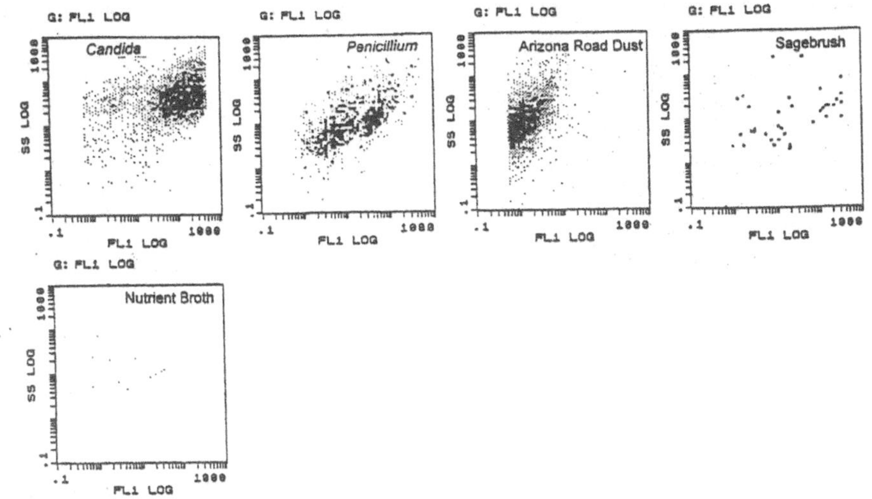

Figure 3. Dual parameter histograms of log side-scatter versus log green fluorescence of YOYO-1 Iodide stained materials. The results show that the materials that mold spores (*Candida* and *Penicillium* sp.) will stain with YOYO-1, but have a higher intensity than bacteria. Other materials show either weak fluorescence (Arizona road dust) or no fluorescence (Sagebrush pollen). Although Sagebrush pollen does have DNA, the nature of the outer coat of the pollen may make it impermeant to the dye, even after alcohol fixation.

Another phase of this study deals with the use of linear mode fluorescence. Linear mode is often used to resolve peaks that lie close together in a histogram. It is primarily used in DNA analysis of mammalian cells. In order to investigate this in bacteria, YOYO-1 was used because it appears to be specific for double stranded nucleic acid and it has a high signal to noise ratio. Therefore, the brighter or more intense the signal, the more nucleic acid the cell should have. The changes in the intensity of the brightness could indicate differences in genome length among the bacteria.

Table 4. The Results Obtained for Several Strains of Bacteria When Stained with YOYO-1 and Evaluated Using the Linear Mode of the Flow Cytometer. The Mean Channel Number (MCN) denotes differences in the signal intensity and may be indicative of the nucleic acid content of the organism.

Bacteria	Mean Channel Number (MCN)
M. luteus	8.1
B. globigii	12.2
B. megatherium (vegetative)	54.3
S. aureus	55.3
S. marcescens	75.3
B. subtilis (vegetative)	80.0
E. coli	85.7
Y. enterocorulitica	94.1
Ps. aeginosa	140.0

Table 4 shows the results obtained with several different bacteria. The bacteria were at 10^5 CFU/ml and stained with the YOYO-1. The Mean Channel Number (MCN) denotes differences in the intensity of YOYO-1dye that is observed when it stains bacteria. This intensity or MCN may be indicative of the ability of the dye to enter the cell or may also indicate the concentration of double-stranded nucleic acid within the cell.

The data also suggests that one might be able to get an idea of the viability of the bacteria. The *M. luteus* that is described is from the Sigma powder and is dead; therefore, it occurs to the left (MCN 8.1) of the BG spores (MCN 12.2), suggesting that it has little double-stranded nucleic acid. The other bacteria used were from live stocks and were fixed in IPOH. The basis for these observation is being correlated to nucleic acid electrophoretic patterns.

Figure 5 shows that a bacterial mixture is resolvable by the linear mode. The auto-gate function of the cytometer was evaluated to see if the cytometer itself could detect different among a mixed population of the bacteria. It was able to successfully sense the different components within the sample. This data shows that YOYO-1 can be an effective dye to determine the presence of bacteria in the sample and can possibly provide a presumptive identification of the bacterium.

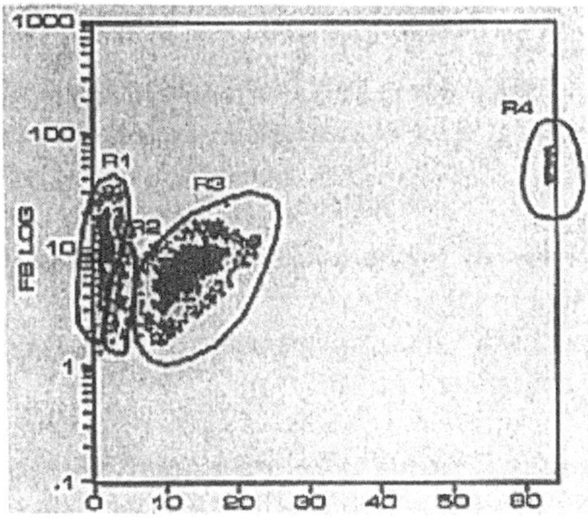

Figure 5. Histogram of a mixture of *M. luteus*, *B. globigii*, and *E.coli* in linear mode. There are distinct clusters in each of the regions that are easily discernible. Region R1 is *Micrococcus luteus*; Region R3 is BG spores; and Region R4 is *Pseudomonas aeruginosa*. Region R2 is debris or possibly an artifact from the "Autogate" function of the cytometer.

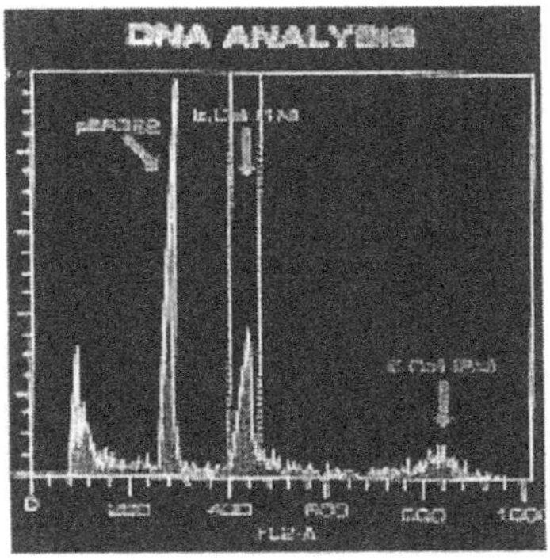

Figure 6. Data obtained from *E. coli* with the pBR 322 plasmid when analyzed by flow cytometry.

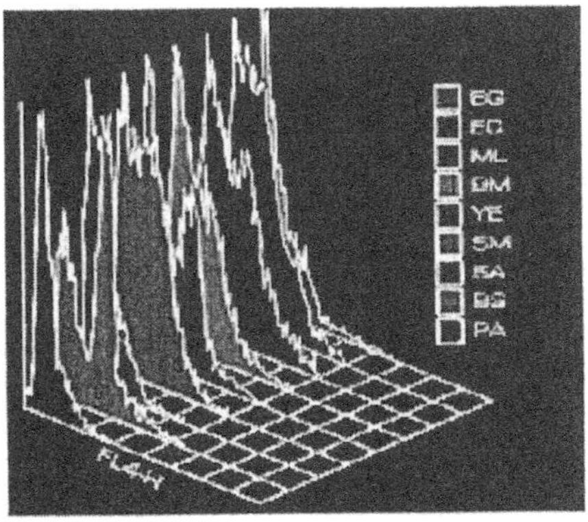

Figure 7. Data obtained from a variety of bacteria when stained with TOTO-3.

Figure 6 is data obtained with the Becton-Dickinson FACSCalibur flow cytometer in pulse processing mode. This mode is used to measure apoptosis of cells and is routinely used in cancer diagnosis. In this use, it can easily detect the presence of a plasmid within *E. coli*.

Figure 7 shows the data obtained when a variety of bacteria is stained with a red fluorescent dye. This study suggests that cytometers can by made even smaller through the use of red diode lasers. The data shows that there are discernible differences among the bacteria when they are stained with this dye.

Figure 8 shows data obtained with the *Sytox* ™ kit from Molecular Probes. The data suggests that limited viability information can be obtained with this kit.

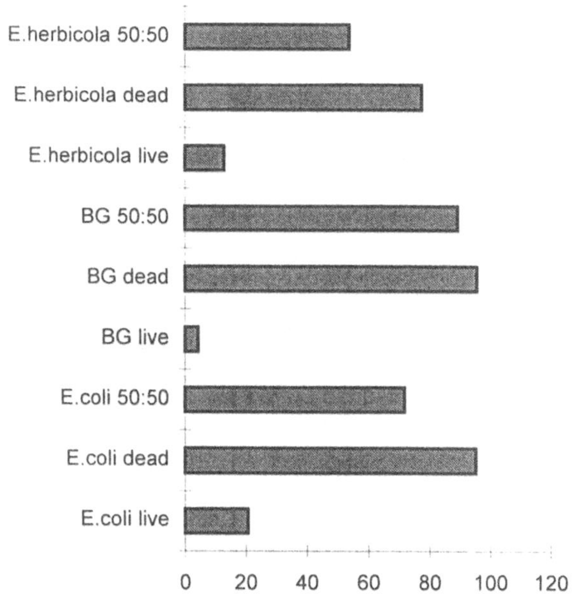

Figure 8. Data obtained with *SYTOX* Live/Dead Kit on the Coulter flow cytometer. The data shows that one can obtain limited data on the viability of bacteria in a sample using this kit.

3.3. IMMUNOASSAYS FOR BACTERIA

An example of a typical histogram of an immunoassay for a bacterium utilizing an antibody labeled with fluoroscein is presented in Figure 9. Figure 10 represents the dose-response curves that were generated for BG spores, *E. herbicola*, *B. anthracis* spores, and *E .coli* (0157:H7). The data show that the immunoassay approach can be extended to the detection of several different types of bacteria and that a sensitive immunoassay with detection limits of 10^3 CFU/ml can be achieved.

The results are reported as "Percent Positive Fluorescence" (PPF) rather than "Counts" per channel. This was done since the desired result was to develop an assay that can be visually interpreted. The PPF was a way of capturing this qualitatively.

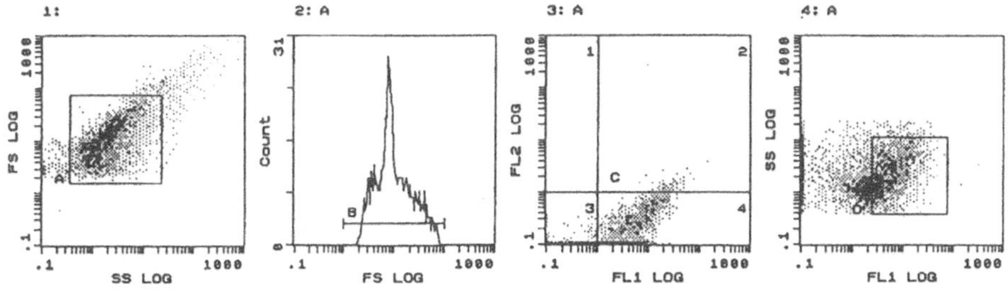

Figure 9. Example of a typical histogram used in an immunoassay. A peak in region D of Histogram and a cluster pattern in region B1 of Histogram 2 denote a positive result.

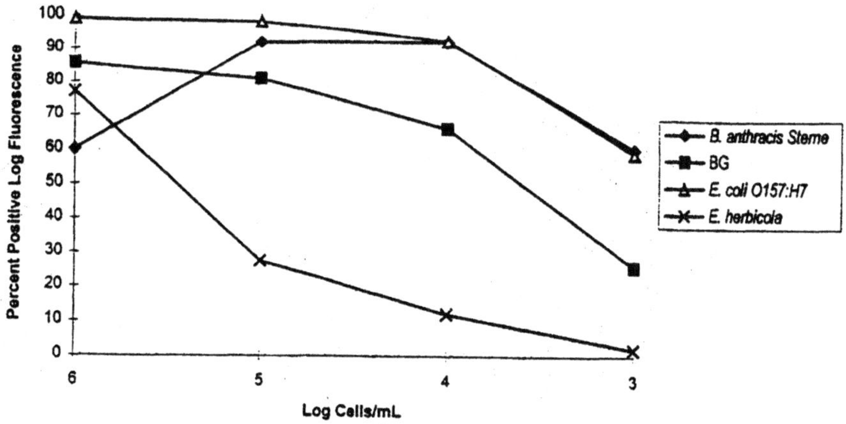

Figure 10. Dose-response curve generated for immunoassays to detect *B. anthracis*, *B. globigii*, *E. herbicola*, and *E. coli* (0157:H7). Detection limit for these materials is 10^3 CFU/ml.

Strategies have also been developed to use these immunoassay techniques as a confirmatory identification system. In this case, antibodies against plasmid specific products of *B. anthracis* are used to measure the appearance of these materials in various strains of bacteria. The accompanying table shows the results of these studies. Note the specificity of the antibody. Other approaches have also been used for immunoassay techniques. Figure 11 shows that simultaneous two colour assays are possible so that one can screen for more than moiety in a sample. Soluble antigens can also be detected through the use of bead-based immunoassays. Figure 12 is an example of such an assay. It is positive for 2 different types of toxins. Notice how the negative beads stay clustered while the positive ones change significantly.

Antibodies Used in *B. anthracis* Immunoassays and Evaluated with Various Strains of *Bacillus*

Bacillus species	MAb 6G6-FITC (*B. anthracis* cell wall)	MAb FDF1B9-FITC (*B. anthracis* capsule)	PCAb *Bacillus* sp.
Δ *B. anthracis* Sterne (pX01⁺, pX02⁻) cells	+	-	+
Δ *B. anthracis* Vollum 1B (pX01⁺, pX02⁺) cells	-	+	+
B. cereus NCTC #9946	-	-	+
B. thuringiensis ATCC #33646	-	-	+
B. subtilis var. *niger*	-	-	+

\+ denotes a positive result.
\- denotes a negative result.

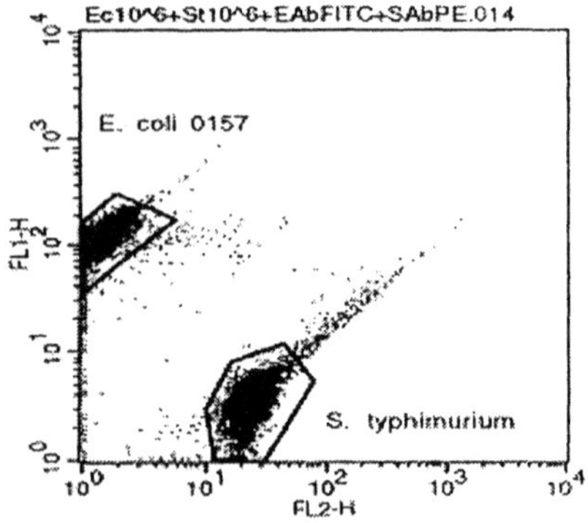

Figure 11. Example of a two color simultaneous immunoassay for *E. Coli* and S. typhimurium

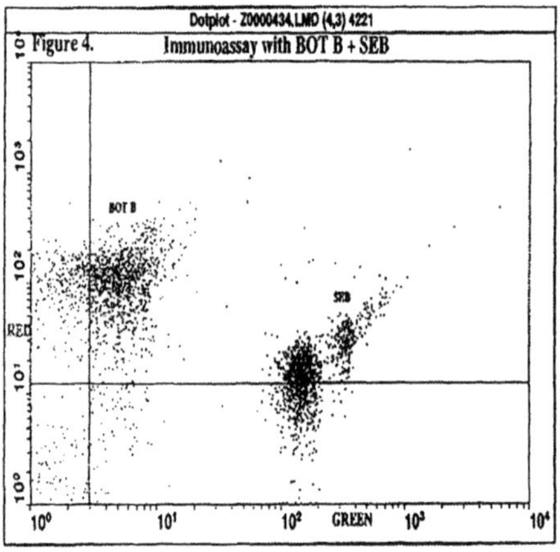

Figure 12. Example of a bead-based assay system for the determinations of soluble toxins in a sample. In this case, the sample is positive for SEB and BOT B toxins.

3.4. AEROSOL STUDIES

The Coulter flow cytometer was used to evaluate aerosol samples in real time during Joint Field Trial II. These samples consisted of liquid impinger fluids collected during 40 releases of 4 different simulant materials and blanks. The simulants included *Bacillus subtilis* var. *niger* (BG- spores), *Erwinia herbicola* (vegetative bacteria), MS2 coliphage virus, and ovalbumin. During each trial, samples were analyzed according to the "Sample" protocol for CPO determinations. In addition, some samples of MS2 were identified as bacteria, which is technically a false positive; but, the MS2 virus is a coliphage and there could have been intact *E.coli* cells or fragments that were mixed in with the material that was disseminated. The FCM was capable of detecting this with the YOYO-1 dye (data not shown). Since there was no referee data regarding this, it was not included in the table. Despite whether or not the material was properly identified with the reagents given by the field trial administrators, the flow cytometer correctly detected the presence of a change in the type and/or concentration of biological materials in the air.

Table 5 summarizes the results as a function of the concentration of the aerosol. The data shows that CPO dye has a high degree of reliability with samples that had a concentration of 10^4 CFU/ml or greater while YOYO-1 had a high degree of reliability with aerosols that had a concentration of 10^3 CFU/ml or greater.

Table 5. Performance of YOYO-1 and CPO Dyes as a Function of Bacterial Concentration in Sampler Fluids Collected during Joint Field Trial II.

Concentration	CPO	YOYO-1
10^1	0 of 1	0 of 1
10^2	0 of 2	0 of 2
10^3	0 of 4	3 of 4
10^4	3 of 3	3 of 3
10^5	4 of 5	5 of 5
Totals	7 of 15	11 of 15

Table 6. Results of Immunoassay Testing of Sampler Fluids Collected During Joint Field Trial II.

Trial Number	Simulant Disseminated	Avg. Sample Concentration (CFU/ml)	Immunoassay ResultS
1	Dry Bg	1.05x10^5	Y
2	Wet Bg	1.37x10^5	Y
6	Dry Bg	3.30x10^1	N
8	Dry Bg	3.02x10^5	Y
9	Wet Bg	6.51x10^4	Y
10	EH	6.08x10^2	Y
12	EH	3.33x10^2	N
15	EH	1.07x10^3	Y
16	MS2	1.00x10^5	FP for EH
19	EH	4.83x10^3	N
23	Dry Bg	5.80x10^4	Y for Bg&EH
24	Wet Bg	1.37x10^3	Y
32	EH	3.16x10^3	N
33	EH	1.59x10^4	N
36	Wet Bg	4.46x10^4	Y
40	EH	3.25x10^5	Y

The results of the immunoassay determination are given in Table 6. In this table, Y means that positive results were obtained; N means that the material was not detected; and FP means that the there was a false positive reaction for that particular trial.

The results of the aerosol studies show that one can use the flow cytometer to determine the biomass content in an air sample; however, further work needs to be performed not only to refine the technique, but also to fully characterize the disseminated sample. In the field trial efforts, we were able to successfully detect the presence of biological materials approximately 50% of the time with the CPO dye and approximately 70% of the time with the YOYO-1. This difference is attributable to the increased sensitivity that the YOYO-1 dye has for the bacteria since it allows detection of bacteria at about an order of magnitude lower than does CPO.

3.5. WATER SAMPLES

Well water samples were also analyzed by this technique. In this study, the samples were stained with CPO and YOYO-1. Typical histograms are shown in Figure 13 for CPO and in Figure 14 for YOYO-1. Table 7 shows a summary of the data. Although there were some differences observed when the sample was stained with CPO, the change in the histogram was not appreciable above the background. On the other hand, there was a dramatic change in the YOYO-1 histograms. The data shows that there was a correlation when compared to classical tests for the presence of enteric bacteria.

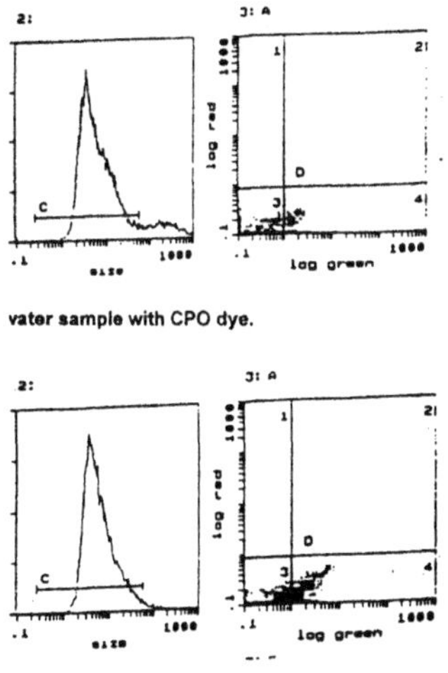

Figure 13. (Top) A negative water sample and a positive (Bottom) water sample stained with CPO dye. Note that there is no appreciable difference between the two.

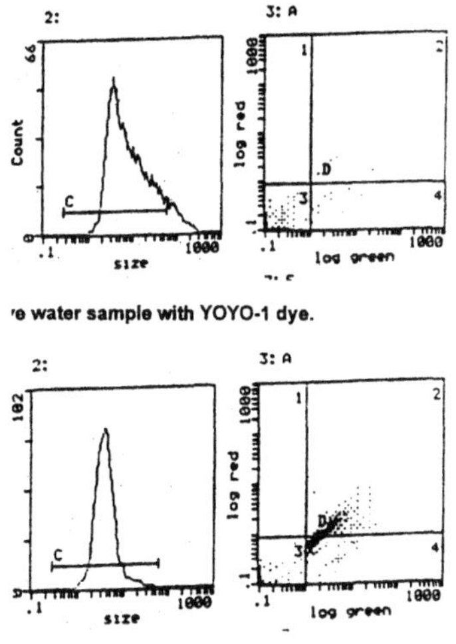

Figure 14. (Top) A negative water sample and a positive (Bottom) water sample stained with YOYO-1 dye. Note that there is the appearance of a cluster in region 2 of histogram 3.

Table 7. Results obtained from 10 well water samples analyzed by flow cytometry.

10 Well Water Samples
Analyzed by Flow Cytometry and
Classical Microbiological Techniques.

Sample Number	YOYO-1 Result	CPO Result	Growth on Nutrient Agar	Growth on EMB Agar	Citrate Test
1	Negative	Inconclusive	+	+	*Turbid*
2	Negative	Inconclusive	+	-	Not Tested
3	Negative	Inconclusive	-	-	Not Tested
4	Negative	Inconclusive	-	-	Not Tested
5	Positive	Inconclusive	+	+	Clear
6	Negative	Inconclusive	-	-	Not Tested
7	Positive	Inconclusive	+	+	Slightly Turbid
8	Positive	Inconclusive	+	+	*Clear*
9	Positive	Inconclusive	-	+	*Clear*
10	Negative	Inconclusive	-	-	*Not Tested*

+ denotes growth.
- denotes no growth.

4. Conclusions

Detection and identification of microbiological agents in the field has been hampered by the lack of adequate techniques that are amenable to non-laboratory use (11,12,13). In addition, detection of any particular material is, by definition, difficult since it will reside on a background of naturally occurring biological flora.

In these studies, two complementary strategies were examined. First, utilization of organic dyes that detect specific properties of nucleic acids indicated that limited microbiological characterization is achievable through the appropriate combination of dyes and instrumental parameters. This has been applied to aquatic systems in the past (14), but this is one of the first applications to aerosol systems. The use of dyes that detect specific parameters of DNA, such as double-strandedness or are specific for particular sequences in the DNA, could be coupled with linear modes of fluorescence to characterize the bacteria in the sample to the genus level. This capability has been previously described using UV dyes (15), but has not been described for blue excitable dyes.

The second approach described uses antibody-based flow cytometric analyses. The materials chosen for this investigation were those which have been shown to simulate pathogenic agents. However, the literature also shows that this approach can be extended to other bacteria that may be of interest (16,17,18).

The application of flow cytometry to microbiological systems is still in its infancy. We are still trying to develop optimal strategies for the use of this technology. Our study shows that flow cytometry is a promising new technique for the characterization of biological materials in aerosols.

5. References

1. Gucker, F.T., O'Kunski, C., Pickard, H.B., and Pitts, J.N., " A photoelectric counter for colloidal particles", *Am. J. Chem.*, **69,** 2422-2431, 1947.

2. Sharpless, T.K., Bartholdi M., et al., "Size and refractive index dependence of simple forward angle scattering measurements in a flow system using sharply focussed illumination", *J. Histoche,. Cytochem.*, **25**, 845, 1977.

3. Steen, H.B. and Lindmo, T., " Flow Cytometry: A high resolution instrument for everyone", *Science*, **204**, 403-404, 1979.

4. Davey, C.L., Kell, D.B., et al., "SKATFIT: A program for determining the mode of growth of individual cells from flow cytometric data", *Binary*, **2**, 127-132, 1990.

5. Boddy, L., morris, C.W., et al., " Neural network analysis of flow cytometric data for 40 marine phytoplankton species', *Cytometry*, **15**, 283-293, 1994.

6. Davey, H.M. and Kell, D.B., " Flow cytometry and cell sorting of heterogenous microbial populations: the importance of single cell analysis, *Microbiological Reviews*, **60(4)**, 641-696, 1996.

7. Shapiro, H. (1992). Practical Flow Cytometry, 2nd edition. New York, Alan R. Liss, In.

8. Back, J.P., and Kroll, R.G., " The differential fluorescence of bacteria stained with acridine orange and the effects of heat", *J. Appl. Bacteriol.*, **71**, 51, 1991.

9. McFeters G.A., Singh A., et al., "Acridine Orange staining as an index of physiological activity in *Eschericia coli*", *J. Microbiol. Methods*, **13**, 87, 1991.

10. Lebaron, P., Parthuiost N., et al., " Comparison of blue nucleic acid dyes for flow cytometric enumeration of bacteria in aquatic systems", *Appl. and Env. Microbiol.*, **64**, 1725, 1998.

11. Lange, J.L., Thorne, P.S., et al., "Application of flow cytometry and in situ hybridization for assessment of exposures to airborne bacteria", *Appl. And Env. Microbiol.*, **45**, 1557, 1997.

12. Tervzieva, S., Donnelly, J., et al., "Comparison of methods for detection and enumeration of airborne microorganisms collected by liquid impingement", *Applied Environmental Microbiology*, **62(7)**, 2264-2272.

13. Bernander, R., Stokke, T., and Boye, E., "Flow cytometry of bacterial cells: comparison between different flow cytometers and different DNA stains", *Cytometry*, **31**, 29, 1998.

14. Robertson, B.R., and Button, D.K., "Characterizing aquatic bacteria according to population, cell size, and apparent DNA content by flow cytometry"" *Cytometry*, **10**, 70, 1989.

15. Van Dilla, M.A., Langlois, R.G., at al., "Bacterial characterization by flow cytometry", *Science*, **220**, 620, 1983.

16. Saunders, G.C., Jett, J.H., and Martin, J.C., "Amplified flow cytometric separation-free fluorescence immunoassay", *Clinical Chemistry*, **31**, 2020, 1985.

17. Seo, K.H., Brackett, R.E., et al., "Immunomagnetic separation and flow cytometry for rapid detection of *E. coli* (0157:H7)", *J. Food Prot.*, **61**, 812, 1998.

18. Clarke, R.G., Pinder, A.C. et al., " Improved detection of bacteria by flow cytometry using a combination of antibody and viability markers", *J. Appl. Microbiol.*, **84**, 577, 1998.

A PORTABLE FLOW CYTOMETER FOR THE DETECTION AND IDENTIFICATION OF MICROORGANISMS

HAZEL M. DAVEY AND DOUGLAS B. KELL
Institute of Biological Sciences, Edward Llwyd Building, University of Wales, Aberystwyth, Ceredigion, WALES, U.K.

Flow cytometry can be used for the detection and identification of microorganism [1]. However traditional instrumentation is bulky and complicated to operate and align. These factors combine to make flow cytometry less than ideal for the field detection of biological agents.

Recently a new flow cytometer, the Microcyte [2], has been developed. The Microcyte is a battery-operated, portable instrument that is simple to use and needs no operator alignment and thus is well suited to field use. The Microcyte uses a 635 nm laser diode as the light source and we present data for a number of microorganisms stained with a variety of fluorescent stains amenable to excitation at this wavelength.

1. Introduction

Cytometry refers to the measurement (-metry) of cells (cyto-). These measurements may be of the cell's physical properties (length, volume, etc.) or of its biochemical properties (protein content, lipid content etc). Traditionally these measurements have been made by light or fluorescence microscopy. Microscopy is a labor-intensive task, prone to error and operator fatigue. It is also slow to perform and consequently conclusions are drawn from measurements of, at best, a few hundred cells. An alternative method of measuring protein content for example, would be to perform bulk biochemical measurements on a population of cells. However in doing this one makes the assumption that all of the cells in a population are behaving in a similar manner. The extent to which this assumption is true will vary with the cell type under investigation but in many cases the heterogeneity of a population of cells will be large, not least because of changes that occur as the cells progress *through* the cell cycle [3]. When samples are collected from the environment a variety of organisms and other particles will be present, increasing the heterogeneity still further.

The termflow *cytometry* refers to these same physical and biochemical measurements [4], but the measurements are made on *individual* cells as they flow past an array of detectors. The cell sample is introduced into the center of a stream of sheath fluid. The

P.J. Stopa and M.A. Bartoszcze (eds.),
Rapid Methods for Analysis of Biological Materials in the Environment, 159–167.

sheath fluid is pumped much more quickly than the sample and so the cells are constrained to the center of the sheath fluid. This process, known as hydrodynamic focusing, allows the cells to be delivered reproducibly to the center of the measuring point.

At the measuring point in a typical microbial flow cytometer [4-6], individual particles pass through an illumination zone, typically at a rate of some 1000 cells.s^{-1}, and appropriate detectors, which may be gated electronically, measure the magnitude of a pulse representing the amount of light scattered. The magnitudes of these pulses are sorted electronically into "channels", permitting the display of histograms of the number of cells possessing a certain quantitative property vs. channel number. Although many of the purposes for which one might use flow cytometry, such as microbial discrimination, require only a qualitative output, the essential character of the flow cytometric approach is strictly quantitative in nature.

The angular-dependence of scattered light provides further information on the nature of the scattering particles [7] but more importantly, appropriate fluorophores may be added to the cell suspension. These may be stains which bind to (or react with) particular molecules such as DNA, RNA or protein [8], fluorogenic substrates which reveal distributions in enzymatic activity [9], indicators which change their property as a function of internal pH [10] or which are taken up in response to membrane energization or, increasingly, antibodies [12] or oligonucleotidcs [13]tagged with a fluorescent probe. Flow cytometry thus represents a powerful method for the rapid analysis of heterogeneous microbial populations [3, 14, 15].

However, flow cytometry has historically possessed four major disadvantages that have greatly decreased its utility for the routine analysis of microbial cell suspensions. The first is the substantial cost: a typical laser-based flow cytometer capable of analyzing but not sorting might work out at *£75,000.* The second disadvantage is that because these are sophisticated instruments, skilled operators are usually required in order to obtain the optimum (and sometimes even any kind of acceptable) performance. Thirdly, in many models the sample flow rate is variable and not measured, so no *absolute* counts are possible. Fourthly, the majority of commercial instruments have been designed primarily for the analysis of animal (particularly human blood) cells and consequently bacteria are on the limit of detection [16, 17]. When one is considering instrumentation for the field detection of biological agents one must also be aware of other disadvantages such as the size (typically 1 to $5\ m^3$), lack of robustness for use outside of a laboratory, large power requirements (especially in laser-based models) and the fact that aerosols of the sample are produced in the analysis chambers of some instruments [18-20].

Recently a new flow cytometer [2] - the Microcyte - has been developed by M/s Optoflow A'S (Oslo, Norway), and commercialized by Aber Instruments Ltd, Science Park, Ceffi Llan, Abetystwyth, Wales, U.K. The Microcyte has been designed primarily for the analysis of microorganisms and detects light scattering and fluorescence from particles as small as 0.5 μm diameter in the standard configuration (and this can be adjusted to allow the detection of even smaller particles if required). The instrument is small with dimensions of 33 x 43 x 16 cm and a weight ~12 kg. The light source is a low-power 635 nm laser diode allowing the instrument to be powered from internal batteries. In addition, the instrument is factory aligned and so does not require highly-trained operators. The sample flow rate in the Microcyte is carefully controlled via feedback, and so the instrument can be used to provide an accurate particle count for concentrations between $1x10^3$ and $1x10^7$ cells.ml^{-1} with an analysis time of less than 10 seconds. The sample analysis area is enclosed and no aerosols are generated during analysis, and after analysis bacteria are trapped in a disposable 0.2 μm filter.

Because almost all existing flow cytometers exploit the argon ion laser line exciting at 488 nm, relatively few studies have been carried out with 635 nm-excitable dyes. However, for most cell types (the main exception being plant cells) the autofluorescence background is significantly lower as one moves to the red end of the spectrum. However there are a growing number of suitable stains for this wavelength and data for some of these are presented.

2. Methods and Methods

Microorganisms

Bacillus subt His var niger spores: Spores from a dry preparation (CBDE, Porton Down, U.K.) were resuspended in K-HEPES buffer (see below).

Escherichia coil: (Gram-negative rod) *E. coli* was grown on L-broth (1% tryptone, 1% yeast extract and 70 $mg.l^{-1}$ $MgSO_4$, pH 6.8) at 37°C for 24 hours. The flask was then left unstirred at room temperature for 24 hours.

Micrococcus luteus (NCIMB 13267): (Gram-positive coccus) *M luteus* was grown on E-broth for 24 hours. The flask was then left unstirred at room temperature for 24 hours.

Baker's Yeast: An isolated strain was grown on YPG medium (5% glucose, *0.5%* yeast extract and 0.5% bacteriological peptone, pH *5)* at 30°C for 3 days.

Au media components were obtained from BDH or LabM.

Buffers

A K-HEPES buffer was used to dilute and/or resuspend cell samples for flow cytometric analyses. The same buffer was also used as a sheath fluid in the Coulter Epics Elite. The buffer contained 150 mM KC1 and 10mM HEPES. The pH of the buffer was adjusted to 6.8 before filtering through a 0.1 μm Whatman WCN filter.

The sheath fluid for the Microcyte was prepared in accordance with the manufacturer's manual and contained 0.1 mM sodium azide to inhibit microbial growth. The pH of the sheath fluid was adjusted to 3.5 by the addition of sulphuric acid and filtered in the same way as the HEPES buffer described above.

Sample preparation

All cultured microorganisms were diluted in K-HEPES to give appropriate concentrations for flow cytometric analyses (~$1x10^6$ $.ml^{-1}$). These organisms thus provided samples of high (but not 100%) viability. Dead samples were prepared by the addition of 70% ethanol to the above samples. After fixing for ~1 minute the fixative was removed by centrifugation and the samples were resuspended in sheath fluid.

Flow Cytometers

All samples were run on the Microcyte flow cytometer in accordance with the manufacturer's manual. In addition, all stained samples were analyzed on a Coulter Epics Elite flow cytometer using a 633 nm HeNe laser for excitation.

Particle Count

Samples containing a range of numbers of particles between 6 x 10^3 and 4 x 10^6 were prepared by dilution from a stationary phase culture of Baker's yeast. Particle counts were performed using the Microcyte in accordance with the manufacturers manual. A region of interest was set around the sample peak on the light scatter display and the analysis time was set to 2 seconds (1 μl). The sample concentration (in particles.ml^{-1}) was read directly from the integral display.

The particle concentration was also determined using a haemocytometer. Sample concentrations of greater than 10^6 particles.ml^{-1} are required for accurate counts by this method and so only the highest two concentrations could be enumerated directly by this method. The concentrations of particles in samples with lower concentrations were determined from their dilution factors.

Fluorescent Stains

The following dyes were obtained from Molecular Probes Ltd, P0 Box 22010, Eugene, Oregon 97402, USA [21].

TO-PRO-3 was obtained as a 1 mM solution in DMSO. It was added to samples to give a final concentration of 1 μM.

Oxonol V was dissolved in ethanol at a concentration of 1 mg.ml^{-1}. This was added to samples to give a final concentration of 1 μg.ml^{-1}.

3,3'-diethyl-thiadicarbo-cyanine iodide (DiSC$_2$(5)) was dissolved in filtered, distilled water at a concentration of 1 mg.ml^{-1}. This was added to samples to give a final concentration of 1 μg.ml^{-1}.

SYTO-17 was obtained as a 1 mM solution in DMSO. It was added to samples to give a final concentration of 1 μM.

3. Results and Discussion

There are many instances in microbiology where a reliable count of the number of particles present in a sample is required. This is true in an industrial or research setting where one may wish to monitor the progress of a fermentation. In the case of bio-warfare agent detection the particle count can be used to give an indication of bacterial load in an air sample and may serve as a useful trigger for other more specific tests. Figure 1 shows a comparison of counts of yeast performed using the Microcyte and by haemocytometry. Only the two most concentrated samples could be enumerated by haemocytometry. In this case haemocytometry took 10-15 minutes per replicate, while 3 counts were done on the Microcyte in less than 30 seconds.

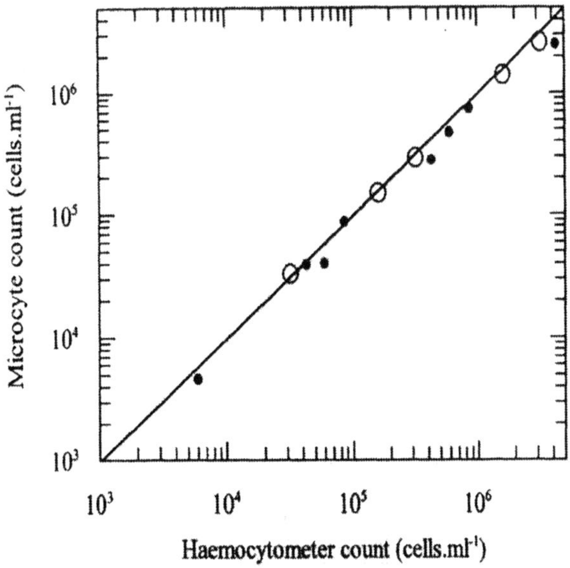

Figure 1 : Comparison of haemocytometer and Microcyte counts. • = Yeast, O = 5 μm latex beads. The best fit line by linear regression has a slope of 0.970, an intercept of 0.079 and a correlation coefficient of 0.996.

Fluorescent stains such as TO-PRO-3 which are excluded by the membranes of intact cells can be used as the basis of flow cytometric viability measurements. Figure 2 shows the results of such an assay, performed with Baker's yeast. Two separable populations can clearly be seen in the stained, unfixed sample; according to the dye exclusion method of viability determination those with lower fluorescence are "viable", while those that are more brightly fluorescent are dead.

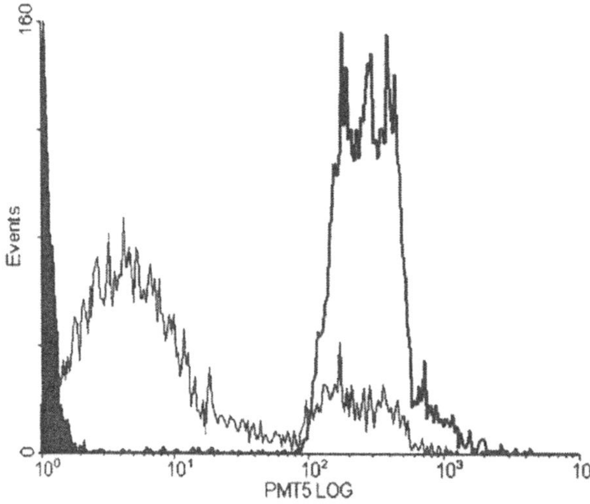

Figure 2 : Flow cytometric analysis of Baker's yeast performed on the Coulter Epics Elite. The 633 nm HeNe laser was used for excitation and fluorescence was collected via a 675 nm band-pass filter. The amplifier gain was logarithmic. ■ Unstained yeast, thin line = unfixed yeast, thick line = fixed yeast. TO-PRO-3 is excluded by the membrane of viable cells but freely enters dead (fixed) cells. In the example shown 18.7% of the unfixed yeast suspension were brightly stained and according to this method would be recorded as dead.

Some fluorescent stains also offer a level of discrimination between microorganisms. Figure 3 shows the results of flow cytometric analysis of a range of microorganisms stained with a variety of fluorescent stains. The data are displayed as collected on the Coulter Epics Elite. However, the results obtained on the Microcyte were qualitatively the same.

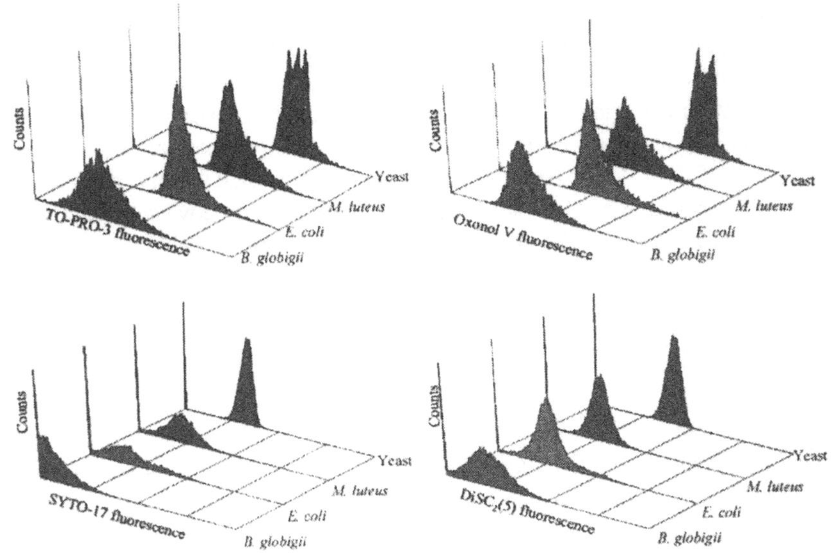

Figure 3: Flow cytometric analysis of *B. globigii*, *E. coli*, *M. luteus*, and yeast with four red-excitable fluorescent stains. The data were collected using the Coulter Epics Elite flow cytometer as described in the legend to Figure 2. In most cases the spores of *Bacillus globigii* are somewhat less fluorescent than the vegetative cells of the other organisms measured.

Figure 4 shows the broad distribution of light scattering properties exhibited by *Bacillus globigii* spores. This is in contrast to the tight distribution observed in 1 μm beads. This is due to the fact that spores form clumps with each other and with cell debris [22].

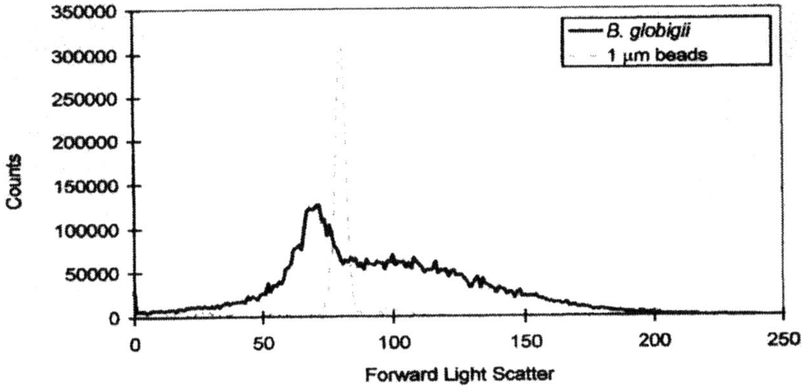

Figure 4: Flow cytometric analysis of *Bacillus globigii* was performed on the Microcyte as described in the materials and methods. The logarithmic amplifier was used for both samples. A broad distribution of light scattering from the spores is observed, however the tight distribution observed for the 1 μm diameter beads shows that the Microcyte is functioning optimally.

In conclusion, flow cytometry has many advantages for the detection of biowarfare agents, not least because of the fact that single cells are analyzed and consequently identification of these agents against a background of other particulates is possible. New instrumentation is beginning to overcome many of the historical disadvantages of the technique and we can expect this to continue in the future. There is a constantly expanding range of fluorescent dyes and these encompass a wide variety of excitation wavelengths. Many of these offer some level of discrimination between particle types. Where the presence of a particular organism is suspected fluorescently labeled antibodies will offer a higher level of specificity, but initial experiments with broad-specificity dyes will continue to be an important first step in the detection and identification of microorganisms.

4. Acknowledgements

The authors would like to acknowledge the U.S. Army through its European Research Office for financial support, Aber Instruments Ltd for the loan of the Microcyte flow cytometer and Roger Laine of Anomeric Inc. and Oddbjorn Gjelsnes of Optoflow A/S for useful discussions.

5. References

1. Davey, H.M. and Kell, D.B. *Rapid flow cytometric detection and identification of microbial particles using multiple stains and neural networks,* in *Scientific Conference on Chemical and Biological Defense Research.* 1996. Edgewood, Aberdeen Proving Ground. MD.

2. Gjelsnes, 0. and Tangen, R., *Liquid flow cytometer.* 1994: Norway.

3. Kell, D.B., Ryder, H.M., Kaprelyants, A.S., and Westerhoff. H.V., *Quantjfying heterogeneity: flow cytometry of bacterial cultures.* Antonie Van Leeuwenhoek. 1991. 60: p. 145-158.

4. Shapiro, H.M., *Practical flow cytometry (3rd edition). 1995.* New York: Alan R. Liss Inc.

5. Steen, H.B. and. Lindmo, T., *Flow cytometry: a high-resolution instrument for everyone.* Science, 1979. 204: p. 403-404.

6. Davey, H.M., Davey, C.L., and Kell, D.B., *On the determination of the size of microbial cells using flow cytometry,* in *Flow cytometry in microbiology,* Lloyd. D., Editor. 1993. Springer-Verlag: London. p. 49-65.

7. Salzman, G.C., Mullaney, P.F., and Price, B.J., *Light-scattering approaches to cell characterization,* in *Flow cytometry and sorting,* Melamed, M.R., Mullaney, P.F., and Mendelsohn, M.L., Editors. 1979, Wiley: New York. p. 105-124.

8. Waggoner, A.S., *Fluorescent probes for cytometry,* in *Flow cytometry and sorting (2nd edition),* Melamed, M.R., Lindmo, T., and Mendelsohn, M.L., Editors. 1990, Wiley-Liss Inc.: New York. p. 209-225.

9. Dolbeare, F.A. and Smitfr R.E., *Flow cytoenzymology: Rapid enzyme analysis of single cells,* in *Flow cytometry and sorting,* Melamed, M.R., Mullancy, P.F., and Mendelsohn, M.L., Editors. 1979, Wiley: New York. p. 3 17-333.

10. van Em, P.E.J., Jansen, M.J.J.M., dc Jongh, G.J., Boezeman, J.B.M., and Schalkwijk, J., *Radiometric measurement of intracellular pH in cultured human keratinocytes using carboxy-SNARF-1 and flow cytometry.* Cytometry, 1991. 12: p. 127-132.

11. Kaprelyants, A.S. and Kell, D.B., *Rapid assessment of bacterial viability and vitality using rhodamine 123 and flow cytometry.* Journal of Applied Bacteriology. 1992. 72: p. 410-422.

12. McClelland, RAIL, *Detection of Salmonella typhimurium in dairy products with flow cytometry and monoclonal antibodies.* Applied Environmental Microbiology. 1994. 60(12): p. *4255-4262.*

13. Wallner, H., Amann, R., and Beisker, W., *Optimizing fluorescent in situ hybridization with rRNA targeted oligonucleotide probes for flow cytometric identification of microorganisms.* Cytometry, 1993. **14:** p. 136-143.

14. Edwards, C., Porter, J., Saunders, J.R., Diaper. J., Morgan, J.A.W., and Pickup, R.W., *Flow cytometry and microbiology.* S.G.M. Quarterly. 1992. November: p. 105-108.

15. Lloyd, D., *Flow cytometry in microbiology.* 1993. London: Springer-Verlag. 188.

16. Jernaes, M.W. and Steen, H.B., *Staining of Escherichia colifor flow cytometry: influx and efflux of ethidium bromide.* Cytometry, 1994. 17: p. 302-309.

17. Davey, H.M. and Kell, D.B., *Flow cytometry and cell sorting of heterogeneous microbial populations: the importance of single-cell analyses.* Microbiological Reviews. 1996. 60(4): p. 641-696.

18. Ferbas, J., Chadwick, K.R., Logar, A., Patterson. A.E., Gilpin. R.W., and Margolick, J.B., *Assessment of aerosol containment on the Elite flow cytometer.* Cytometry, *1995.* 22(1): p. *45-47.*

19. Schmid, I., Nicholson, J.K.A., Giorgi, J.V., Janossy. 0., Kunkl. A., Lopez, P.A., Perfetto, S., Seamer, L.C., and Dean, P.N., *Biosafety guidelines for sorting of unfixed cells.* Cytometry, 1997. 28(2): p. 99-117.

20. Schmid, I. and Dean, P.N., *Introduction to the biosafety guidelines for sorting of unfixed cells.* Cytometry, 1997. 28(2): p. 97-98.

21. Haugland, R.P., *Molecular probes handbook flourescent probes and research chemicals (6th edition)*, ed. Spence, M.T.Z. 1996: Molecular Probes Inc. 679.

22. Davey, H.M. and Kell, D.B. *Rapid flow cytometric detection and identification of microbial particles.* in *Scientific Conference on Chemical and Biological Defense Research.* 1994. Edgewood, Aberdeen Proving Ground, MD.

BIODIVERSITY OF BACTERIAL POPULATIONS MONITORED BY FLOW CYTOMETRY

ELENA STABNIKOVA
Department of Microbiology and Biotechnology, Ukrainian University of Food Technologies, 68 Vladimirskaya Str., Kyiv 252017, Ukraine

Abstract

Biodiversity of population of *Bacillus thuringiensis, B. licheniformis,* and *B. megaterium* was studied by flow cytofluorometry and other methods. The distributions in size, DNA content and RNA and protein contents of the cells were measured by FACStarPlus flow cytofluotometer (Becton Dickinson). The bacterial neuston contains more cells from the beginning stage of cell division cycle and DNA replication cycle that the cells from the bulk of water. Anabiotic and endotrophic subpopulations of bacilli are accumulated in the water-air interface. The distributions of the cells by DNA content have the different forms depending on the share of C-period within the division cycle. Flow cytofluorometry analysis of the synchronous cultures of bacilli shown that the curves have the patterns with one maximum, two maxima or with one maximum and plateau. The share of cells in B-period and share of exotrophic cells of population are suitable characteristics of the physiological state of populations and can be rapidly determined by the flow cytofluorometry. Their determination may be used for identification of the source and term of biohazard's release into the environment by the comparance of physiological state in the release point and in the environment; for the determination of the physiological state of biohazard in the natural ecosystems which is important for the risk assessment; and for the determination of the physiological state of biohazard before release into the environment by the comparance of the changes of physiological state in the sampling point(s) of the environment.

Key words: flow cytofluorometry, *Bacillus,* population, biodiversity, environment, monitoring.

1. Introduction

The heterogeneity of microbial populations is originated from the temporal, physical, chemical and biological relations between the elements of microbial system [3,4]. One type of the heterogeneity of microbial populations is the distribution of the cells of one population by DNA content, which is depending on the position of a cell within the cell cycle. An analysis of the bacterial populations is complicated by the absence of the synchronization between the cell division cycle and DNA replication cycle [1]. The share of the cells with the different content of DNA depends on the duration of B, C,

P.J. Stopa and M.A. Bartoszcze (eds.),
Rapid Methods for Analysis of Biological Materials in the Environment, 169–175.

and D –periods. Period B of bacterial cell cycle is the time from the end of cell division and the beginning of chromosomal DNA replication, period C is the duration of chromosomal DNA replication and period D is the time between the termination of the chromosomal DNA replication and the initiation of cell division.

Flow cytometry is a powerful analytical technique that can be used to yield the information about the physiological diversity of bacterial population. The analysis of DNA content and use of specific probes to enumerate the viable bacteria and to measure the RNA and protein content, membrane potential, and to identify the bacterial species, give the possibility to study the physiological heterogeneity of population not in laboratory but in the natural ecosystems also [7].

The aim of the work is to discuss the possibility of flow cytofluorometry for rapid analysis of physiological state of bacterial populations in the environmental monitoring.

2. Materials and Methods

The objects of our study were the strains of *Bacillus megaterium, Bacillus thuringiensis,* and *Bacillus licheniformis* from the Collection of Bacilli, Institute of Microbiology and Virology, National Academy of Sciences of Ukraine. Bacteria were cultivated in the fermenter with the working volume 1L, at the 30°C and pH of medium 7.0. Bacilli were cultivated on media of the following composition (g/L): glucose, 10.0; Na_2PO_4, 3.0; K_2HPO_4, 3.0; NaC1, 2.0; $MgSOx7H_20$, 0.1; peptone, 0.5 (for *B. thuringiensis)*; sucrose, 6.5; KH_2PO_4, 1.5; K_2HPO_4, 3.5; NaC1, 01; NH_4C1, 1.0; KC1, 0.04; Na_2SO_4, 0.3; $MgC1_2x5H_2O$, 4.3 (for *B. licheniformis)*; ethanol, 0.5; NH_4NO_3, 2.0; $MgC1_2$, 0.1; KH_2PO_4, 3.0; K_2HPO_4, 7.0 (for *B. megaterium*). Synchronization of bacterial proliferation was provided by separation from the population of quickly flotating cells [11]. This fraction was used as an inoculum. The technique ensures the index of synchronization from 0.55 to 0.70.

To determine the DNA content in the cells by flow cytofluorometry the known method of dyeing by propidium iodide solution with prehydrolysis of RNA and protein was used [2]. The curves of the distribution of 20 thousand cells by DNA content (intensity of red fluorescence) or by cell size (intensity of forward scattered light) were obtained using FACStarPlus flow cytofluorometer (Becton Dickinson). The fluorescence of the cells was excited by argon laser with wavelength of 488 nm. To measure the red fluorescence of propidium iodide- DNA complex, a 590 nm, barrier filter scatter was used in front of the red channel of photomultiplier . Forward light scatter was measured at angles between 1.5° and 19° to the illuminating beam.

To determine the share of exotrophic cells in the population of *Bacillus megaterium*, the microbial suspension, grown in the medium with ethanol as sole carbon source, was incubated 10 minutes in the solution of amyl alcohol (final concentration was 8 g/L), and dead cells were determined by the luminescence microscopy after the dyeing in the solution of acridine orange (final concentration was 0.2 g/L [5].

3. Age Structure of Population: Theory and Results

Bacterial cells occupying different periods of cell division cycle and DNA replication cycle may be determined by flow cytofluorometry. Figure 1 shows the theoretical variants of the distribution of DNA content in the cells for the cases of slow, medium, and fast rate of DNA replication

FIGURE 1. Structure of bacterial call cycle and distribution of DNA content.

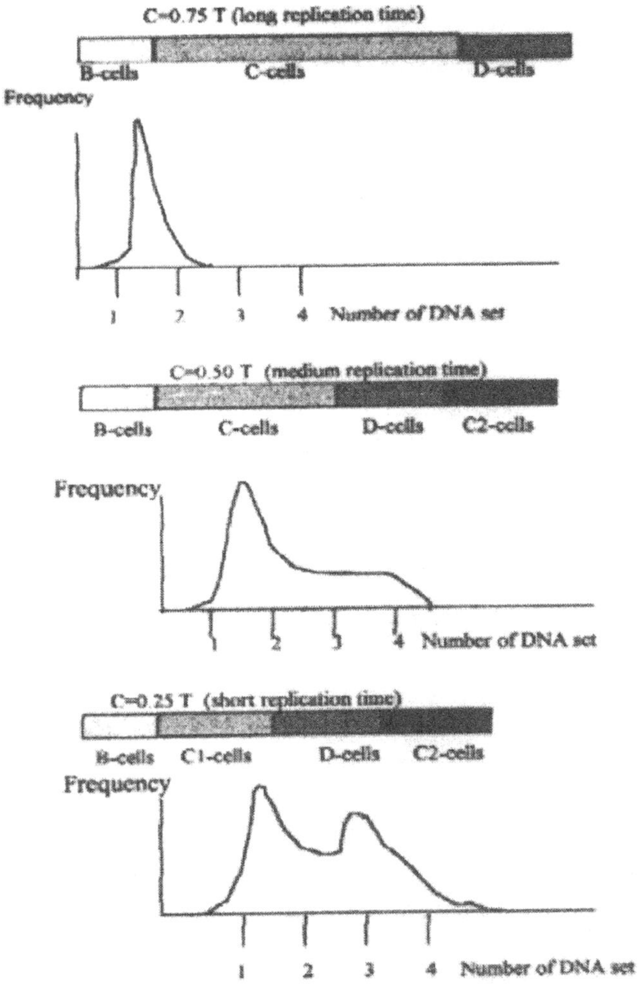

The curves have the pattern with one maximum, two maxima or with one maximum and plateau. All these theoretical variants of the distributions of DNA content were shown by the experimental flow cytofluorometry analysis of synchronous populations of *Bacillus thuringiensis, Bacillus licheniformis* and *Bacillus megaterium* [6]. The share of the cells in B-period determined from the DNA content distribution may be used to determine the physiological state of the bacterial populations.

4. Biodiversity of Neuston and Plancton

The monitoring of biohazards in the aquatic systems by flow cytofluorometry must account the difference between the bacterial plancton and neuston. Bacterial neuston is microbial cenosis of the biofilm of air-water interphase. The population of *Bacillus thuringiensis* from the bacterial neuston contains more small cells (Figure 2) and cells within the B-period of DNA replication cycle (Figure 3) than the population from the bulk of water. Similar data were obtained for another specia of bacilli [16]. The spores of bacilli and cells of R-forms are also accumulated within the bacterial neuston [12,13]. The ability of the vegetative cells of bacterial specia to be accumulated within the neuston is not a stable genetic character [14].

Figure 2. Size distributions of the cells of *Bacillus thuringiensis* from bacterial neuston (dark lines) and from the bulk of water (dotted lines).

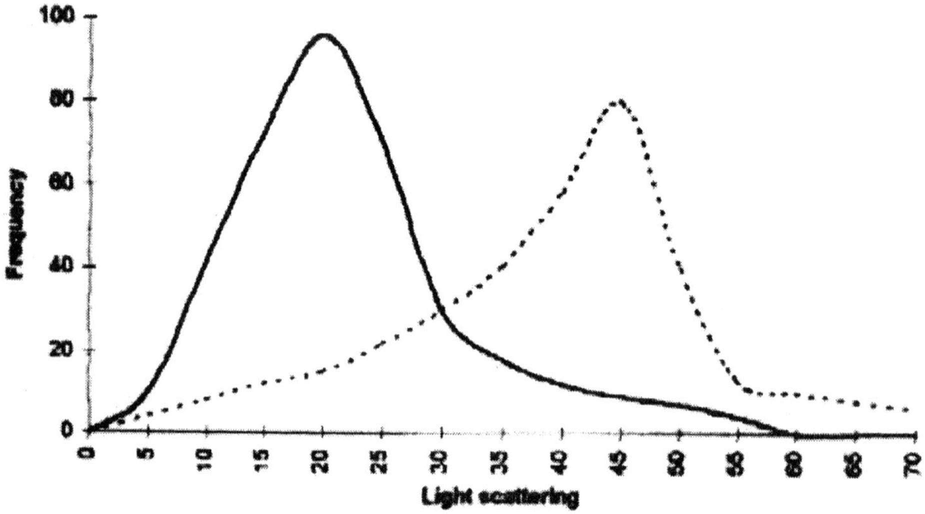

Figure 3. DNA content in the cells of *Bacillus thuringiensis* from bacterial neuston (dark lines) and from the bulk of water (dotted lines).

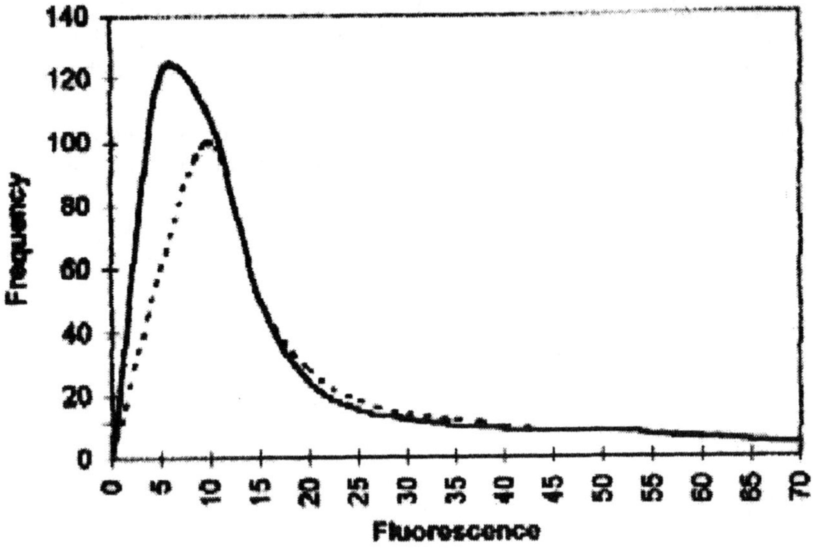

5. Sub-population of Exotrophic Cells

Another aspect of biodiversity of one population is the presence of exotrophic and endotrophic cells. To explain the oscillations of energetic metabolism during the cell cycle a hypothesis was proposed by Vladimir Ivanov [4]. Main point of this hypothesis is that in some periods of the cell cycle a cell utilizes an energy, which is accumulated in the previous periods of the cycle. We studied the synchronous and asynchronous populations of some yeast and bacterial specia by physiological and cytofluorometrical methods. The data of research may be summarized in the following conclusions [3-5, 15]. A cell utilizes the carbon and energy sources from medium during the G1- and G2- phases (mytotic cycle) or B- period (bacterial cell cycle). Such cells may be revealed cytologically as exotrophic ones. The method of determination is based on the toxic affect of the substrate analog and further determination of the dead cells by their specific fluorescence after dyeing [5]. Another known fluorometric techniques for the determination of the survival and mortality of the bacterial cells may be used in this analysis [7, 8, 10]. The share of exotrophic cells in the asynchronous populations of bacteria and yeasts positively correlates with the specific growth rate and is diminishing sharply during starvation of the cells [4, 5, 15].

6. Applications

A pattern of DNA content distribution or share of exotrophic cells determined for the cells dyeing by *specific fluorescent-labeled oligonucleotide probes* or *fluorescent-labeled immunochemical probes* may be used for the rapid monitoring of biohazards in the natural aquatic systems. The main applications may be such as the following:

*Identification of the source of biohazard's release into the environment by the comparison of physiological state in the release point and in the environment.

*Determination of the term of biohazard's release into the environment by the comparison of physiological state in the release point and in the environment.

*Determination of the physiological state of biohazard in the natural ecosystem. This determination is important for the risk assessment. The examples of such determinations is given in [9, 10, 17].

*Determination of the physiological state of biohazard before release into the environment by the comparison of the changes of physiological state in the sampling point(s) of the environment. It can be used for the characterization of the suspected source of biohazard. The non-cultivated biohazards, wastes as example, will be characterized by the distribution of DNA content with the peak in the region of one set of DNA. The cultivated biohazards, as the biohazards released from the water treatment facilities or bioweapons, will have the patterns of distribution of DNA content which are specific one for the proliferating cells.

7. References

1. Cooper, S. (1990) The Escherichia coli cell cycle, *Res. Microbiol.*, **141**, 685-693.
2. Hutter, K.-J. and Eipel, H.E. (1979) Microbial determinations by flow cytometry, *J. Gen. Microbiol.*, **113**, 369-375.
3. Ivanov, V.N. and Ugodchikov, G.A. (1984) Cell cycle of microorganisms and heterogeneity of their populations, Naukova Dumka, Kiev (in Russian).
4. Ivanov, V.N. (1990) Exo- and endotrophy of a cell, Naukova Dumka, Kiev (in Russian).
5. Ivanov, V.N. and Nudga, A.Y. (1990) Exo- and endotrophic cells in population of Bacillus megaterium, *Mikrobiologichesky Zhurnal (Kiev)*, **52**, 17-21.
6. Ivanov, V.N., Svechnikova, T.A., Stabnikova, E.V., and Gregirchak, N.N. (1995) Structure of cell cycle and age structure of bacterial populations, *Mikrobiologichesky Zhurnal (Kiev)*, **57**, 3-11.
7. Lloyd, D. *Flow Cytometry in Microbiology* (1995) Springer-Verlag.
8. Lopez-Amoros, R., Comas, J., and Vives-Rego, J. (1995) Flow cytometric assessment of Escherichia coli and Salmonella typhimurium starvation-survival in seawater using rhodamine 123, propidium iodide, and oxonol, *Applied and Environmental Microbiology*, **61**, 2521-2526.
9. Marie, D., Partnesky, F., Jacquet S., and Valuot D. (1997) Enumeration of cell cycle analysis of natural populations of marine picoplankton by flow cytometry using the nucleic acid stain SYBR Green I, *Applied and Environmental Microbiology*, **63**, 186-193.
10. Servais, P., Vices-Rego, J., and Billen, G. (1992) Survival and mortality of bacteria in natural environment, in J.C. Fry and M.J. Day (eds.), *Release of genetically engineered and other microorganisms*. Cambridge University Press., Cambridge, pp. 100-119.
11. Stabnikova, E.V., Zimberg, E.A., and Gregirchak, N.N. (1990) Synchronization of the bacterial proliferation by flotation, *Mikrobiologia (Moscow)*, **59**, 278-282 (in Russian).
12. Stabnikova, E.V., Gregirchak, N.N., and Taranenko, T.O. (1991) Peculiarities of flotation of bacillar cells and spores, *Mikrobiologichesky Zhurnal (Kiev)*, **53**, 38-42 (in Russian).

13. Stabnikova, E.V., Milko, E.S., and Gregirchak, N.N. (1991) Flotative fractionation of cells o R- and S-variants of some bacteria. *Mikrobiologichesky Zhurnal (Kiev)* **53**, 52-54 (in Russian).
14. Stabnikova, E.V., Gregirchak, N.N., Taranenko, T.O., and Nudga, A. Yu (1991) Selection of neuston forms of bacteria. *Mikrobiologichesky Zhurnal (Kiev)*, **53**, 33-37 (in Russian).
15. Stabnikova, E.V., Gordienko, A.S., Ksenofontov, B.S., Poberezhniy, V.Y., Stavskaya, T.Z., and Ivanov, V.N. (1991) *Interaction between the cells and water-air interface*, Naukova Dumka, Kiev (in Russian).
16. Stabnikova, E.V., Gregirchak, N.N., and Ivanov, V.N. (1992) Age specificity of the bacilli cell interaction with water-air interface. *Mikrobiologia (Moscow)*, **61**:110-114 (in Russian).
17. Stabnikova, E.V., Ivanov, V.N., Gregirchak, N.N., and Dulgerov, A.N. (1993) Utilization of neuston forms of bacilli for treatment and decontamination of surface waters, *Mikrobiologichesky Zhurnal (Kiev)*. **55**. 88-94 (in Russian).

MEASUREMENT OF BIOLOGICAL AEROSOL WITH A FLUORESCENT AERODYNAMIC PARTICLE SIZER (FLAPS): CORRELATION OF OPTICAL DATA WITH BIOLOGICAL DATA

JIM HO, MELVIN SPENCE AND PETER HAIRSTON*
National Defence Canada, Defence Research Establishment Suffield, Box 4000, Medicine Hat, Alberta, T1A 8K6, Canada
**TSI Incorporated, P.O. Box 64394, St. Paul, MN 55164 USA*

Abstract

Biological aerosol measurement in real time is an urgent military requirement that also has many potential non-military applications. Such detection capabilities will be useful in environmental monitoring, for example, in gathering information in perceived hazardous areas like housing developments downwind of sewage treatment plants.

Experience gained from measuring fluorescence signals of single bacterial spores under flow cytometry using UV excitation at 340-360 nm, was applied to concept testing of a prototype instrument, built to do the same for aerosols. This machine was capable of resolving particle size as well as fluorescence intensity of each particle under laboratory and field conditions; it was called the fluorescent aerodynamic particle sizer (FLAPS). A second generation FLAPS (FLAPS2) was designed to be smaller, power efficient and field portable. FLAPS2 was challenged under refereed conditions in blind trials to determine if it could detect biological aerosols in natural field environment. This paper describes practical aspects of measuring biological aerosols when the results must be compared to reference samplers that provide culturable or "live" data. Treatment of particle size and fluorescence information is discussed with respect to FLAPS and reference data fidelity. Finally, an objective method is introduced to evaluate FLAPS data correlation to reference data. The measurements suggest that there is positive correlation between FLAPS measurements and live biological aerosol particles.

1. Introduction

Microorganisms are naturally aerosolized in the atmosphere, often becoming a biological burden to downwind communities. For example, municipal wastewater aeration tanks have been studied and found to produce, by the minute, over 130 viable particles per m^3, about 10% being coliforms (Sawyer et al. 1993). Hosing down of dairy processing plants was shown to produce a biological aerosol that remained above ambient concentrations for as long as 40 minutes (Kang and Frank, 1990). Diffusion modeling showed that aerosolized microorganisms could spread infection over a 150 km^2 area among 10 swine herds (Grant et al., 1994). Indoors microbiological air quality

P.J. Stopa and M.A. Bartoszcze (eds.),
Rapid Methods for Analysis of Biological Materials in the Environment, 177–201.

in homes and offices has been reported for bacteria and fungi (Lis et al. 1997). The bacterial numbers ranged from 200-800 (homes) and 100-500 (offices) viable particles/m^3 while those for fungi were about half as abundant. Fairly similar findings were reported by the US Environment Protection Agency which has recently taken an interest in measuring indoor and outdoor biological aerosol quality (EPA, 1998). Before meaningful health assessments can be made, it is clear that extensive biological aerosol measurements have to be carried out over long periods of time. The job can be executed more efficiently if it can be done in real time without having to resort to using the conventional but resource intensive microbiological growth assessment methods.

Measurement and detection of biological aerosols in real time has been a difficult task to achieve. Conventional approaches involving growth of biological samples take days to produce results. Nevertheless, as an initial attempt, we developed a system based on a particle sizing instrument that measured some unique size characteristics of biological particles in air to differentiate them from ambient material. Its development was based on work done at Defence Research Establish Suffield, DRES, (Ho, 1994) in the 1980's using commercial technologies coupled with a knowledge of biological aerosol characteristics from artificial sources. This system employed particle sizing algorithms to distinguish between observed physical characteristics of biological aerosols and background material. The particle sizing instrument estimated particle size by time-of-flight information derived from scattered light. This scheme was based solely on fundamental physics of light scattering and did not measure any intrinsic biological characteristics. Thus dust clouds with similar scatter properties to biological material could cause undesirable interference. The system was deployed as a biological agent detection system and served its purpose at the unique locations of deployment primarily due to predicted well behaved background aerosol characteristics.

Measurement and detection of biological aerosols in real time has been a difficult task to achieve. Conventional approaches involving growth of biological samples take days to produce results. Nevertheless, as an initial attempt, we developed a system based on a particle sizing instrument that measured some unique size characteristics of biological particles in air to differentiate them from ambient material. Its development was based on work done at Defence Research Establish Suffield, DRES, (Ho, 1994) in the 1980's using commercial technologies coupled with a knowledge of biological aerosol characteristics from artificial sources. This system employed particle sizing algorithms to distinguish between observed physical characteristics of biological aerosols and background material. The particle sizing instrument estimated particle size by time-of-flight information derived from scattered light. This scheme was based solely on fundamental physics of light scattering and did not measure any intrinsic biological characteristics. Thus dust clouds with similar scatter properties to biological material could cause undesirable interference. The system was deployed as a biological agent detection system and served its purpose at the unique locations of deployment primarily due to predicted well behaved background aerosol characteristics.

We showed that FCM provided unequivocal evidence of single spore intrinsic fluorescence when excited by 365 nm light (Ho and Fisher, 1993). At this wavelength, the most probable biological molecule that emits fluorescence is reduced nicotinamide adenine dinucleotide (NADH), a key metabolic intermediate of all cellular life systems (Ichinose, 1991). Davey and Kell (1996) summarize other important FCM contributions to bacterial studies. Of special interest is their work on describing viability and vitality of single bacterium in a population using FCM (Kaprelyants and Kell, 1992, Kell et al. 1998).

From this observation it was clear that measuring fluorescence from individual spores in air could provide a means of enhancing signal-to-noise over background material. Indeed, other workers have used 266-280 nm UV excitation to study characteristics of protein related fluorescence in biological particles (Bronk and Reinisch, 1993). More recently, Chang's group at Yale applied pulsed UV at 266 nm on single bacterial particles with fast electronics that permitted excitation selectivity (Chen et al. 1996). In contrast, our approach emphasized the use of 340-360 nm as the excitation wavelength of choice. Numerous reports suggest this wavelength region provide fluorescence information closely associated with biological molecules linked to critical metabolic activity (Kell et al. 1991). Even more meaningful and potentially more definitive was that the fluorescing molecules excited at 340-360 nm could be related to living microorganisms (Harrison and Chance, 1970). If fluorescent property is related to living organisms, this connection would suggest that an aerosol detector performing under similar characteristics could potentially distinguish live from dead or inanimate particles.

In conventional quantitative analytical biochemistry, it has been standard practice to use short UV wavelengths between 260-280 nm to elicit protein fluorescence as a standard method for concentration estimation (Groves et al. 1968). However, in the biological detection context, fluorescence excitation at this wavelength is not specific to living organisms, as recently acknowledged by Pinnick et al. (1998). In addition there is suggestion that other molecules such as organic pollutants in air interact with ozone to form particles that fluoresce broadly at 300-800 nm when excited with short wavelength UV (Day and Lee, 1983). Employing 340-360 nm as the selective NADH excitation wavelength and measuring the fluorescence at 400-550 nm would provide the sought after selectivity for living organisms.

We have reported on the first generation fluorescence aerodynamic particle sizer (FLAPS1); we demonstrated single biological spore particles in an air stream could be detected (Hairston et al. 1997). It was effective in field trial detection of biological aerosol where no false alarms were encountered (Boulet et al. 1996). Background biological aerosol characteristic measurement using the FLAPS has been described (Ho and Spence, 1998). This paper describes performance of FLAPS2, a field portable instrument optimized for rapid biological aerosol detection and environmental monitoring.

2. Methods and Materials

2.1. FLUORESCENT AERODYNAMIC PARTICLE SIZER (FLAPS)

With DRES specifications, TSI (St. Paul, MN 555164-0394) built FLAPS1 using a HeCd CW laser coupled to their existing aerodynamic particle sizer (APS). The success of this project led to a second contract to construct FLAPS2 based on their APS II. This instrument is now designated as model 3312a; it is also called "Ultraviolet Aerodynamic Particle Sizer" (UV-APS). By itself, UV-APS produces raw data without analysis or display capabilities.

Proprietary display and alarming software was developed by DRES with Dycor (Edmonton, Alberta) as programming contractor. The original code has been in development since mid 80's when the APS was used in DRES field biological detection. It evolved into a functional tool, used successfully during international field trials (Ho et al. 1995). Further development work brought the software up to date and in its present form is used to control the FLAPS. The name "FLAPS" describes a functional package consisting of the fluorescence enabled APS hardware and DRES software combination.

The FLAPS2 configuration was equipped with a pulsed UV light source ("StableLight" model S349-30Q, Uniphase, 163 Baypointe Parkway, San Jose, CA 95134, 800-644-8674). It was powered by a diode-pumped Nd:YLF laser frequency tripled to produce 349 nm. Energy per pulse was rated at 30 μJ running at pulse width of about 5 to 8 nsec.

As the pulse energy was higher than optimal, an optical density 1 filter was used to pass only 10% of the laser output to the measurement chamber. This resulted in lower background noise and better signal-to-noise characteristics. The laser was triggered when a particle was detected with a minimum time between triggers of 200 μsec.

To compensate for different flight times due to particle size, a "skeet shooting" look-up table was used to synchronize laser trigger times. The table provided information that relates particle size to time delay for firing the laser. Small particles that travel faster through air than larger ones have proportionally shorter delay times. This table was generated from calibration data using standard fluorescent latex beads. In addition, by using firmware commands, selective firing could be exercised to excite only particles of desired size characteristics. This feature not only offered a "smart" technique for interrogating the biological characteristics of a single particle, it also avoided unnecessary firing of the laser at background material (<1 μm size).

During real time sampling, the instrument presents size and fluorescence intensity information for each particle. Data derived from a sampling period, typically 3 seconds, can be reduced to a fractional number (gated % fluorescent); this represents the percent of particles that exhibited fluorescent signal above a preset size and intensity threshold. A threshold level can be determined for each environmental condition by observing the prevailing background conditions. An alternate method to present fluorescence data is to derive a "signal-to-noise" ratio. This is done but comparing the mean of background gated % fluorescent with new "unknown" or "biological" data.

As a way to enrich the population of background fluorescent particles for better statistical counting, a prototype aerosol concentrator (model XMX, SCP Engineering, St. Paul, MN) was used as a front end to the FLAPS2 intake. This was a modified version of the original XM2 used for concentrating biological aerosol, as reported previously (Brenner et al. 1988). The present work employed a smaller version, designed for optimal size, weight and power consumption. It operated at 400 liter/min concentrating to 1 liter/min delivered to the FLAPS intake. Improved particle throughput provided by this setup facilitated a rapid sampling time of 3 seconds. In practice, an additional 1 second penalty was incurred for computational and data handling overhead. By this protocol, aerosol data could be collected continuously every 4 seconds over long periods unattended by the operator.

2.2. FIELD TRIAL PROTOCOLS

Field biological aerosol dissemination (spores of *Bacillus subtilis* var niger, BG) and reference sample collection were services provided by Dugway Proving Grounds, DPG, (US Army, Dugway, Utah). Biological aerosol or tracer smoke dissemination took place about 800 m upwind of the detection grid. Andersen and Mohr (1996) have described the procedures in detail and only brief comments will be stated here where warranted.

Critical to understanding the characteristics of the test aerosol, culturable particles in air must be captured and grown to determine the presence of "live" content. Biological aerosol particles were impacted onto s 15 cm nutrient agar plate situated in a slit sampler (model STA 203, New Brunswick Scientific, Edison, NJ 08818-4005). The air flow rate was 28.3 liter/min. Each sampler was electronically modified to rotate at 1 revolution/min. A sampler array consisted of 10 devices serially connected to function as a continuous 10 minute collector. The sampling station was located on an elevated platform, set longitudinally adjacent to the FLAPS under test. The close proximity of these devices was critical to generating good results. However, position optimization was compromised by the distance created by the width of the array of slit samplers (30 cm each). Air intake height for the reference samplers was about 1 meter higher than that of the FLAPS. Most trials took place with the wind from a south to north direction.

On passage of a test cloud, the reference samplers were switched on manually by the trial operators. Switch timing was estimated from the moment of cloud passage, determined by a tracer gas (propylene) and detector system. The tracer was disseminated simultaneously with the biological aerosol. A series of gas detectors provided real time information to determine the optimal moment for triggering the slit samplers. Precise trigger was critical for effective operation of the reference system. There was a maximum usable collection time of 10 minute and generally, the operators hit the on switch fairly close to biological cloud passage. During this time, biological particles, if present in the aerosol, were impacted on to the surface of a nutrient agar plate. After over night incubation, live particles were counted as bacterial colonies by means of a flat bed scanner driven by custom software developed jointly by DRES, DPG and Spiral Biotech (model "STAR" version 1.5, Spiral Biotech, Bethesda, MD 20814). Reference data were expressed as agent containing particles per liter of air (ACPLA). In this report, fluorescent data were compared to reference data to illustrate timing relationships whenever biological hits occurred.

2.3. STATISTICAL ANALYSIS AND DATA TRANSFORM

In wanting to know if two variables are interdependent or vary together, the common approach is to perform a correlation analysis (Sokal and Rohlf, 1981). A typical assumption is that the two variables are both effects of a common cause. Translated to our present problem, we wish to estimate the degree to which FLAPS measurements vary together with live reference data and the common cause is the presence of live biological aerosols. Before applying this test, there is one precaution to note- the data must be tested for distribution normality. Non-normally distributed data like FLAPS measurements and slit sampler data (shown in results) require non-parametric methods. Spearman's non-parametric correlation coefficient was determined by SigmaStat version 2 (SPSS Inc. 444 N. Michigan Avenue, Chicago, Illinois, 60611) as described by Sokal and Rohlf (1981). This test requires 2 data sets of equal size and timing intervals. However, raw slit sampler and FLAPS data sets derived from two fundamentally different measurement approaches, have different timing intervals (2 and 4 sec. respectively). As a consequence, they are structurally dissimilar and unsuitable for this type of analysis.

To overcome this discrepancy, each data set consisting of time and the variable was transformed by first converting the time format into decimal time (e.g. 12:00:00 to 0.5) using the "timevalue" function in Excel (Microsoft Office 97, Microsoft Inc., Redmond, Washington 98052-6399 U.S.A). Then x and y data columns were imported into TableCurve 2D (version 4, SPSS Inc. 444 N. Michigan Avenue, Chicago, Illinois, 60611) where non-parametric "local regression" estimation was done using Gaussian weighting. Three other non-parametric methods were tested to determine the optimal approach (local regression with Loess weighting; spline estimation with non-uniform rational B-spline [NURBS] weighting and smoothing spline estimation with fast fourier transform [FFT] and NURBS weighting). For generating new interpolated data sets, increment decimal time value of 0.000023148 or about 5 sec. was used along with specified beginning and end times. After performing the conversion procedure for both FLAPS and reference data, just the y portions of each set were grouped, constituting the variable pair suitable for Spearman's test.

2.4. RESULTS AND DISCUSSIONS

2.4.1. *Location of Samplers*

Existing open air measuring devices recommended for sampling biological aerosols are essentially point detectors (Chatigny et al. 1989). When attempting to compare performance of two or more samplers in an environment of unpredictable wind characteristics, it is important to co-locate these devices as closely as possible to avoid sampling errors. Significant errors can occur if two samplers have air intakes of different heights relative to the ground as illustrated in Figure 1.

In this experiment, optically visible red smoke was disseminated to provide a test for determining if interfering aerosols would degrade detector performance. Unlike biological aerosol, which is invisible, one advantage to using red smoke was that the experimenters could actually see the aerosol passage through the test site. At the time of day shown in figure 1, there was an apparent temperature inversion creating differential air layers at various heights. As a result, the aerosol particles could be observed to move horizontally downwind as a linear layer (to the naked eye) at a distance of several meters above ground level. A distinct clear air/aerosol interface was easily visible, with the particle-laden layer on top. For about 3 minutes, the air below this interface remained clear. This unique observation was confirmed by analyzing particle data collected by two FLAPS instruments with air intakes set at two different heights as indicated in this figure. On this occasion, particle numbers were plotted versus time to illustrate the effects of environmental conditions on measurement parameters. Also, there was no detectable fluorescence from red smoke particles. The important lesson to learn here is that due to unusual environmental conditions, particle detection times appear to vary significantly if aerosol samples were extracted at different heights.

Apart from detecting the smoke at different times, the lower instrument also measured concentrations at almost an order of magnitude higher than the second instrument at around 02:22:00. It is possible that this difference was partially due to complex uneven mixing or interaction of the stratified aerosol layers. It is clear that from this observation, much remains to be learned from open air aerosol measurements under unique kinds of environmental conditions.

These data also provide an extra dimension of awareness to those whose interest is in modeling of aerosol transport. A fairly common practice in aerosol modeling is the expedient use of simplistic environmental assumptions that differ from reality. Grave errors in computational predictions may occur if the lessons shown here are not carefully assimilated.

As a result of these observations, we made great efforts at co-locating FLAPS and the reference samplers. This was not as practical as we would have liked as the slit samplers occupied a platform of about 3 meters in width. The closest possible location FLAPS could be positioned was about 1 meter to the right of the platform. Nevertheless, having defined the limitations, we were glad to note that our bioaerosol measurements closely matched the aerosol dynamics of the reference samplers in almost all cases.

2.4.2. *Fluorescence Signal to Noise*

The prototype instrument (FLAPS1) employed a continuous wave HeCd laser as excitation source (Hairston et al. 97). After successfully demonstrating the working concept, we set out to reduce its size, weight and power requirements. Hardware constraints dictated that FLAPS2 had to be designed around a pulsed laser resulting in a package that could be housed in a meter cube enclosure. However, due to this change in laser illumination, it was of interest to demonstrate that the two instruments had comparable performance. Figure 2 shows an experiment where the two instruments were presented with an aerosol cloud in the field under ideal environmental conditions, unlike that encountered in figure 1. In this example, the signal-to-noise ratio was plotted for each showing that the instruments detected the cloud in almost identical fashion. Both instruments detected the leading edge of the cloud at about 21:38:00 where S/N registered about 2. To illustrate a concomitant presence of biological aerosol containing culturable particles, figure 2 also shows the scatter plot for live particles (ACPLA) collected by slit samplers. There is evidence to suggest that in this example, both FLAPS began to register S/N value of 3 in the presence of about 10 ACPLA.

Five minutes after the initial event, a higher concentration cloud of about 20 ACPLA caused both detectors to register S/N of about 5 to 6. In the next 5 minutes there was apparently a bimodal cloud that was resolved by both instruments as having a "dip" in S/N levels at around 21:45:00. Of significance, the reference samplers concomitantly detected this event. By 21:49:00, both detectors reported background levels after exhibiting a steeply declining slope from a peak S/N of about 10 to 11. This peak was equivalent to about 40 ACPLA, obtained by inspecting referee data.

The time course data in Figure 2 appear to indicate that fluorescent signals from low as well as high biological aerosol concentrations can be unambiguously resolved. The data suggest that the instrument design conferred a useful dynamic range of at least from 5 to 50 ACPLA. It is also of importance to note the rapid drop in optical signal levels after passage of a cloud. It is possible that the APS sheath flow may have played a role in contributing to good aerosol clearance characteristics after passage of a fairly heavy biological cloud. In contrast, there are other optical designs that emphasize large surface area exposure of particles in the optics (Jeys et al. 1997) and thus may account for difficulties with slow clearance of residual material long after the passage of the aerosol cloud.

2.4.3. *Particle Number Measurement versus Fluorescence Measurement*

One disadvantage in using particle light scatter signals for detection triggering is the presence of ambient non-biological particles that may cause errors. To reduce the potential for false positive detection events, a detector should be designed to measure real biological entities rather than particle-based "noise". Figure 3 illustrates the benefit of using fluorescence-based measurements as compared to those based solely on light scatter. It is shown that of the 3 particle based events (01:12:30, 01:22:00 and 01:32:00), only the middle one corresponded to fluorescence material and by inference, biological content.

When this data set was compared to reference biological data (figure 4), it can be seen that there was a close to perfect temporal match. Note that the time of day scale has been expanded to focus in on the details of the middle peak. It can be seen that the reference data set began at 01:20:49 whereas FLAPS2 indicated the presence of fluorescent signals at about 01:20:00, suggesting an apparent discrepancy. However, as was mentioned earlier, the reference samplers were manually activated and occasionally the starting cue could be delayed due to timing errors. In this case, a lag of about 49 seconds resulted in the reference samplers capturing the aerosol cloud well after the initial up slope. Another reference data anomaly can be seen at 01:24:45 where there was an apparent spike registering 22 ACPLA. This data was not confirmed by any of the 4 other reference sampling locations, suggesting that it might be due to a "noise" spike rather than a real event that was not detected by FLAPS.

The rest of the reference data set gave a remarkably good visual match with respect to FLAPS2 fluorescent data. From a subjective interpretation viewpoint, the coincidence of these two data sets would lead us to speculate that the two measurement systems were detecting the same phenomenon. It is possible to be convincing using this graphical approach when the data sets show little to no time event discrepancies. However, there were instances where imperfect matches existed (figure 6), and so a more objective analytical approach was required to assess detector performance.

Although test aerosol dissemination was a continual process, occasionally, most likely due to wind and ground related effects, two distinct biological clouds could be detected. Figure 5 shows an example of two successive hit events. Between these two hits, just before 17:06:00 and only lasting for a few sampling intervals, the reference samplers measured no live particles. This coincided with a similar interval where FLAPS also registered background fluorescent levels. This is a good illustration of the resolving power of the FLAPS optical system. It can be seen that the clean air content between the two clouds was clearly resolved as less than 10-15 seconds and immediately after that the instrument responded to the second cloud with no obvious lag.

Unlike what happened at the initial aerosol hit point shown in figure 4, here in figure 5 the reference aerosol sampling was perfectly executed. This is depicted in a series of clear background data sets representing clean air at 17:01:00. Half a minute later, in the presence of 3 to 5 ACPLA biological aerosols, FLAPS detection was positively registered. Similarly, the rise and fall of aerosol content throughout the two hits were optically tracked with high fidelity.

Background aerosol clouds normally consist of a constant proportion of fluorescent particles mixed in with other particles of non-biological origin. However, under high background conditions gated % fluorescent values may become distorted when compared with reference data. This can be especially critical if a high background level of non-fluorescent particles masks a detection event involving a low level of fluorescent particles. Fortunately, this condition has been encountered only rarely. However, figure 6 shows an instance where slight to moderate distortion did occurred. In this example, it is visually less obvious that the two raw data sets describe the same aerosol event. Thus it becomes necessary to employ statistical tools to help evaluate such "difficult" data sets.

2.5. STATISTICAL APPROACHES

Some basic descriptive statistical analysis was done to the particle data sets to determine the correct approaches to take with respect to correlation testing procedures. Figure 7 shows typical example of a test outcome, in this case, using total fluorescent particle data for illustration. With a skewness coefficient of 1.12, the data set was not normally distributed (normal = 0). This is not surprising; the observation that biological data are not normally distributed has been commonly reported before (Tillett and Carpenter, 1991). Similar observations were obtained with the other data sets. For this reason, non-parametric local regression methods were selected for interpolation estimations. As mentioned previously, we needed matched data sets to perform comparison tests and the first step was to select the best transform method. However, since there are a variety of techniques to chose from, a survey study was required to determine the best data transform model.

Figure 8 shows the results of subjecting raw gated fluorescent particle data to local regression treatment (with gaussian weighting). The new predicted or transformed data compared very well with the original, as indicated by an excellent regression coefficient ($r^2 = 0.99$). Other methods of regression transform also yielded good coefficients (data not shown). When this transformed data set was compared with the transformed reference data using the non-parametric Spearman's test (Cleveland, 1993), a correlation coefficient of 0.813 was obtained. This high value can be interpreted to mean the two measured parameters are positively correlated. In plain terms, for this data set, the FLAPS measurement can be considered statistically equivalent to the live biological data. To take this inference one step further, it implies what the FLAPS measured was indeed that of the live biological aerosol.

Expanding on what has been shown in Figure 8, we proceeded to repeat this treatment on other data sets from the same experiment. The resultant matrix of Spearman's coefficient values is summarized in table 1. The obvious observation is that in all treatments, gated % fluorescent data gave the poorest results. Again we emphasize that since this value is derived from prior numeric manipulation, (it is a fractional ratio number derived from dividing gated fluorescent numbers by gated particle numbers) the lower coefficients obtain are not surprising. Its utility can be appreciated in that it gives an objective evaluation to the same data shown in figure 6 where visual inspection for a fit is difficult to make. A coefficient value of 0.646 (table 1, column 2, last row) can be considered as mildly positively correlated. Its usefulness is apparent when the detectors are used in an environment with wildly fluctuating ambient non-biological content.

By inspecting Table 1, local regression with Loess weighting may appear to give better coefficient results. But the numbers did not stand up to statistical (difference) testing, most probably due to insufficient data items. So for practical purposes, any of the four estimation methods will work well. Similarly, gated fluorescent counts may appear to give better test data but the advantage is marginal.

To summarize, we have submitted biological aerosol data to rigorous statistical analysis and learned that non-parametric techniques are most appropriate. Accordingly, Spearman's correlation coefficient method was used to determine if FLAPS measurements were a function of live biological aerosol clouds that hit the samplers. The results certainly suggest positive correlation, regardless of which method of data transform was used.

Table 1 indicates that under ideal test conditions, it is possible to positively correlate the measured particle parameters with the presence of viable biological aerosol. Although we did not have sufficient replicates to statistically conclude that local regression with Loess weighting was better than the other transform methods, it became our choice. If we have to make a recommendation as a guide for other workers to treat their biological data, then the following would be fairly representative:

1. test raw data sets for normality distribution
2. use a data transform method to obtain time base compatibility between optical data and biological data
3. if data sets fail normality test, use Spearman's method to measure correlation
4. derive conclusions from positively correlated observations

By following these steps, we would have a uniform approach to objectively make sense of the optical measurements. This is especially crucial when there are numerous optical parameters that can be derive from a single particle. For example, the FLAPS instrument can also measure side scatter. At present, we do not have the transmission speed to cope with another data stream. But soon enough, this bandwidth hurdle will be solved, then it will be relevant to ask if side scatter characteristics can be correlated with biological aerosols.

2.6. CONCLUSIONS

We have shown that FLAPS technology permits measurement of fluorescence signals from single particles in an aerosol. A graphical technique has been employed to permit visual inspection of optical data in comparison to data obtained from culturable or living organisms in artificially generated aerosol. The instrument signal quality is relatively unaffected by inanimate stray particles and also is unaffected by varying background particle concentrations. Also shown is that sampling of aerosols in the field can be influenced by unexpected environmental conditions. Placement of detectors with respect to referee systems can be crucial to obtaining high fidelity data. There is evidence to show that the two generations of FLAPS detectors, built with different laser designs, have similar performance characteristics. This suggests that the basic concept of measuring single particle characteristics to isolate biological material from background noise is sound. Evidence has been provided to indicate particle number analysis by itself is insufficient to characterize a biological cloud because it is susceptible to interference by dust or non-biological particles. By measuring fluorescent emission on UV excitation at 340-360 nm, we have demonstrated that the measured signals closely correlate with live or culturable biological particles. Statistical analyses can be used to enhance raw data sets for improved performance in difficult aerosol conditions.

There have been previous reports of optical fluorescence measurements of biological aerosol with bench top instruments using the wavelengths that excite NADH and related biomolecules (Hairston et al. 1997, Pinnick et al. 1998). An important contribution made here, for the first time, is that a laboratory concept has been taken out to the field to measure ambient air. In addition, evidence is presented to demonstrate that this field instrument is capable in detecting the presence of culturable particles making it potentially useful in monitoring naturally occurring biological particles in air. We speculate that this instrument, coupled to the concentrator will be useful in gathering detailed background biological aerosol profiles at rapid sampling intervals. This instrument, if deployed as a network, will make it possible to map atmospheric biological content over a broad territorial range. With such a networked system, we envision new areas of investigation will become feasible, for example, prediction of phyto-pathogenic fungal spore migration over broad terrain (Davis and Main, 1984).

Figure 1. Demonstration of Air Stratification Effects on Particle Distribution. Note Detection Hit Time Disparity of 3 Minutes. Two FLAPS instruments collected particle size and number information, each at different heights. The air intake of each measured material delineated by a distinct air layer that was visible by the naked eye. The instrument confirmed this observation.

Figure 2. Signal-to-noise Ratio Analysis. Comparison of FLAPS1 and 2 Performance. A broadly distributed biological cloud was measured by the two instruments simultaneously. Inspection of the normalized data show similar characteristics in their performance.

Figure 3. Fluorescent vs. Particle Numbers Analysis. The instrument logged particle numbers sorted by their fluorescence brightness. Particles that do not fluoresce can be assumed to have no biological characteristics. The experiment showed that of the three particle clouds that were detected, only one had biological properties.

Figure 4. Comparing Fluorescent Data and Biological Reference Data. Inspection of the time plot for fluorescence compared well with that from a slit sampler that measured culturable particles. Visually, it appeared that the two samplers were indeed tracking the same biological aerosol cloud.

Figure 5. Optically Resolving Two Distinct Biological Clouds. The usefulness of a biological detector depends on how rapidly it can resolve biological aerosols with high fidelity. This test shows detection of two clouds each of about 5 minutes in duration. Biological aerosol concentration is represented by slit sampler data. Note the particle fluorescence characteristics showed remarkably similar fluctuation dynamics.

Figure 6. Example of Imperfectly Matched Data Sets. Low concentration clouds are difficult to measure. In this example, there were fewer than 5 ACPLA measured in the cloud according to the reference sampler. Visual inspection of the two data sets provide uncertainties as to their relationship. Some objective analytical technique is required to provide a better assessment.

Figure 7. Normality Test (Kolmogorov-Smirnov) Failed
Skewness = 1.12 (normal = 0)
K-S Distance = 0.145 P = <0.001
Tested with total fluorescence data set. A standard approach in using statistical tools is to first determine if the data set is normally distributed. Non-normally distributed data require the use of non-parametric techniques. Most biological derived data are not normally distributed and this example is no different.

Figure 8. Gated Fluorescent Particle Data Non-Parametric Estimation
Local Regression & Gaussian Weighting r2 = 0.99
Spearman's Coefficient for Gated Fluorescent Numbers vs ACPLA = 0.813 For non-normally distributed data, the standard method to determine if two data sets are related is the Spearman's test. It requires equal numbers of data points for each set. Since the two measuring instruments ran on different time scales, the resultant data sets had unmatched sample numbers. One way to overcome this is to make them equal by interpolation. In this example, local regression with gaussian weighting was the method used for generating the new data sets. The resultant Spearman's test of 0.813 can be considered to represent good relationship.

Table 1. Summary of Spearman's correlation with referenced to ACPLA data tested with different estimation methods.

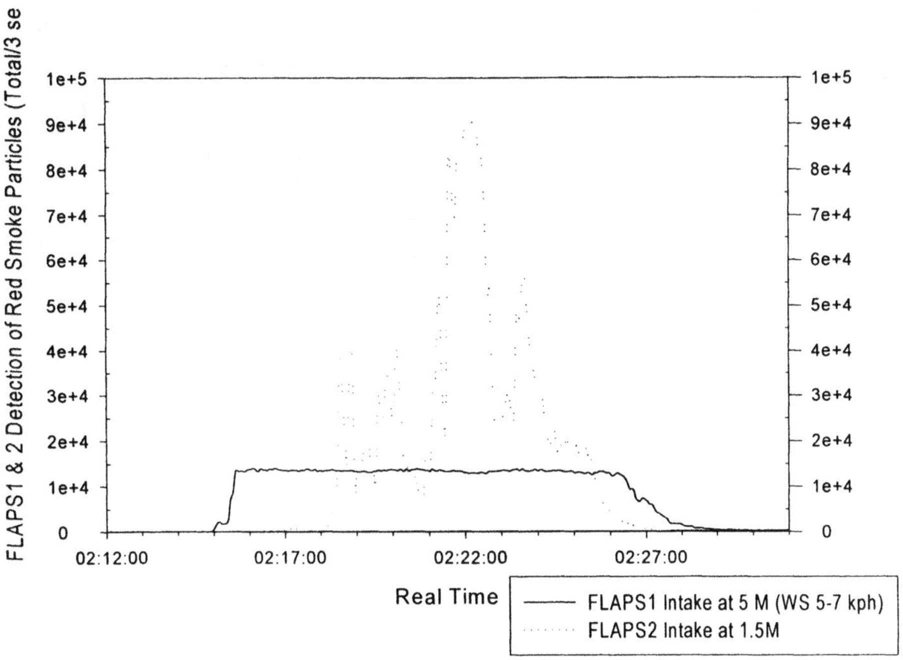

Figure 1. Demonstration of Air Stratification Effects on Particle Distribution
Note Detection Hit Time Disparity of 3 Minutes

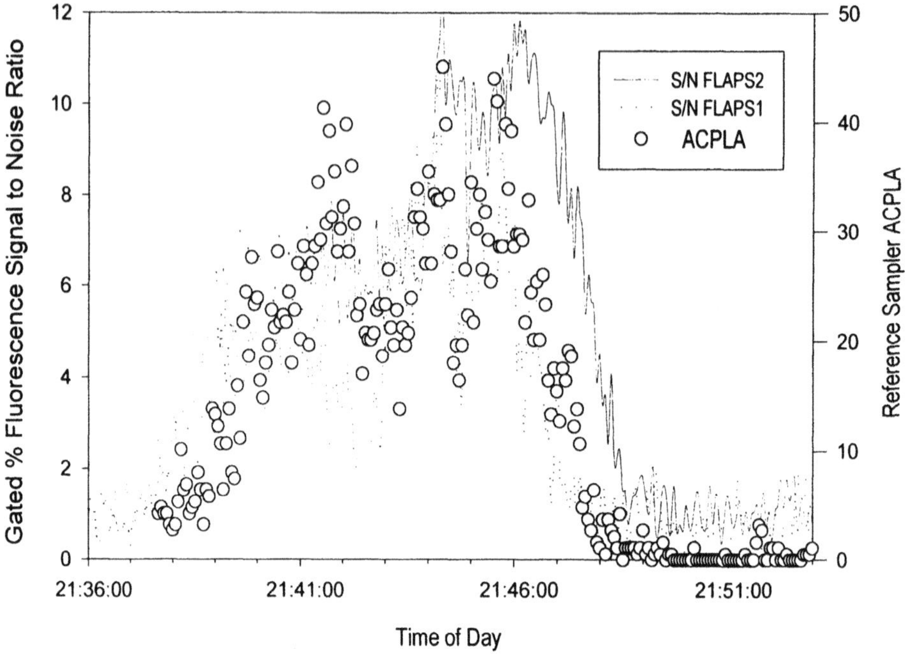

Figure 2. Signal-to-noise Ratio Analysis
Comparison of FLAPS1 and 2 Performance

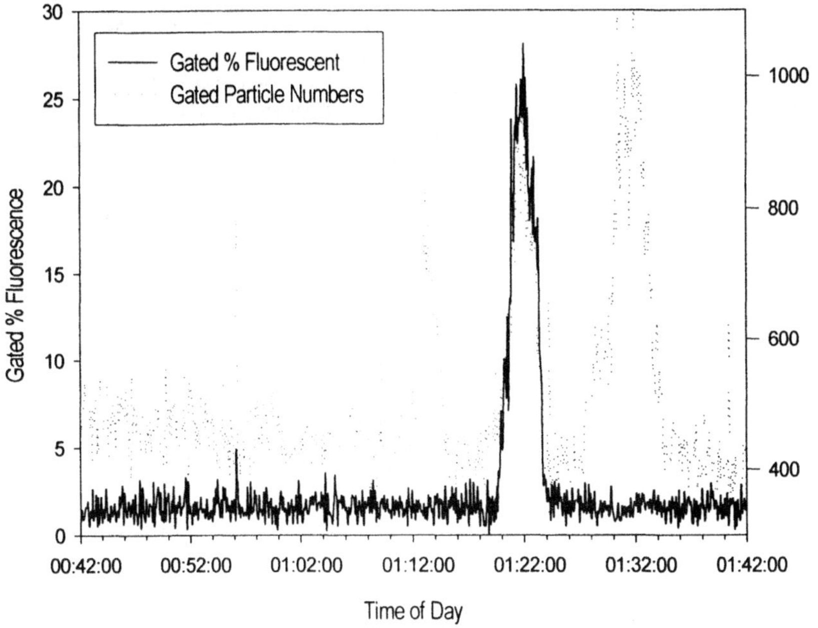

Figure 3. Fluorescent vs Particle Numbers Analysis

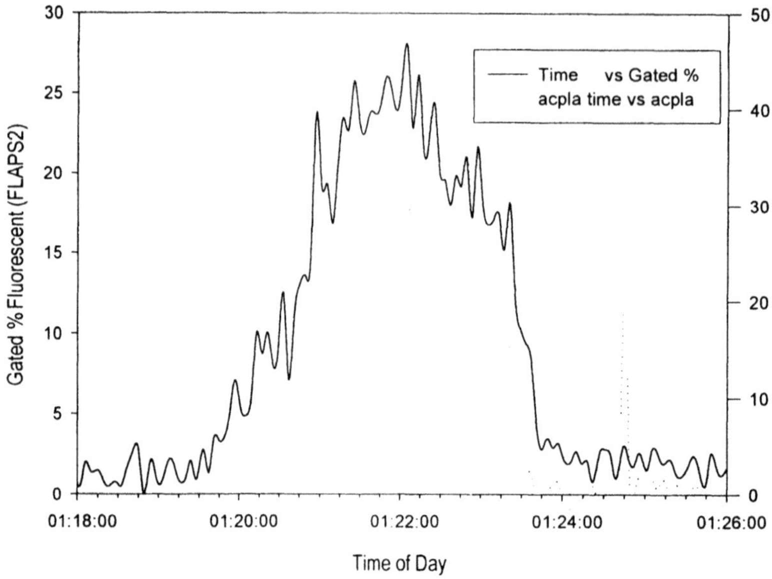

Figure 4. Comparing Fluorescent Data and Biological Reference Data

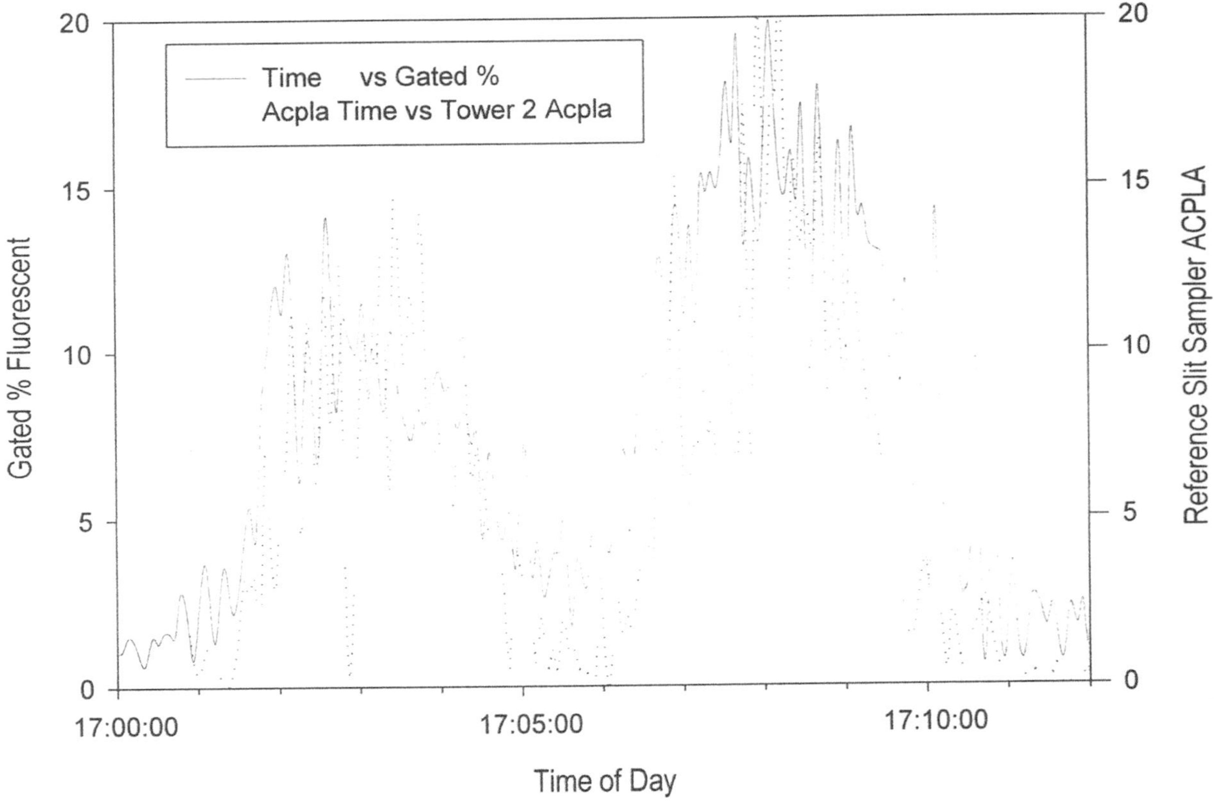

Figure 5. Optically Resolving Two Distinct Biological Clouds

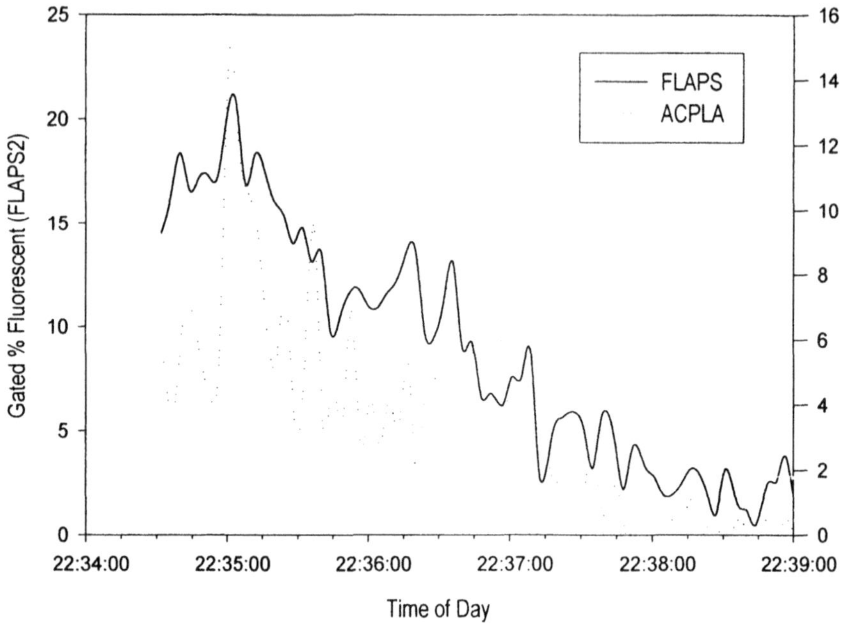

Figure 6. Example of Imperfectly Matched Data Sets

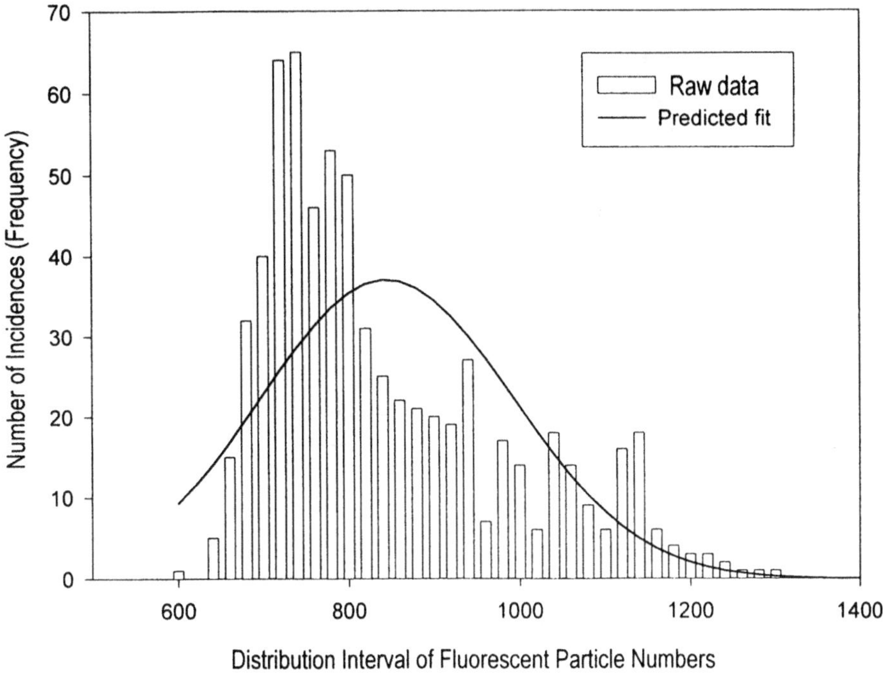

Figure 7. Normality Test (Kolmogorov-Smirnov) Failed
Skewness = 1.12 (normal = 0)
K-S Distance = 0.145 P = <0.001
Tested with total fluorescence data set

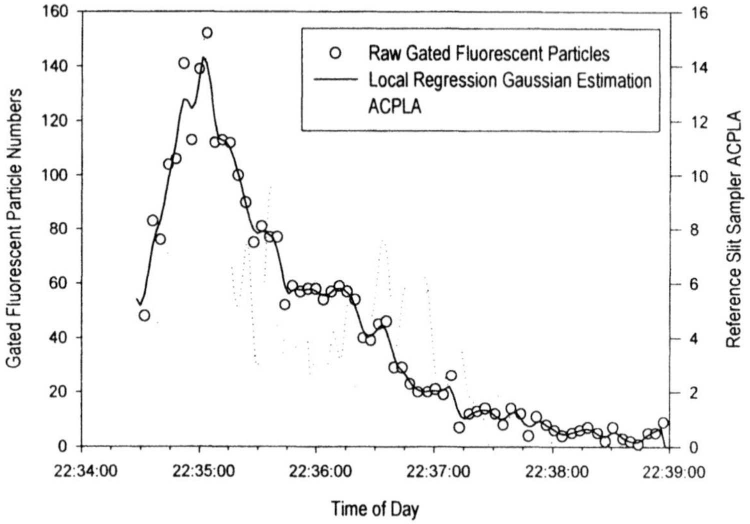

Figure 8. Gated Fluorescent Particle Data Non-Parametric Estimation

Local Regression & Gaussian Weighting $r^2 = 0.99$

Spearman's Coefficient for Gated Fluorescent Numbers vs ACPLA = 0.813

Table 1. Summary of Spearman's correlation with referenced to ACPLA data tested with different estimation methods. The different methods to interpolate data were tested to determine which is better. Each method was used to generate the two data sets (FLAPS & slit samplers) which were then subjected to Spearman's test. The FLAPS is able to produce different data types and these were also tested. Based on these limited observations, it was difficult to score one method over the other.

	Local Regression Gaussian	Local Regression Loess	Spline Estimation NURBS	Smooth Spline Estimation FFT & NURBS
Total Particle Count	0.806	0.8822	0.812	0.798
Gated Particle Count	0.813	0.830	0.819	0.803
Total Fluorescent Count	0.806	0.817	0.774	0.800
Gated Fluorescent Count	0.813	0.831	0.821	0.803
Gated % Fluorescent	0.646	0.640	0.655	0.623

3. References

Andersen, D.O. and A.J. Mohr. 1996. Abbreviated test plan for outdoor field trial (JFT III) of prototype biological detection components and systems. DPG document No. DPG-TP-96-041, JPO for Biological Defense, 5201 Leesburg pike, Skyline #3, Suite 1200, Falls Church, VA 22041-3203.

Boulet, C.A., J. Ho, L. Stadnyk, H.G. Thompson, M.R. Spence, G.A. Luoma, R.E. Elaine and W.E. Lee. 1996. Report on the Canadian Integrated Biological Detection System Trial Results of the Combined Joint Field Trials II, for Biological Detection. Defence Research Establishment Suffield SR 652 (Unclassified). Canadian defense documents are available from: Information Holdings-Operations Department of National Defence Ottawa, Ontario K1A 0K2, Canada.

Boyle, E. and H. Steen. 1993. The Physical and Biological Basis for Flow Cytometry of Escherichia coli. In Flow Cytometry in Microbiology, D. Lloyd ed. Springer-Verlag, New York.

Brenner, K.P., P.V. Scarpino and C.S. Clark. 1988. Animal viruses, coliphages, and bacteria in aerosols and wastewater at a spray irrigation site. Appl. Environ. Microbiol. 54:409-415.

Bronk, B.V. and L. Reinisch. 1993. Variability of steady-state bacterial fluorescence with respect to growth conditions. Appl. Spectr. 47:436-440.

Chatigny, M.A., J.M. Macher, H.A. Burge and W.R. Solomon. 1998. Sampling airborne microorganisms and aeroallergens pp 199-220, in Air Sampling Instruments for Evaluation of Atmospheric Contaminants, 7th edition ed. S.V. Hering, publ. American Conference of Governmental Industrial Hygienists, Inc.

Chen, G., P. Nachman, R.G. Pinnick, S.C. Hill and R.K. Chang. 1996. Conditional firing aerosol fluorescence spectrum analyzer for individual airborne particles with pulsed 266 nm laser excitation. Opt. Lett. 21:1307-1309.

Cleveland, W.S. 1993. Visualizing Data. Publ. Hobart Press, Simmit, NJ. ISBN: 0963488406

Davey, H.M. and D.B. Kell. 1996. Flow cytometry and cell sorting of heterogeneous microbial populations: the importance of single-cell analysis. Microbial. Rev. 60:641-696.

Davis, J.M. and C.E. Main. 1984. A regional analysis of the meteorological aspects of the spreading and development of the blue mold on tobacco. Boundary-Layer Meteorol. 28:271-304.

Day, R.L. and L.C. Lee. 1983. Optical emission from photoexcitation of aerosol particles produced by reaction of ozone with 1,3-butadiene. Applied Optics 22:2546-2550.

EPA, 1998. Building Assessment Survey and Evaluation (BASE) Study and the Temporal Indoor Monitoring and Evaluation (TIME) Study. www.epa.gov/iaq/base/baseenvi.html

Grant, R.H., A.B. Scheidt, L.R. Rueff. 1994. Aerosol transmission of a viable virus affecting swine: explanation of an epizootic of pseudorabies. Int. J. Biometeorol. 38:1 33-39.

Groves, W.E., F.C. Davies and B.H. Sells. 1968. Spectrophotometric determination of microgram quantities of protein without nucleic acid interference. Anal. Biochem 22:195-210.

Hairston, P., J .Ho, F R. Quant. 1997 Design of an instrument for Real-time Detection of Bioaerosols Using Simultaneous Measurement of Particle Aerodynamic Size and Intrinsic Fluorescence. J. Aerosol Sc. 28:471-482.

Harrison, D.E.F. and B. Chance. 1970. Fluorimetric technique for monitoring changes in the level of reduced nicotinamide nucleotides in continuous cultures of microorganisms. Appl. Microbiol. 19:446-450.

Ho, J. And G. Fisher. 1993. Detection of BW agents: flow cytometry measurement of *Bacillus subtilis* (BG) spore fluorescence. Defense Research Establishment Suffield memorandum 1421.

Ho, J. (1994) "Detection of Biological Warfare Agents". In Today Science Tomorrow Defence, pp 11-18, ed. C. Boulet, Government of Canada Cat No. D4-1/175E, ISBN 0-662-21570-2, (Unclassified).
Ho, J. M. Spence and G.R. Fisher. 1995. Detection of BW agents: Dugway trinational BW field trial 20-30 October 1993, Dugway, Utah; Mobile aerosol sampling unit data analysis (U). Defense Research Establishment Suffield report 619.

Ho, J. and M. Spence , (1998). "Background Aerosol Characteristics Measured with a Fluorescence Aerodynamic Particle Sizer: Sensitivity of FLAPS Performance". Suffield Memorandum 1504, (Unclassified).

Ichinose, N. 1991. Fluorimetric analysis in biomedical chemistry: trends and techniques including HPLC applications in chemical analysis 109:5-44, ed. N. Ichinose, g. Schwedt, F.M. Schnepel and K. Adachi.

Jeys, T., S. DiCecca, V. Daneu, G. Rowe, N. Newbury, A. Sanchez, C. Primmerman and F. Reyes. 1997. Bioaerosol fluorescence sensor. Scientific conference on Obscuration and Aerosol Research, 23-27 June, ERDEC, APG, MD. Available from NTIS, 5285 Port Royal Road, Springfield, VA 22161.

Kang, Y.J. and J.F. Frank. 1990. Characteristics of biological aerosols in dairy processing plants. J. Dairy Sci. 73:621-626.

Kaprelyants, A.S. and D.B. Kell. 1992. Rapid assessment of bacterial viability and vitality using rhodamine 123 and flow cytometry. J. Appl. Bacteriol. 72:410-422.

Kell, D.B., H.M. Ryder, A.S. Kaprelyants and H.V. Westerhoff. 1991. Quantifying hertrogeniety: flow cytometry of bacteria cultures. Antonie van Leeuwenhoek 60:145-158.

Kell, D.B., A.S. Kaprelyants, D.H. Weichart, C.R. Harwood and M.R. Barer. 1998. Viability and activity in readily culturable bacteria: a review and discussion of the practical issues. Antonie van Leeuwenhoek 73:169-187.

Lis, D.O., J.S. Pastuszka and R.L. Górny. 1997. The prevalence of bacterial and fungal aerosol in homes, offices and ambient air of Upper Silesia. Preliminary results. Rocz Panstw Zakl Hig. 48:59-68.

Pinnick, R.G., S.C. Hill, P. Nachman, G. Videen, G. Chen and R. Chang. 1998. Aerosol fluorescence spectrum analyzer for rapid measurement of single micrometer-sized airborne biological particles. Aerosol science and Technology 28:95-104.

Sawyer B. G. Elenbogen, K.C. Rao, P. O'Brien, D.R. Zenz, C. Lue-Hing. 1993. Bacterial aerosol emission rates from municipal wastewater aeration tanks. Appl Environ Microbiol. 59:3183-3186

Sokal, R.R. and F.J. Rohlf. Biometry: The principles and practice of statistics in biological research. Page 607. 2nd ed. 1981. Freeman and Co. New York.

Steen, H.B. 1983. A microscope-based flow cytometer. Histochem. J. 15:147-160.

Tillett, H.E. and R.G. Carpenter. 1991. Statistical methods applied in microbiology and epidemiology. Epidemiol. Infect. 107: 467-478.

MEASUREMENT OF BIOLOGICAL AEROSOLS IN AN AEROSOL CHAMBER

Z. KROCOVA, J. KREJCI, M. KROCA, AND A. MACELA
Institute for Immunology
Purkyne Military Medical Academy, Hradec Kralove
561 66 Techonin
Czech Republic

KEYWORDS/ABSTRACT: aerosol / aerosol chamber / *Francisella tularensis* LVS/ mice

A standard system for the aerosol exposure of mice with viable *Francisella tularensis* live vaccine strain (LVS) in an aerosol chamber was developed. The pneumatic nebulizer and the dynamic "nose only" aerosol chamber for exposure of 12 mice were used. Pressure of the orifice and flow rate through the atomizer, flow rate of the air, temperature, and humidity in the chamber were measured. Density, viscosity, consumption of the aerosolized suspension, and the number of cultivable bacteria in the suspension were measured. The correlation between the technical parameters and the number of the cultivable bacteria in the initial suspension, aerosol, and the murine lungs were determined.

1. Introduction

The standardization of aerosol application of the intracellular pathogen the *Francisella tularensis* live vaccine strain (LVS) was developed within the part of project "Vaccines against the intracellular pathogens - *F. tularensis* model". The immune response against the standard vaccine *F.tularensis* LVS and a new vaccine, named *F. tularensis* ghost, after different ways of the exposure, including aerosol, will be studied. *F. tularensis* ghosts are based on the expression of the gene of the E-phage ΦX 174 in the bacterial cell. *Francisella tularensis*, a facultative intracellular bacterium, initiates fatal human disease called Tularemia. Disease manifestation depends upon the route of inoculation and ranges from ulceroglandular infection (entry through the skin) to respiratory tularemia (inhalation) or to a typhodial disease (unknown route of entry). Regardless of the initial site of infection, pneumonia is a common sequel of tularemia. A relatively avirulent *Francisella* strain LVS is used as a vaccine in humans, albeit with only variable effectiveness, to protect against the virulent *F. tularensis*. This vaccine strain itself, however, is fully pathogenic for certain animals and causes a lethal infection in mice that is indistinguishable from human disease. The susceptible (BALB/c) and resistant (C57Bl/10) murine strain to tularemic infection were used.

P.J. Stopa and M.A. Bartoszcze (eds.),
Rapid Methods for Analysis of Biological Materials in the Environment, 203–217.

For the standard application of the biological aerosol into the aerosol chamber it was necessary to develop a system for the exposure of mice in the aerosol chamber and the measurement of biological aerosol. Mice were exposed in the dynamic aerosol chamber, "nose only" type. The syringe-pump pneumatic atomizer was used for the generation of bioaerosol.

The deposition of bioaerosol in the murine lungs is influenced by a number of technical and non-technical parameters. The technical factors include the method of aerosol generation that results in size of aerosol particles and the preparation, storage and consumption of the suspension. Major factors governing the effectiveness of the deposition mechanisms are the size of the inspired particles, ventilatory flow, geometric disposition of airways, inhaled volume, and duration of ventilatory cycle. The forces involved correspond to mechanical inertia in the case of particle diameters higher than 1 mm, and Brownian movements of thermodynamic diffusion for particle diameters smaller than 0.1 μm. Particles of a 0.1- to 1-μm diameter produce deposition through the action of both the mechanisms. For the particles with a diameter smaller than 1 μm, the influence of nasal or oral breathing is low. Deposition in the terminal respiratory units (slow-clearing airways) is also depicted by a bell-shaped curve: particles larger than 15 μm cannot penetrate them and only particles of the 2 μm and under can deposit there. However occupation of drops by *F. tularensis* microbes that are approximately 0.8 μm large rise with increasing diameter of particles. For instance aerosol with a mass median diameter (MMD) = 3.5 μm and geometric standard deviation (GSD) equal two contain 88% of noninfectious aerosol. The rest of the particles show 70% by one microbe only. The influence of particle size on deposition in the lung was fixed.

Temperature, humidity and a flow rate of the inlet air and air going through the chamber were measured. The MMD and GSD were determined by the May Cascade Impactor and by microscopic analysis.

Common methods for sampling of microorganism in aerosol employ collection onto a filter, impaction onto an agar surface, or impingement into a liquid. The agar collector, the Andersen Six-Stage Viable Particle Sizing sampler, and the All-Glass Impinger with liquid collection medium were used in proposed work. From the determination of the concentration of the cultivable *Francisella* in the initial suspension, impinger, and the viable concentration of aerosol in the lungs, stability of microbial viability after nebulization, and the correlation between the concentration of microbes in the impinger and the lungs were established.

2. Materials and Methods

2.1. ANIMALS

Specific-pathogen-free BALB/c and C57Bl/10N female mice were purchased from Charles River Company, Germany and used at 6 to 8 weeks of age for experiment. Mice were housed in a barrier environment.

2.2. BACTERIA

F. tularensis LVS (ATCC 29684, American Type Culture Collection, Rockville, USA) was cultured on McLeod agar supplemented with hemoglobin (Difco, Detroit, USA) and IsoVitaleX (Becton Dickinson, USA) for 24 hr. at 37°C. Suspension for aerosol application was prepared in 1% bacto-pepton and 1% lactose (Difco, Detroit, USA).

Enumeration of bacteria. The number of cultivable bacteria was determined by colony forming units (CFU) on McLeod agar plates. The samples from aerosolized suspension and impingers were cultivated on agar plates for 48 to 72 hours at 37°C. Bacterial CFU's from organs were made from aseptically removed organs, the tissue was disrupted through a sterile mesh, homogenized and resuspended in PBS, serially diluted, and plated on McLeod agar plates for 48 to 72 hours at 37°C.

2.3. AEROSOL EXPOSURE

The dynamic aerosol chamber "nose only" (Techonin, Czech Rep.) was used for exposure of 12 mice in one experiment (Figure 1). Flow rate through the chamber was 6 l/min, velocity 0.054 m/sec, temperature of air inside the chamber was 22 ± 2°C and humidity ~ 60%. The syringe-pump atomizer (Techonin, The Czech Rep.), as described in, was used for the generation of aerosol containing *F. tularensis* (Figure 2). The air flow through the generator was 6 l/min. Different pressure on orifice and secondary air going against the aerosol falling to closed reservoir were applied for achievement of particle size optimization. The viability of the circulated suspension was determined by plating and was not significantly changed for the 6 minute interval of exposure. The concentration of the suspension was according to requirement of the number of microbes caught in the lungs. The consumption of the suspension varied from 80 to 170 μl/min under the influence of physical conditions. The sampling of flow rates, consumption of the medium and bioaerosol sampling in the chamber were made during each exposure of the mice. Determination of the particle size of the aerosol in the chamber was determined before the exposure of mice. Deposition of microbes in the lungs was determined by CFU's from the lungs immediately after the exposure. The correlation between the concentration of bacteria in aerosol and lung was made. For a standard application of *F. tularensis,* a logging system was created where the following data fields were defined: the flow rate through the chamber; the consumption of the suspension; time of the exposure; amount of the impinger solution; flow rate through the impinger; MMD; GSD; "live" concentration of the suspension; "live" concentration in the impinger; number of bacteria in the lungs; number of live bacteria in one liter of air; "live" concentration of aerosol; stability of viability after the nebulization and deposition of *F. tularensis* in the lungs.

2.4. PARTICLE SIZE

The size of aerosol particles was measured by the May four-stage cascade impactor (Casella London). Operating flow rate of this impactor is 17.5 l/min, orifice dimensions are from 0.027 to 0.65 cm, jet velocities from 2.37 to 77 m/sec, and the effective cut aerodynamic diameter (ECAD) 0.5 - 12.5 μm. The glass slides, on which the aerosol was captured, were coated with a micro-layer of cedar oil. Immediately after an exposure to aerosol, slides were analyzed under a microscope with a screen delivered by the manufacturer of the impactor. At each scale 200 drops were measured and 6 measurements were provided for determination of a result. Evaluation was provided according to the valid internationally used formulas for a calculation of CMD, MMD, and GSD.

2.5. BIOAEROSOL SAMPLING

Two types of bioaerosol samplers were used in this study. The Andersen Six-Stage Impactor Sampler (Graseby Andersen, Smyrna, Ga.) consists of six stages with a decreasing nozzle diameter (0.118 - 0.025 cm) such that successive stages collect progressively smaller particles (ECAD is from 7.1 to 0.6 μm). Thus, the six-stage sampler measures the cultivable bioaerosol concentrations in specific particle size ranges. McLeod agar was used for filling of the Petri dishes placed inside the impactor and the cultivation of the collected bacteria under the conditions described below. The second sampler was the All-Glass Impinger (Techonin, Czech Rep.). The flow rate of the inlet air is 6 l/min, incoming flow velocity 127.3 m/sec, nozzle diameter 1 mm, cross-section area 0.79 mm^2, stopping distance S_{50} was 0.125 mm and calculated "cutoff size" $d_{50} = 0.35$ μm. The impinger was filled by 5 ml of collection medium (0.5% gelatin in PBS).

3. Results

The "nose only" chamber for murine exposure was used. Animals were restrained in holders and their noses were placed in the air stream carrying the aerosol. Aerosol produced by the syringe-pump atomizer was equally distributed to each of the 12 mice exposed. The pressure on the orifice was between 250 and 350 kPa. Aerosol was transferred to the atomizer space and moved to the chamber at a flow rate of 6 l/min. The larger particles impacted on the opposite wall of the generator and dropped back into the closed reservoir.

Experiments were performed, varying different parameters. The pressure on the orifice was usually between 250 and 350 kPa and one experiment was made at pressure 250 kPa and used the secondary air that was going against falling particles in the atomizer. The size of the particles, count median diameter (CMD), mass median diameter, and geometric standard deviation were designated for three different technical parameters. An aerosol designated as "**A**" was generated at 250 kPa and with using of the secondary air; an aerosol "**B**" was produced at 250 kPa and without the secondary air; and an aerosol "**C**" was made at 350 kPa and without the secondary air. The smallest CMD = 1.22 μm and MMD = 2.58 μm were found for aerosol **A** (TABLE 1). Aerosol **B** had CMD = 1.46 mm and MMD = 2.63 μm. The biggest size of particles was measured for the aerosol **C**, CMD = 2.81 μm and MMD = 4.8 μm. From the distribution of aerosol particles according to number of particles (Figure 3), it is evident that aerosols **A** and **B** have the majority of particles smaller than 3.5 mm while aerosol **C** also has a considerable amount of particles lying between 3.5 and 6.5 mm, with all the particles having this size creating the major percentage of the volume (MMD = 4.8 μm).

Although the suspension was recycled, the decay of viability of *Francisella tularensis* in the suspension during one 6-minute cycle was only about 10 % in average (TABLE 2).

The Andersen Six-Stage Cascade Impactor was used for measurement of aerosol carrying bacteria of *F. tularensis*. From the curve of the CFU distribution (Figure 4) it is possible to see that in aerosol **C** the bacteria predominantly occurred at larger sized particles. The curves of the dependence between the numbers of the particles or CFUs on particle size are shown in Figure 5. The figure shows that the occupation of particles with 1 µm diameter is approximately half the occupation of particles with diameter 3 µm.

The influence of deposition of *F. tularensis* in the murine lungs on particle size was evaluated. Deposition is the ratio between the number of bacteria captured in the lungs and cultivated from the lung suspension immediately after the exposure and the theoretically counted number. Percentage of the deposition was the same at both the murine strains (data not shown). Deposition of the aerosol **A** with MMD = 2.58 µm was 19%. For aerosol **B** (MMD = 2.63 µm), the deposition was 25%. In the case of aerosol **C,** which had the highest mass median diameter, the deposition was 41% (Figure 6). The correlation between the measured concentration of the cultivable bacteria in the impinger and the deposited bacteria in the lungs of animals was made (Figure 7). The concentration curve has a linear trend. It is possible to determine from it the number of bacteria captured in the lungs. This is especially important for the lowest doses of infectious agents. By comparing the change of the suspension concentration required with the amount of bacteria in the lungs, it is possible to obtain the whole range of usable infection doses (TABLE 3). Data of the bacterial concentration in the impinger are also shown in this table.

4. Discussion

A functional system of the aerosol application of the infection *F. tularensis* LVS on mice working under the standard condition (TABLE 1) was developed. In this system the aerosol with standard parameters can be created repeatedly in the sufficient amount and with the required particle distribution.

The system is also applicable of creating aerosols containing infectious agents, such as *F. tularensis*. However, in our case, a low decrease of bacterial viability during experiments (TABLE 2) was observed.

Aerosol particles are delivered uniformly to each of twelve positions for experimental animals, which was verified by comparison of the deposited amount of bacteria in mice according to their position in the chamber (data not shown). Our results from deposition of aerosol with different particle properties in the murine lungs (Figures 5 and 6) are in good accordance with previous data obtained on another systems.

Concentration of aerosolized infectious particles in the impinger after exposure is linearly dependent on the nebulized suspension concentration and also on the deposited dose in the lungs. For the very low doses of the deposited bacteria, it is thereby possible to calculate this dose from the numbers of bacteria caught in the impinger (Figure 7).

5. Conclusion

The standard system for creating and application of biological aerosols on mice was developed with the possibility to challenge mice by defined doses of infectious agents in controlled conditions (TABLE 4). This system will be used for the future immunological studies in the field of anti-microbial immunity.

This work was supported by Ministry of Defense, The Czech Republic.

TABLE 1. The influence of technical parameters on properties of generated aerosol.

Type of the aerosol	A	B	C
Pressure on the orifice (kPa)	250	250	350
The secondary air (l/min)	2.5	without	without
The consumption of the suspension (ml/min)	0.170 ± 0.025	0.078 ± 0.011	0.098 ± 0.010
CMD (μm)	1.22 ± 0.26	1.46 ± 0.23	2.81 ± 0.83
MMD (μm)	2.58 ± 0.76	2.63 ± 0.81	4.80 ± 0.63
GSD	1.79 ± 0.05	1.83 ± 0.15	1.96 ± 0.07

TABLE 2. The viability of *F. tularensis* suspension measured before and after aerosol exposure attached by plating.

The number of the cultivable bacteria *F. tularensis* / ml		The decay of the visibility (%)
Before exposure	After exposure	
5.63×10^9	5.27×10^9	7.0
4.80×10^9	4.40×10^9	8.4
1.70×10^9	1.45×10^9	14.3
4.40×10^9	3.83×10^9	13.0
4.40×10^9	3.97×10^9	9.8

TABLE 3. The dependence of *F. tularensis* deposition in lung on suspension concentration. The concentration of suspension in impingers is also showed.

The concentration of the suspension *F. tularensis* / ml	The concentration of the impinger *F. tularensis* / ml	The concentration of the lungs *F. tularensis* / ml
4.4×10^9	1.1×10^7	3.3×10^4
4.4×10^8	1.1×10^6	4.0×10^3
3.7×10^7	4.1×10^4	2.1×10^2
3.3×10^6	4.8×10^3	1.5×10^1
4.3×10^5	3.8×10^2	3.3×10^0

TABLE 4. Parameters for production of standard biological aerosol carrying bacteria of *F. tularensis* LVS intended for infection of mice with tularemia and following immunological studies.

Q	the flow rate through the chamber	6 l/min
f	the consumption of the suspension	0.078 ± 0.011 ml/min
t	time of the exposure	6 min
V	breath volume of mouse (weight $\approx$ 20 g)	0.024 l/min
a	the amount of the impinger solution	5 ml
CMD	count median diameter	1.46 ± 0.23 μm
MMD	mass median diameter	2.63 ± 0.81 μm
GSD	geometric standard deviation	1.83 ± 0.15
c_s	"live" concentration of the suspension	$3.5 \pm 2.2 \times 10^9$ *F.tul*/ml
c_i	"live" concentration in the impinger	$5.7 \pm 2.8 \times 10^6$ *F.tul*/ml
c_{lung}	the number of bacteria in the lungs	$2.2 \pm 1.2 \times 10^4$ *F.tul*/lungs
c_{disp}	the number of live bacteria in 1 l of the air $c_{disp} = f \, . \, c_s / Q$	$4.6 \pm 2.9 \times 10^7$ *F.tul.*/lml
c_v	"live" concentration of aerosol	$7.5 \pm 4.7 \times 10^5$ *F.tul*/ml
η	the stability of the viab. after the nebuliz. $\eta = c_v / c_{disp} \, . \, 100$	$3.6 \pm 3.5\%$
D	the deposition of *F.tul* in the lungs $D = c_{lung} \, . \, V. \, 100 / c_v \, . \, Q \, . \, t$	$24.5 \pm 5.5\%$

FIGURE 1. Aerosol chamber "nose only" for mice. a) The side view on the chamber. The holders with the fixed mice are shown. b) The upper view of the chamber

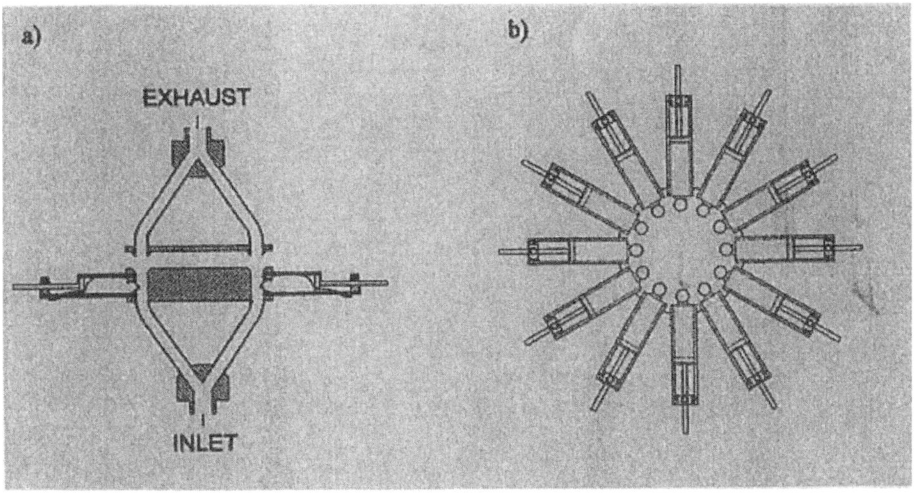

FIGURE 2. The principle of atomizer that was used for the generalization of biological aerosol.

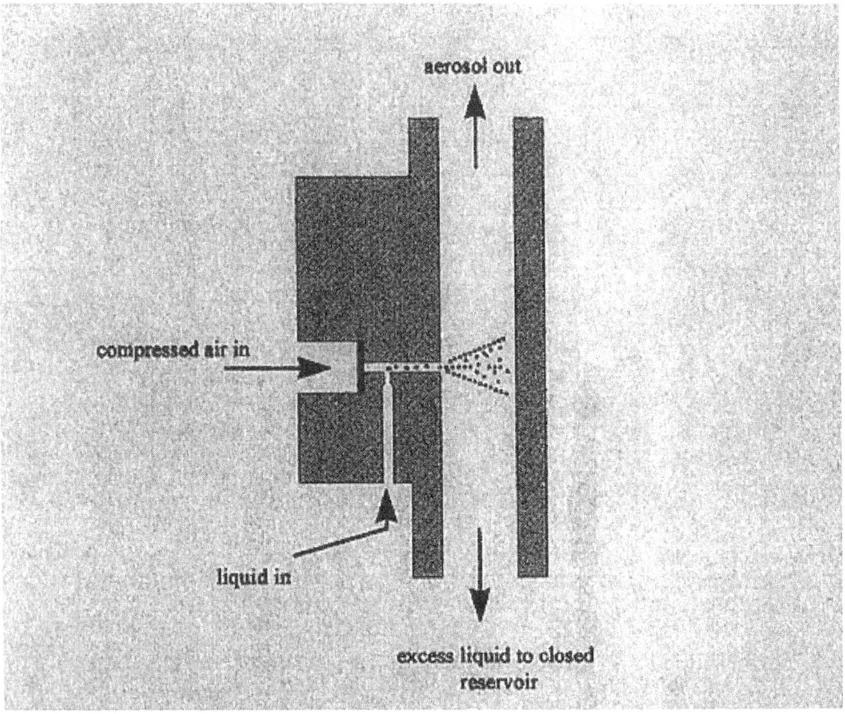

FIGURE 3. Distribution of particles of aerosol **A** (250 kPa, secondary air), **B** (250 kPa, without sec. air) and **C** (350 kPa, without sec. air) according number of particles.

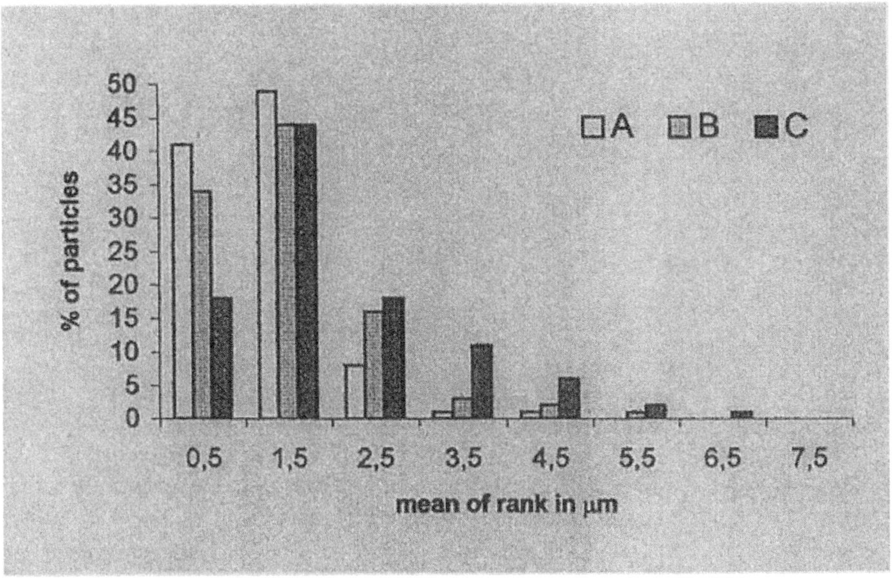

FIGURE 4. Distribution of particles in aerosol **A, B** and **C** constructed from number of CFUs cultivated on the plates of Andersen cascade impactor in dependence on particle size.

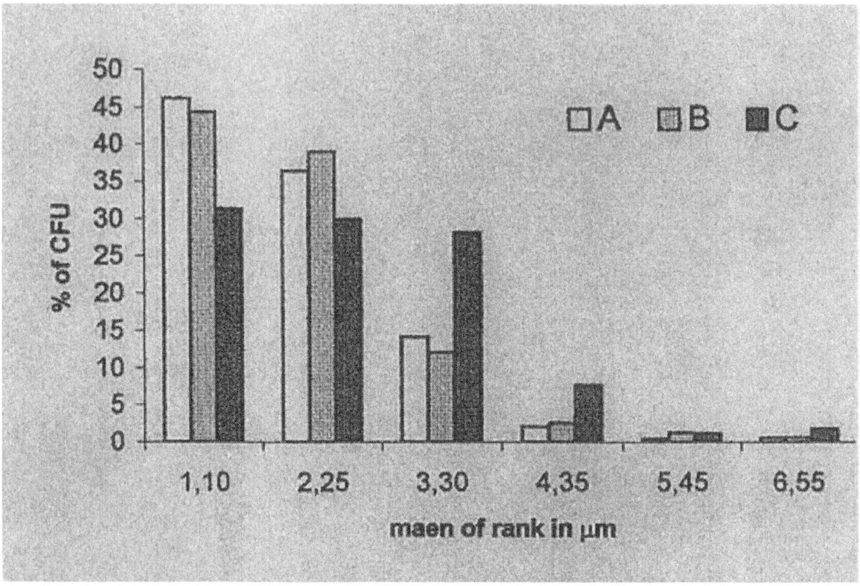

FIGURE 5. Comparison of the dependence of the number particles and number CFUs on particle size for aerosol **A, B** and **C.**

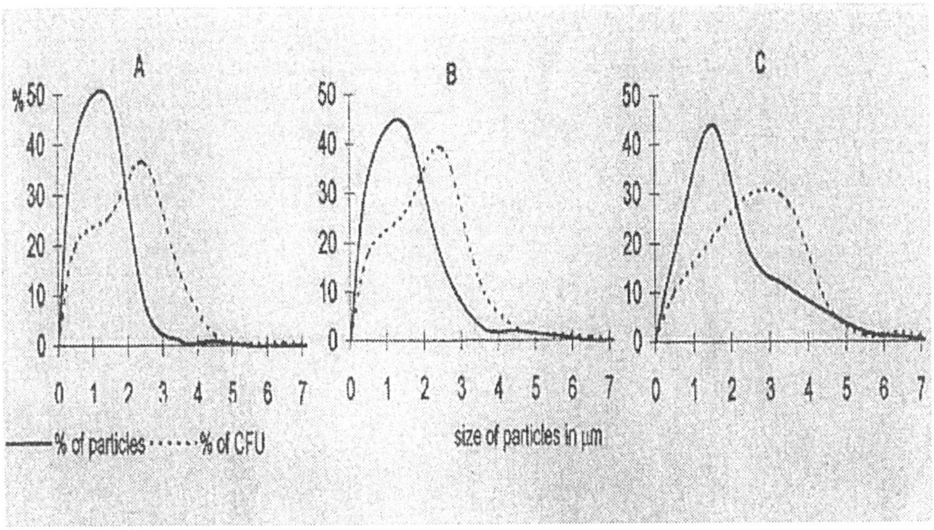

FIGURE 6. The dependence of *F. tularensis* deposition in murine lungs on the aerosol mass median diameter.

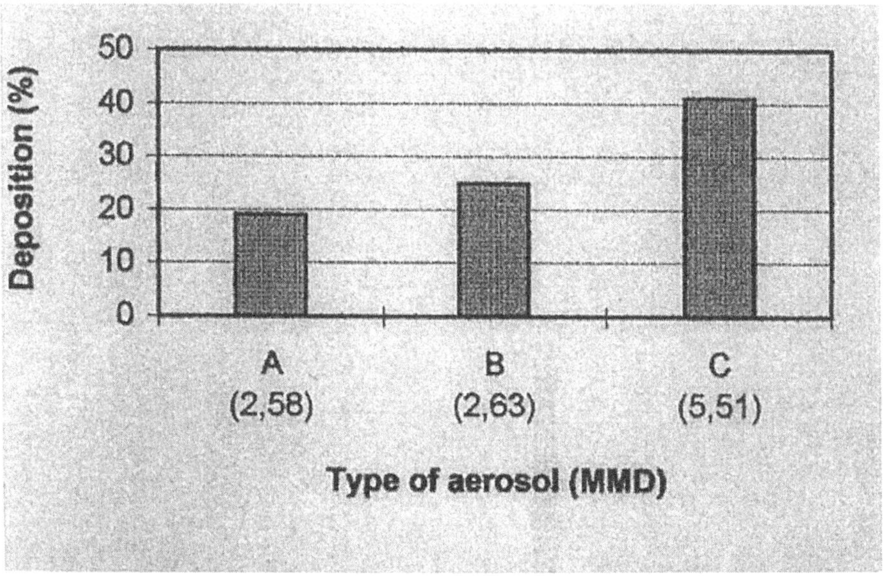

FIGURE 7. Correlation of bacterial CFUs in impingers and lung of infected mice. The results were achieved under following parameters: pressure on atomizer orifice 250 kPa, flow rate through atomizer 3.4 l/min, consumption of aerosolized suspension 0.078 ml/min, MMD 2.63 μm, GSD 1.83, flow rate through chamber and impinger 6 l/min, time of exposure 6 min, breath volume of mice with weight app. 21 g was 0.024 l/min.

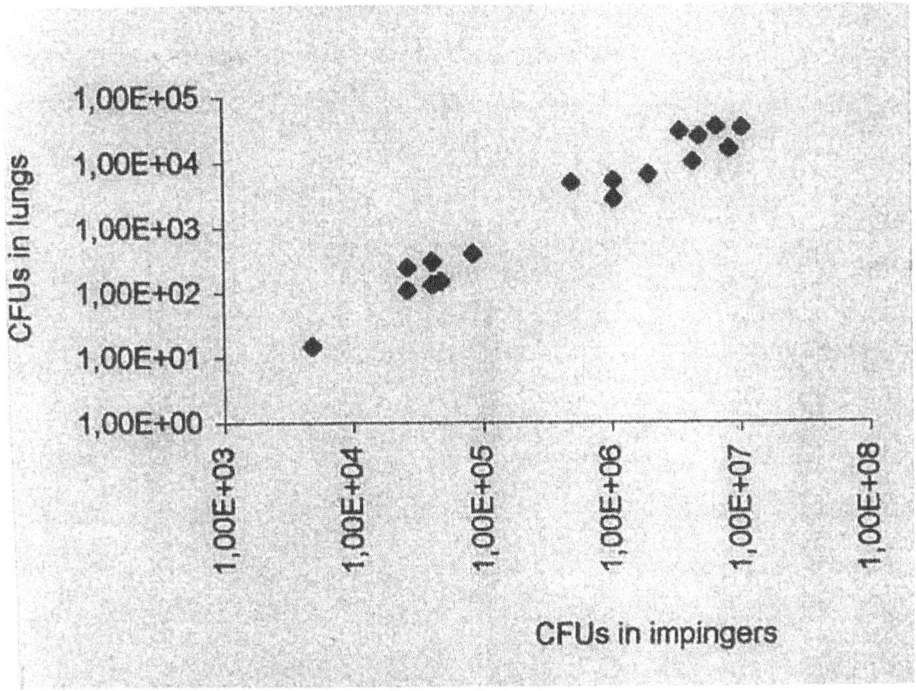

DEVELOPMENT OF PCR-BASED ASSAYS FOR THE DETECTION AND MOLECULAR GENOTYPING OF MICROORGANISMS OF IMPORTANCE TO BIOLOGICAL WARFARE

VITO G. DELVECCHIO, RAJENDRA REDKAR, SHUQUI CHENG, AND MIKE GRESS
Institute of Molecular Biology and Medicine, University of Scranton, Scranton, PA 18510

Keywords/Abstract

PCR, Fingerprinting, *Enterococcus, Franciscella tularensis*

Species- or strain-specific polymerase chain reaction assays offer an extremely sensitive and accurate detection method for biological warfare agents. Several different strategies can be used for identifying and cloning unique nucleic acid sequences which can detect one type of microorganism from others found in mixed samples or the environment. Specific PCR probes can be obtained from plasmid and chromosomal libraries, ribosomal DNA genes, rep-PCR libraries, and gene sequences in DNA databases. Such probes should be conserved in the genome of the organism in question.

Conventional molecular genotyping or DNA fingerprinting is often time-consuming, laborious, and demands the full attention of a dedicated technician. Repetitive element sequence-based PCR (Rep-PCR), which amplifies regions of DNA found between repetitive elements, has been exploited as a means of genotyping. The seemingly random distribution of these repeats permit complementary oligonucleotide primers to generate fragment patterns that are specific for individual strains of a microorganism. Rep-PCR is relatively rapid, not labor-intensive, accurate, and reproducible. It can readily be adapted to any laboratory setting without great expense.

1. Introduction

Polymerase chain reaction (PCR) has revolutionized the detection of pathogenic microorganisms. PCR assays are highly specific once a signature sequence is found for a particular microorganism. Specificity is a primary requisite when developing such assays. Another concern is that the signature or target sequence is conserved in all members of a pathogenic strain or species. PCR is extremely sensitive since it can rapidly detect a few copies of a target sequence, not time consuming, and not labor intensive. Thus, liberation of nucleic acid targets from pathogens should be a relatively simple process. These attributes have made PCR attractive for the detection of a variety of biological warfare agents.

P.J. Stopa and M.A. Bartoszcze (eds.),
Rapid Methods for Analysis of Biological Materials in the Environment, 219–229.

PCR has also streamlined the process of molecular genotyping or fingerprinting of microorganisms. Repetitive element sequence-based PCR (Rep-PCR), which amplifies regions found between repetitive segments of chromosomal DNA, has offered a relatively simple means of genotyping and identifying specific target sequences for species-specific probes. This report presents an outline of the various strategies our laboratory has used in the development of DNA probes for the detection of pathogenic microorganisms as well as those of importance to biological warfare. These strategies include the use of chromosomal libraries, ribosomal genes, Rep-PCR fragments, and gene sequences obtained from the scientific literature or various gene databases.

2. Chromosomal Libraries

Enterococcus faecium accounts for 5 to 16% of enterococci isolated from human nosocomial or hospital acquired infections. It is one of the most physiologically heterogeneous groups of all enterococcal species and can easily acquire multiple antimicrobial agent-resistance. Strains have been reported to possess high-level resistance to vancomycin, aminoglycosides, and penicillin. These strains have not only been isolated from nosocomial but also from different environmental samples (1). Since species identification and antimicrobial susceptibility profiles are necessary for defining colonization, infection patterns, and analysis of epidemiological data, an *E. faecium*-specific PCR assay was developed as described below (2).

DNA was isolated as reported (3) except mutanolysin was used to disrupt the bacterial call wall. The DNA was digested with *Eco*RI, and a genomic library was created using the cloning vector pUC18. The library was transformed into *E. coli* DH5*a* cells, and recombinant plasmid DNA was isolated and digested with *Eco*RI. *E. faecium*-derived insert DNA were electrophoresed through 1.0% SeaKem GTG agarose, and size was calculated by comparison with concurrently run DNA standards. Southern blots (4) of insert DNA were probed with *E. faecalis Eco*RI-digested chromosomal DNA. Clones containing insert DNA that hybridized with the *E. faecalis* DNA were eliminated from the investigation. However, clones that did not hybridize with the *E. faecalis* DNA were selected for further study, and their inserts were used to probe blots of *Eco*RI-digested genomic DNA of enterococcal reference strains. Figure 1 illustrates the specificity of one of the clones examined.

It can be seen that the insert DNA hybridized only with DNA fragments found in *E. faecium*. DNA sequencing of this insert DNA was accomplished with the DyeDeoxy Terminator Cycle Sequencing Kit (Applied Biosystems, Foster City, CA) on the 373A Automatic DNA sequencer. A pair of primers (EM1A:5'-TTG AGG CAG ACC AGA TTG ACG and EM1B:5'-TAT GAC AGC GAC TCC GAT TCC) was selected for the PCR assay using the Oligo ver 4 Primer Software (National Biosciences, Inc., Plymouth, MN). A two-temperature PCR amplification was performed in a total volume of 25 μl which consisted of 10 mM Tris HC1 (pH 8.3), 75 mM KC1, 3.5 mM $MgC1_2$, 200 μM of each deoxynucleoside triphosphate, 0.2 μμM of each primer, 25 ng of template DNA, and 0.5 U of *Taq* polymerase. After a 4 min preamplification step at 94°C and 1 min at 72°C. An extension step of 8 min at 72°C was included for both assays. The PCR assay was observed to be specific for *E. faecium* as can be seen in Figure 2.

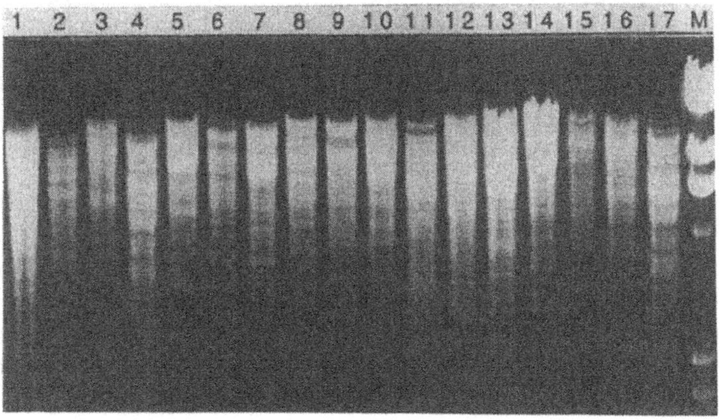

B

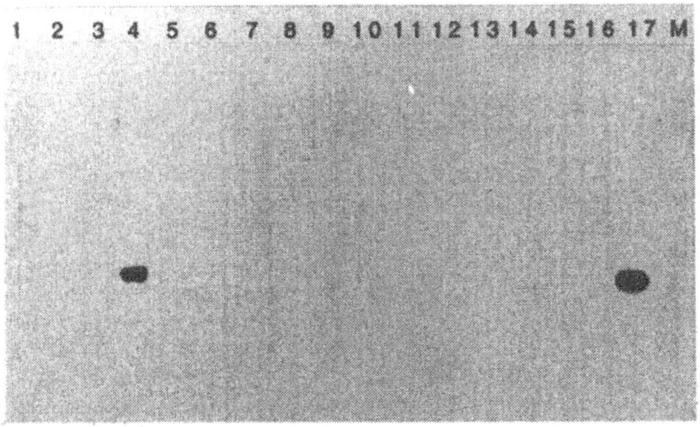

Figure 1. Agarose gel electrophoresis of *Eco*R1-digested genomic DNA of *Enterococcus* species (A) and corresponding Southern blot probed with *E. faecium*-specific probe (B). Lanes; 1, *E. durans*; 2, *E. avium*; 3, *E. faecalis*; 4, *E. faecium*; 5, *E. casseliflavus*; 6, *E. saccarolyticus*; 7, *E. malodoratus*; 8, *E. cecorum*; 9, *E. seriolicida*; 10, *E. pseudoavium*; 11, *E. raffinosus*; 12, *E. gallinarum*; 13, *E. faecalis*; 14, *E. hirae*; 15, *E. faecalis;* 16, *E. faecalis*; 17, *E. faecium*; M, 100-bp ladder.

Figure 2. Agarose gel electrophoresis of PCR products from DNA of *Enterococcus* species. Lanes: 1, *E. durans*; 2, *E. avium*; 3, *E. faecalis*; 4, *E. faecium*;5, *E. casseliflavus*; 6, *E. saccharolyticus*; 7, *E. malodoratus*; 8. *E. cecorum*; 9, *E. seriolicida*; 10, *E. pseudoavium*; 11. *E. raffinosus*; 12, *E. gallinarum*; 13, *E. faecalis*; 14, *E. hirae*; 15, *E. faecalis*; 16, *E. faecalis*; 17, *E. faecium*; M, 100-bp ladder.

3. Cloning of Rep-PCR Fragments

Many investigators have recognized the importance of detailed epidemiological studies, and molecular genotyping methods have contributed to our understanding of infectious outbreaks. Rep-PCR employs oligonucleotide primers that are complementary to segments of repetitive elements. These rep elements vary in position and number on the chromosomes of strains of microorganisms (5). The characteristic distribution of such elements allows primers to amplify segments of chromosomal DNA sequences found between the interspersed rep elements. Agarose gel electrophoresis of the amplified products provides banding patterns of sufficient complexity to distinguish species and strains of microorganisms. Thus, the banding pattern can act as a bar code for a specific strain of microorganism.

The first step in the development of a Rep-PCR genotyping system for enterococci (6) was to find a rep element on the chromosome of different strains and species of enterococci. In a previous study on the development of Rep-PCR of methicillin-resistant *Staphylococcus aureus*, it was found that is shared homology with a repetitive sequence present in *Mycoplasma pneumoniae* (7). This similarity was exploited for Rep-PCR fingerprinting of *S. aureus* (3). Since *M. pneumoniae* was thought to have evolved from gram-positive bacteria by genome reduction (8), it was found that the same primer could be used for Rep-PCR fingerprinting of enterococci. This primer was labeled RW3A, and its nucleotide sequence is 5'TCG CTC AAA ACA ACG ACA CC. Figure 3 is a gel image of Rep-PCR patterns generated by various species of enterococci. It is obvious that the RW3A primer hybridizes with a repetitive element that is highly conserved in enterococci, and Rep-PCR fingerprinting distinguishes the diverse species.

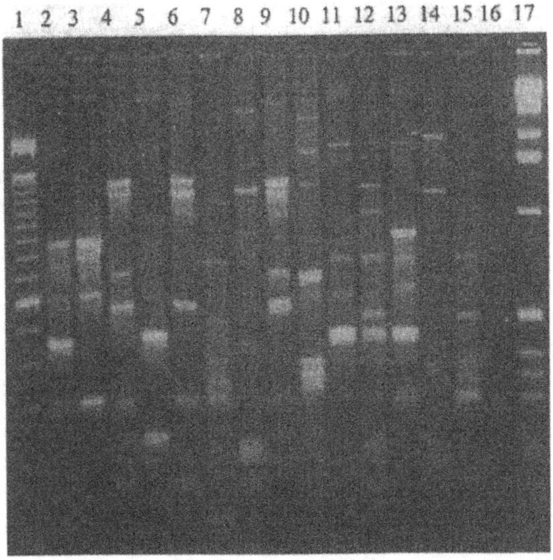

Figure 3. Rep-PCR fingerprints of *Enterococcus* species. Lanes: 1, 100-bp ladder; 2, *E. hirae*; 3, *E. durans*; 4, *E. faecalis*; 5, *E. avium*; 6, *E. faecalis*; 7, *E. faecium*; 8. *E. casseliflavus*; 9, *E. faecalis*; 10, *E. saccharolyticus*; 11, *E. malodoratus*; 12, *E. cecorum*; 13, *E. raffinosus*; 14, *E. gallinarum*; 15, *E. faecium*; 16, negative control; 17, 1-kb ladder.

Rep-PCR was accomplished by using 200 ng of chromosomal DNA as template for amplification. A PCR amplification reaction consisted of 2.5 mM of each of the four deoxynucliotides (dATP, dCTP, dGTP, and dTTP), 75 picomoles of the primer RW3A, 2.5 U of *Taq* polymerase, 10 mM Tris-HCl (pH 9.2), 25 mM KCl, and 3.5 mM $MgCl_2$. A preamplification step of 95°C for 3 min was used to denature the template DNA. This was followed by the additional of *Taq* polymerase and 30 cycles of denaturation at 94°C for 1 min, annealing at 54°C for 1 min and extension at 72°C for 2 min. A final extension step at 72°C for 5 min was included.

Primer RW3A is especially useful for Rep-PCR genotyping of clinical isolates of *E. faecium* and *E. faecalis* (Figure 4). Our laboratories have fingerprinted more than 400 isolates from blood, urine, wounds, and various environmental sites including geographic locations from around the world. Rep-PCR patterns of *E. faecium* had from 10-18 bands while those of *E. faecalis* consisted of 7-18 bands. The amplified bands ranged in size from 50 to 1450 bp and had varying intensities of staining. The banding patterns obtained were observed to be highly reproducible and discriminatory. We have been able to determine the strain of enterococcus responsible for five nosocomial outbreaks. Reproducibility was also verified by the fact that similar genotypes have been observed in strains originating from distant geographic locations. An additional benefit of Rep-PCR analysis is that it is useful as a means of distinguishing *E. faecalis* from other enterococcus species. Two bands of 1,200 and 1,300 bp have been observed in every strain of *E. faecalis* analyzed to date. The entire procedure from isolation of DNA to electrophoresis of the PCR products can be accomplished in an 8-hour workday.

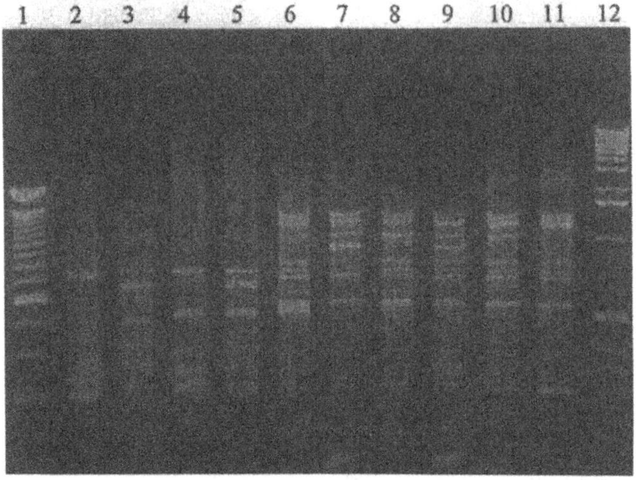

Figure 4. Rep-PCR fingerprints of clinical isolates of *E. faecium*, lanes 2-5 and *E. faecalis* (lanes 6-11). The 1,200/1,300-bp doublet is indicated on the right. Lanes 1 and 12 are DNA size standards.

The presence of the 1,200- and 1300-bp bands in *E. faecalis*, and not in other enterococcus species, facilitated the development of a species-specific probe for this bacterium (9). The 1,200-bp fragment was removed from the agarose gel, cloned into *E. coli*, and specificity of the fragment was determined by probing Southern blots of *Eco*RI-cut genomic DNA of the enterococcus species listed in Figure 3. The nucleotide sequence of the fragment was determined, and a PCR assay was developed. The assay generated the expected amplicon only when *E. faecalis* DNA served as the template as seen in Figure 5.

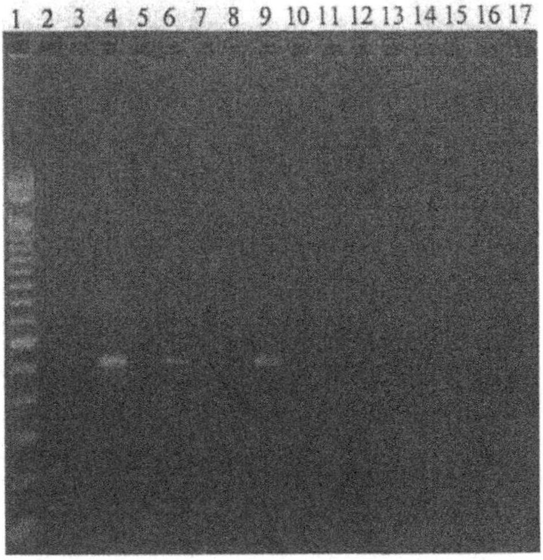

Figure 5. Specific PCR amplification of *E. faecalis.* . Lanes: 1, 100-bp ladder; 2, *E. hirae*; 3, *E. durans*; 4, *E. faecalis*; 5, *E. avium*; 6, *E. faecalis*; 7, *E. faecium*; 8. *E. casseliflavus*; 9, *E. faecalis*; 10, *E. saccharolyticus*; 11, *E. malodoratus*; 12, *E. cecorum*; 13, *E. raffinosus*; 14, *E. gallinarum*; 15, *E. faecium*; 16, negative control; 17, 1-kb ladder. The 513-bp amplicon was generated only if *E. faecalis* was the source of the template DNA.

DNA isolation is not necessary when applying the PCR assays for *E. faecium* and *E. faecalis* since successful amplification can be performed with single colonies of clinical isolates scraped from the surface of agar plates. The colony is suspended in 40 μl of 5% Chelex (BioRad, Richmond, CA) and boiled for 5 min to disrupt the cells. The suspension was then centrifuged at 14,000 x g for 2 min. Five μl of the supernatant was added to the PCR amplification mix. The assay generated the expected amplicon only when a specific enterococcus species DNA served as the template. The assays have been tested with hundreds of clinical samples and their accuracy has been in total agreement with standard bacteriological tests. In some cases, the PCR assay has proven to be more accurate than standard tests that involve culturing of the bacteria.

4. Ribosomal DNA Cloning

Defining the diversity of microbial populations is challenging because microorganisms live in consortia and the majority of bacteria cannot be grown and identified by traditional microbial techniques. An alternative approach to understanding the complexity of microbial communities involves examining variations in the 16S rRNA encoding DNA (rDNA). Based on the variability of the nucleotide sequence of the variable regions of the rDNA of bacteria, archaebacteria, and eukaryotes, phylogenetic and evolutionary relationships can be predicted. The average bacterial rDNA is 1,500 bp long, and the rRNA operon has many copies present in a genome.

Our laboratory has utilized this strategy to determine what organisms are involved in foot ulceration of diabetic patients. Cloning of the rDNA of microorganisms involves extraction of DNA from all organisms present in a microbial community, amplification of the rDNA genes by PCR, construction of an rDNA library, rapid detection of variations in the clones, determination of the nucleotide sequences of the different clones, and comparative sequence analysis. This data can identify new or known species and also pinpoint phylogenetic status. The following procedure for rDNA cloning is adapted from a procedure outlined by Relman (10). The sample is subjected to lysis in a freshly prepared digestion buffer consisting of 50 mM Tris-HC1 (pH *8.5),* 1 mM EDTA, 1% Tween, and 1 mg/ml lysozyme. The mix is then incubated at 37°C for 1 hour. Following the addition of 10 µl of proteinase K (20 mg/ml), the mixture was placed at 55°C for 1 to 3 hours. After heat inactivation of the proteinase K, DNA was isolated. The forward primer (5'-GCG GAT CCG CGG CCG CTG CAG AGT TTG ATC (C/A)TG GCT CAG) and the reverse primer (5'-GGC TCG AGC GGC CGC CCG GG(T/C) TAC CTT GTT ACG ACT T) were used to amplify a 1540-bp PCR product. The forward primer hybridizes with nucleotides at 8 to 27 in the rDNA gene, and the reverse complements a sequence at position 1492 to 1510. The underlined portion of the primer sequence indicates the *Not*I recognition site, which is utilized for increased efficiency of cloning.

PCR reaction contained 10mM Tris HC1 (pH 8.3), 50mM KC1, 2.0mM $MgC1_2$, 50 pmol of each primer, 0.2mM dNTPs, and 1.0 U LD-*Taq* polymerase in a reaction volume of 100 µl with 10 to 50 ng of template DNA. The amplification profile was: 94°C for 5 min, and then 30 cycles at 94°C for 1 min, 55°C for 30 sec, 72°C for 1 min, and a final extension of 5 min at 72°C. The PCR product was then digested with *Not*I at 37°C for 1 hour and then purified using a QiaQuick (Qiagen, Valencia, CA) column. The amplicon was then precipitated with ethanol and ligated into *Not*I-digested, dephosphorylated pBluescript II SK(-) vector (Stratagene, La Jolla, CA). The recombinant DNA molecules were then used to transform *E. coli* XL1-Blue MRF' competent cells.

The recombinant plasmids of selected clones were checked for insert size, and 22 clones were randomly selected for further analysis. Although the insert size is the same for each clone, the nucleotide sequence varies since the template DNA used in the PCR reaction could have been derived from a mixture of microorganisms. We have used melting point determination to classify insert molecules. Using the LightCycler (Idaho Technology, Inc., Idaho Falls, ID), PCR amplification of a hypervariable portion of the rDNA was targeted to yield a 216-bp amplicon. This was observed in real time by the incorporation of SYBR Green I (Molecular Probes, Eugene, OR). SYBR Green I is a double-strand DNA specific dye. Thus there is an increase in fluorescence upon the incorporation of dye into newly formed amplicons. The primers p201 (5'-GAG GAA GGI GIG GAI GAC GT) and p1389 (5'-AGI CCC GIG AAC GTA TTC AC) were used in this amplification (11). The PCR products were classified by melting curve analysis on the basis of GC/AT ratio as well as the distribution of G and C within the sequence (12). The clones were placed in 3 different groups based on the melting temperatures. Representatives from each group were sequenced, and the sequence data used to search for homology in the GenBank database. Three bacteria, *Fusobacterium nucleatum*, *Bacillus subtilis*, and *Provotella veroralis* were identified. Our laboratory is presently in the process of developing specific probe based rDNA-PCR assays for these organisms. *In situ* PCR on fixed tissue prepared during excision of the clinical samples will be probed and examined with confocal microscopy (13).

5. Gene Sequences from Databases

The exponential increase in the number of genes and DNA sequences, which are being added to databases often, facilitates the direct development of species-specific PCR assays. A PCR assay for *Franciscella tularensis* is a case in point. This bacterium is a gram-negative coccobacillus that ranges in size from 0.2 by 0.2 to 0.7 µm. It causes a plague-like disease known as tularemia. *F. tularensis* is extremely infectious and is transmitted by various blood-sucking insects. Wild rabbits and ticks have been identified as the major sources of human infections. Tul4 is a l7-kDa protein membrane protein of F. *tularensis*. Tu14 is one of the major T cell-reactive proteins found on this organism. The gene which encodes for this protein, *tul4* (GenBank accession # M32059), has been isolated (14) and used to develop a PCR assay specific for *F. tularensis* (15).

Our laboratory has selected the primer Tull (5'-TAT CAA TCG CAG GTT TAG C) which hybridizes with nucleotides *585* to 607 on the *tul4* gene, and primer Tu12 (5'-TCG TTC TTC TCA GCA TAC TTA G) which complements nucleotides 864 to 885 on the opposite DNA strand. Thus the expected size of the amplicon is 283 bp as can be seen in Figure 6. Amplification of the target sequence was accomplished with a 25 µl reaction mixture containing 100 ng of genomic DNA, 10Mm Tris HG (pH 8.8), 3.5 mM $MgC1_2$, *75* mM KC1, 0.4 µM of each primer, 0.2mM of each dNTP, and 1.8 U *Taq* polymerase. After an initial denaturation of the template DNA at 94°C for 3 min, amplification was carried out for 30 cycles of denaturation at 94°C for 1 min, annealing at 59°C for 1 min, and extension at 72°C for 1 min. A final extension step at 72°C for 5 min was also included.

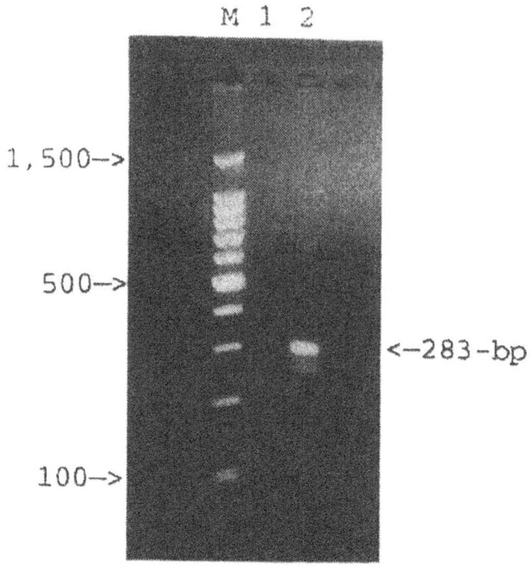

Figure 6 Electrophoresis of *F. tularensis* PCR product. M, DNA size standard; 1, negative control; 2, sample containing *F. tularensis* DNA.

6. References

1. Klare, I., Heier, H., Claus, H., and Witte, W. (1993) Environmental strains of *Enterococcus faecium* with inducible high-level resistance to glycopeptides. FEMS Microbiol. Let 106, 23-30.
2. Cheng, S., McCleskey, K., Gress, M. J., Petroziello, J. M., Liu, R., Namdari, H., Beninga, K., Salmen, A., and DelVecchio, V. G. (1997) A PCR assay for identification of *Enterococcus faecium*. J. Clin. Microbiol. 35, 1248-50.
3. DelVecchio, V. G., Petroziello, J. M., Gress, M. J., McCleskey, K., Melcher, G. P., Crouch, H. K., and Lupski, J. R. (1995) Molecular genotyping of methicillin resistant *Staphylococcus aureus* via fluorophore-enhanced repetitive-sequence PCR. J. Clin. Microbiol.33, 141-5.
4. Southern, E. M. (1975) Detection of specific sequences among DNA fragments separated by gel electrophoresis. J. Mol. Biol. 98, 503-17.
5. Versalovic, J. T., et al. (1991) Distribution of repetitive DNA sequences in eubacteria and application to fingerprinting of bacterial genomes. Nucleic Acids Res. 19:6823-31.
6. Petroziello; J. M., Gress, M. J., McCleskey, F. K., Namdari, H., Del Vecchio, V. G. (1996) Genetic fingerprinting of clinical strains of *Enterococcus faecalis* and *Enterococcus faecalis* by Rep-PCR. Infect. Contr. Hops. Epidemiol. 175,46, abstr no. M34.
7. Wensel, R., and Herrmann, R. (1988) Repetitive DNA sequences in *Mycoplasma pneumoniae*. Nucleic Acids Res. 16, 8337-50.
8. Woese, C. R. (1980) Phylogenetic analysis of the mycoplasmas. Proc. Natl. Acad. Sci. USA 77, 494-8.
9. Gress, M., Keeler, S., Petroziello, J., Cheng, S., Namdari, H., McCleskey, F. K., and DelVecchio, V. G. (1997) PCR detection of *Enterococcus faecalis*. Program Abstr. 97th ASM General Meeting,. abstr. no. C3 1.
10. Relman, D. A. (1993) Universal bacterial 16S rDNA amplification and sequencing, in D. H. Persing, T. F. Smith, F. C. Tenover, and T. J. Wbite (eds), *Diagnostic Molecular Biology, Principles and Application,* American Society of Microbiology, Washington, D.C.

11. Steinman, C. R., Muralidhar, B., Nuovo, G. J., Rumore, P.M., Yu, d., and Mukai, M. (1997) Domain-directed polymerase chain reaction capable of distinguishing bacterial from host DNA at the single-cell level: Characterization of a systematic method to investigate putative bacterial infection in idiopathic disease. Anal. Biochem. 244. 328-39.
12. Rierie, K. M., Rasmussen, R. P., Wittwer, C. T. (1997) Product differentiation by analysis of DNA melting curves during the polymerase chain reaction. Anal. Biochem. 245, 154-60.
13. Wagner, M., Abmus, B., Hartmann, A., Hutzler, P., Amann, R. (1994) *In situ* analysis of microbial consortia in activated sludge using fluorescently labeled, rRNA-targeted oligonucleotide probes and confocal scanning laser microscopy. J. Microsc. 176, 181-7.
14. Sjostedt, A., Sandstrom G., Jaurin, B. (1990) Nucleotide sequence and T cell epitopes of a membrane protein of *Franciscella tularensis.* J. Immunol. 145, 311-7.
15. Long, G. W., Oprandy, J. J., Narayanan, R. B, Fortier, A. H., Porter, K. R., Nany, C. A. (1993) Detection of *Franciscella tularensis* in blood by polymerase chain reaction. J. Clin. Microbiol. 31,152-4.

USE OF TAQMAN, LIGHT CYCLER, AND CONFOCAL MICROSCROPY TO DETECT SPECIFIC PCR

VITO G. DELVECCHIO AND RAJENDRA REDKAR
Institute of Molecular Biology and Medicine, University of Scranton, Scranton, PA 18510 USA

Abstract

Certain amplicon-detection systems can detect target nucleic acid sequences without gel analysis. The TaqMan™ system consists of primers for amplification of specific DNA sequences and a fluorogenic probe tat contains both a reporter and quencher dye. During amplification the probe hybridizes with a target on the amplicon, and the 5'-exonuclease activity of DNA polymerase hydrolyzes the probe. The separation of the reporter from the quencher results in an increase in the fluorescent emission of the reporter. This indicates the presence of the amplicon of interest. The LightCycler™ allows real-time detection of nucleic acid amplification by measuring the increase in fluorescence due to the double-stranded DNA-specific dye SYBR®Green I. The LightCycler uses glass capillary tubes to contain 10 µl of amplification mix. This allows complete amplification and analysis in 10-15 minutes.

The presence of unculturable microorganisms from diverse sources can be investigated by examination of rDNA libraries. Such libraries are formed by the use of primers that will amplify rDNA genes found in all bacteria. Sequencing of each clone type permits identification or phylogenetic categorization of each organism by comparison with other rDNA sequences found in rDNA databases. Fluorophore-labeled primers specific for each rDNA (or bacterial type) can then be utilized to locate the bacteria in fixed histological sections by means of *in situ* hybridization and confocal microscopy.

Extremely rapid, sensitive, and accurate DNA probe assays are of great importance in clinical and environmental investigations because such systems result in faster response to the presence of potentially dangerous microorganisms. Polymerase chain reaction (PCR) and other amplification techniques have contributed immensely to the advancements in nucleic acid-based detection technologies. Prior to the advent of PCR, using time-consuming and labor intensive assays such as the use of dot blot (1) and fluorescent oligonucleotide rRNA probing of single cell (2) usually achieved sensitivity. With the introduction of PCR, the standard method of amplicon detection was agarose gel electrophoresis. More modem methods, which permit real time amplicon detection, have become available, and these include TaqMan and the LightCycler. Since Eric Henscal will present the TaqMan technology in these proceeding, we will concentrate on the use of the LightCycler for amplicon detection.

P.J. Stopa and M.A. Bartoszcze (eds.),
Rapid Methods for Analysis of Biological Materials in the Environment, 231–237.

The Idaho Technology LightCycler permits rapid amplification (10-15 minutes) of target sequences (3). This is possible because extremely small samples are run in glass capillary tubes, and temperature transitions are accomplished by forced air. Thus minimum time is required to achieve desired temperatures. The minimization of denaturation and annealing times has an added bonus, in that the yields and specificity are improved. Along with facilitating rapid heat transfer, glass capillaries are optically clear and can serve as excellent cuvettes for fluorescence analysis. The second major feature of this instrument is the simultaneous differentiation of amplicons by melting curve analysis.

The instrument is equipped with a xenon source that excites fluorescent dyes. The fluorescent excitation then passes through a collection of optics and a series of photomultiplier tubes. The excitation source is then focused on a 5-25 μl of amplification mix, which is positioned by stepper motor control of carousal. The data is collected on a computer equipped to monitor the amplification events. Different dyes can be used to quantify the amount of DNA amplified and to monitor hybridization and denaturation by manipulating the emission of the xenon source. This is achieved by adjusting the iris pinhole, lenses wavelength filter, focusing lens, etc. The double-stranded DNA-specific dye SYBR Green I, which becomes incorporated into forming amplicons, was used in the recording of amplification events. A wavelength range of 520-580 nm is used for measuring SYBR Green I as it is incorporated into the double-stranded amplicon molecule. Fluorescence usually begins to increase at around cycle 15-20 and plateaus at cycle 35.

The LightCycler also provides for data analysis. In addition to monitoring amplification, hybridization and denaturation can be followed during or after the temperature cycling. This is a powerful tool because the melting temperature of amplicons can be determined. Melting point analysis can identify and discriminate PCR products since melting curves are characteristic of the products being denatured. The most significant factors are the size of the amplicon and the GC/AT sequence and contents. It can be used to differentiate amplification products separated by less than 2°C in melting temperature. Thus differences in amplicons can be observed in less than 20 minutes. This is accomplished by plotting fluorescence versus temperature as the thermal cycler heats through the dissociation temperature of the product yielding a DNA melting curve.

Our laboratory has developed a rapid system for the detection of *Franciscella tularensis* using a gene that encodes the Tu14 membrane protein. Tu14 is one of the major T cell-reactive proteins found on this organism. The *tul4* gene (GenBank accession # M32059), has been isolated (4) and used to develop a PCR assay specific for *F tularensis* (5). Tul-1 (5'-TAT CAA TCG CAG GTT TAG C-3) which hybridizes with nucleotides *585* to 607 on the *tul4* gene, and primer Tul-2 *(5'-TCG* TTC TTC TCA GCA TAC TTA G-3') which complements nucleotides 864 to 885 on the opposite DNA strand were selected as primers for the PCR assay. Thus the expected size of the amplicon is 283-bp. Amplification of the target sequence was accomplished with a 10 μl reaction mixture containing 25 ng of genomic DNA, 10 Mm Tris-HC1 (pH 8.8), 3.5 mM $MgC1_2$, 75 mM KC1, 5 pmol of each primer, 0.2 mM of each dNTP, 250 μl/ml

BSA, SYBR Green I at 1:30,000 dilution and 1.0 U *Taq* polymerase. After an initial denaturation of the template DNA at 94°C for 30 sec, amplification was carried out for 30 cycles of denaturation at 94°C for 0 sec, annealing at 59°C for 0 sec, and extension at 72°C for *5* sec. A final extension step at 72°C for 15 sec was also included.

The PCR amplification and simultaneous fluorescence analysis was completed in about 12 mm. The increase in fluorescence for *F tularensis* sample was observed on the computer screen after cycle *15,* which continued to increase exponentially until cycle 30. The negative control or other bacterial DNA's did not exhibit any increase in fluorescence, indicating the specificity of the assay for *F tularensis.* A computer print out of the screen is shown in Figure 1. The Amplicons were recovered and run in a 2% agarose gel to confirm the LightCycler results (Figure 2).

RIBOSOMAL DNA CLONING

Defining the diversity of microbial populations is challenging because microorganisms live in consortia and the majority of bacteria cannot be grown and identified by traditional microbial techniques. An alternative approach to understanding the complexity of microbial communities involves examining variations in the 16S rRNA encoding DNA (rDNA). Based on the variability of the nucleotide sequence of the variable regions of the rDNA of bacteria, archaebacteria, and eukaryotes, phylogenetic and evolutionary relationships can be predicted. The average bacterial rDNA is 1,500 bp long, and the rRNA operon has many copies present in a genome.

Our laboratory has utilized this strategy to determine what organisms are involved in foot ulceration of diabetic patients. Cloning of the rDNA of microorganisms involves extraction of DNA from all organisms present in a microbial community, amplification of the rDNA genes by PCR, construction of an rDNA library, rapid detection of variations in the clones, determination of the nucleotide sequences of the different clones, and comparative sequence analysis. This data can identify new or known species and also pinpoint phylogenetic status. The following procedure for rDNA cloning is adapted from a procedure outlined by Relman (6). The sample is subjected to lysis in a freshly prepared digestion buffer consisting of 50 mM Tris-HC1 (pH 8.5), 1 mM EDTA, 1% Tween, and 1 mg/ml lysozyme. The mix is then incubated at 37°C for 1 hour. Following the addition of 10 µl of proteinase K (20 mg/ml), the mixture was placed at 55°C for 1 to 3 hours. After heat inactivation of the proteinase K, DNA was isolated. The forward primer (5'-GCG GAT CCG CGG CCG CTG CAG AGT TTG ATC (C/A)TG GCT CAG) and the reverse primer (5'-GGC TCG AGC GGC CGC CCG GG(T/C) TAC CTT GTT ACG ACT T) were used to amplify a 1540-bp PCR product. The forward primer hybridizes with nucleotides at 8 to 27 in the rDNA gene, and the reverse complements a sequence at position 1492 to 1510. The underlined portion of the primer sequence indicates the *NotI* recognition site, which is utilized for increased efficiency of cloning.

PCR reaction contained 10 mM Tris-HC1 (pH 8.3), 50 mM KC1, 2.0 mM $MgCl_2$, 50 pmol of each primer, 0.2 mM dNTPs, and 1.0 U *LD-Taq* polymerase in a reaction volume of 100 µl with 10 to 50 ng of template DNA. The amplification profile was:

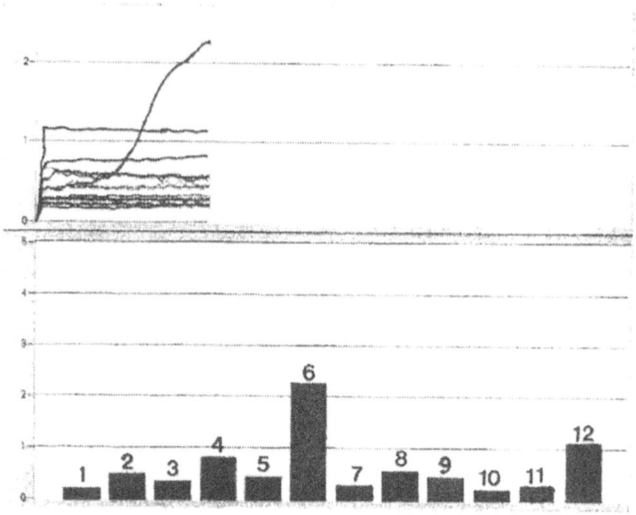

Figure 1. Real-time monitoring of *F. tularensis*-specific assay on LightCycler. Samples: 1, negative control; 2, *Candida albicans*; 3, *Citrobacter diversus*; 4, *Enterococcus faecium*; 5, *Escherichia coli*; 6, *Franciscella tularensis*; 7, *Klebsiella pneumoniae*; 8, *Mycoplasma pneumoniae*; 9, *Pseudomonas fluoroscence*; 10, *Salmonella miami*; 11, *Shigella sonnei*; 12, *Staphylococcus aureus*.

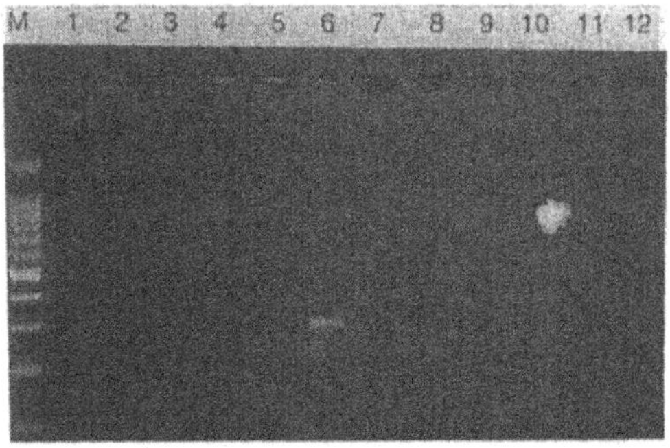

Figure 2. Specificity of *F. tularensis* assay. Lanes: 1, negative control; 2, *Candida albicans*; 3, *Citrobacter diversus*; 4, *Enterococcus faecium*; 5, *Escherichia coli*; 6, *Franciscella tularensis*; 7, *Klebsiella pneumoniae*; 8, *Mycoplasma pneumoniae*; 9, *Pseudomonas fluoroscence*; 10, *Salmonella miami*; 11, *Shigella sonnei*; 12, *Staphylococcus aureus*. The 100-bp molecular size marker is run on the left (M).

94°C for 5 min, and then 30 cycles at 94°C for 1 min, 55°C for 30 sec, 72°C for 1 min, and a final extension of 5 min at 72°C. The PCR product was then digested with *NotI* at 37°C for 1 hour and then purified using a QiaQuick (Qiagen, Valencia, CA) column. The amplicon was then precipitated with ethanol and ligated into *NotI*-digested, dephosphorylated pBlueScript II SK(-) vector (Stratagene, La Jolla, CA). The recombinant DNA molecules were then used to transform *E. coil* XL1-Blue MRF' competent cells.

The recombinant plasmids of selected clones were checked for insert size, and 22 clones were randomly selected for further analysis. Although the insert size is the same for each clone, the nucleotide sequence vanes since the template DNA used in the PCR reaction could have been derived from a mixture of microorganisms. We have used melting point determination to classify insert molecules. Using the LightCycler (Idaho Technology, Inc., Idaho Falls, ID), PCR amplification of a hypervariable portion of the rDNA was targeted to yield a 216-hp amplicon. This was observed in real time by the incorporation of SYBR Green I (Molecular Probes, Eugene, OR). Thus there is an increase in fluorescence upon the incorporation of dye into newly formed amplicons. The primers p201 (5'-GAG GAA GGI GIG GAL GAC GT) and p1389 *(5'*-AGI CCC GIG AAC GTA TTC AC) were used in this amplification (7). A typical 10 µl PCR reaction mixture included: 5 ng of plasmid DNA, 10 mM Tris-HC1 (pH 8.3), 50 mM KC1, 3mM $MgC1_2$, 5 pmol of each primer, 0.2 mM of each dNTP, 250 µg/ml BSA, SYBR Green I at 1:30,000 dilution, gelatin at 0.01% and 0.2 U of LD*Taq* polymerase. The amplification was carried out in glass capillaries using F1/1 filter set and the profile was: initial denaturation at 94°C for 15 sec, followed by 30 cycles of denaturation at 94°C for 5 sec, annealing at 55°C for 5 sec, and extension at 72°C for 10 sec. A final extension step at 72°C for 20 sec was also included. This was followed by melting analysis parameters which were: holding at 95°C for 0 sec, followed by 55°C for 0 sec, and 72°C for 30 sec, and increasing the temperature gradually to 95°C at 0.1°C/sec. The fluorescence was acquired once every cycle after the extension step during cycling for detection of amplification and continuously in the melting analysis. A negative control without DNA was also processed simultaneously.

DNA melting was observed as a sudden decrease in the fluorescence of the dsDNA dye SYBR Green I as the samples were heated through the Tm of the amplicon. Melting curve analysis on the basis of GC/AT ratio, as well as the distribution of GC along the sequence (8) differentiated the PCR products. Selected clones were placed in 3 distinct groups based on their Tm's (Table 1).

Table 1. Melting temperatures and identity of rDNA clones.

Clone#	Tm (°C)	Group	Group Mean Tm (°C)	Identity by sequence comparison	%GC of amplicon
Clone 27	86.11	I		*Fusobacterium nucleatum*	
Clone 26	86.34	I			
Clone 25	86.37	I			
Clone 43	86.47	I		*Fusobacterium nucleatum*	
Clone 42	86.49	I	86.54 ± 0.24		47.93
Clone 39	86.55	I			
Clone 35	86.63	I			
Clone 34	86.78	I		*Fusobacterium nucleatum*	
Clone 30	86.83	I		*Fusobacterium nucleatum*	
Clone 41	86.84	I			
Clone 2	88.97	II		*Bacillus subtilis*	
Clone 7	89.04	II			
Clone 3	89.06	II			
Clone 10	89.08	II	89.08 ± 0.06	*Bacillus subtilis*	53.7
Clone 8	89.08	II			
Clone 13	89.09	II			
Clone 11	89.12	II			
Clone 4	89.13	II		*Bacillus subtilis*	
Clone 6	89.17	II			
Clone 32	89.45	III			
Clone 40	89.85	III	89.73 ± 0.25	*Provotella veroralis*	55.09
Clone 37	89.90	III		*Provotella veroralis*	

Ten clones were placed in Group I having a mean Tm of 86.54 ± 0.24°C. The remaining clones were separated in 2 distinct groups based on their mean Tm's. Representatives from each group were partially sequenced and the sequence information was used to search the GenBank and ribosomal databases to find the identity of the bacteria present in the sample. The nucleotide sequences of representative clones in each group were identical and the GC content wad directly related to mean group Tm values. *Fusobacterium nucleatum, Bacillus subtilis* and *Provetella veroralis* were identified using this approach from a human sample. These results corroborated with cultural findings wherein *F. nucleatum* and *Provetella veroralis* were isolated with conventional techniques. Thus melting peak analysis measures a fundamental characteristic of DNA and offers a product differentiation method that is fully integrated with real-time PCR.

At the present time, we are in a process of identifying specific fluorophore-labeled rDNA primers and probes for detection of bacterial types. These can be utilized in future to locate bacteria in histological specimens by means of *in situ* hybridization and confocal microscopy.

REFERENCES

1. Gallia, G. L., Petroziello. J. M., Brogan. J. M.. McCleskey. F. K., and DelVecchio, V. G. (1995) Development of a diagnostic polymerase chain reaction assay for detection of *Mycoplasma hominis.* Mol. & Cell. Probes 9,415-21.
2. DeLong, E. F. and Shah, J. (1990) Fluorescent. ribosomal RNA probes for clinical application: A research review. Diag. & Clin. Testing 28.41-4.
3. Wittwer, C. T., Ririe, K. M.. Andrew. R. V.. David. D. A., Gundry, R. A., and Balis, U. J. (1997) The LightCycler^{TM}, A microvolume multisample fluorimeter with rapid temperature control. BioTechiques 22,176-181.
4. Sjostedt, A., Sandstrom G., and Jaurin. B. (1990) Nucleotide sequence and T cell epitopes of a membrane protein of *Franciscella tularensis.* J. Immunol. 145, 311-7.
5. Long, G.. W., Oprandy, J. J.. Narayanan. R. B. Fortier. A. H., Porter. K. R., and Nany, C. A. (1993) Detection *of Franciscella tularensis* in blood by polymerase chain reaction. J. Clin. Microbiol. 31,152-4.
6. Relman, D. A. (1993) Universal bacterial 16S rDNA amplification and sequencing, *In* D. H. Persing, T. F. Smith, F. C. Tenover, and T. J. White (eds). *Diagnostic Molecular Biology, Principles and Applications,* American Society of Microbiology. Washington, D. C.
7. Steinman, C. R., Muralidhar. B.. Nuovo. G. J.. Rumore, P. M., Yu, D., and Mukai, M. (1997) Domain-directed polymerase chain reaction capable of distinguishing bacterial from host DNA at the single-cell level: Characterization of a systematic method to investigate putative bacterial infection in idiopathic disease. Anal. Biochem, 244,328-39.
8. Ririe,K.M., Rasmussen, R. P., and Wittwer. C. T. (1997) Product differentiation by analysis of DNA melting curves during the polymerase chain reaction. Anal. Biochem. 245, 154-60.

EVALUATION OF POLYMERASE CHAIN REACTION ASSAYS FOR IDENTIFYING BIOLOGICAL AGENTS

E.A. HENCHAL AND M.S. IBRAHIM
U.S. Army Medical Research Institute of Infectious Diseases
1425 Porter Street, Fort Detrick, MD 21702

Rapid, sensitive, and specific assays for the identification of biological agents in environmental and clinical samples are needed by laboratories within the Department of Defense to protect the health of service members. We evaluated the capability of four different gene amplification formats based upon the polymerase chain reaction (PCR) method to identify biological agents. Microplate-based assays and miniaturization will make these approaches accessible to field laboratories.

1. Introduction

Within the U.S. Department of Defense, the ability to identify some biological agents resides principally at reference and research laboratories, such as the U.S. Army Medical Research Institute of Infectious Diseases (USAMRIID, Fort Detrick, Maryland). Currently, environmental or clinical samples must be collected and shipped hundreds or thousands of miles away for analysis. This labor-and time-intensive process is inconsistent with the delivery of rapid and effective medical countermeasures to protect U.S. forces. With current research approaches, it may take more than 6 to 24 hours after the sample is received by the laboratory to have a preliminary identification. More rapid (less than 60 min) approaches are needed for prompt identification of potential disease threats and treatment of casualties. Because licensed, commercial assays for many biological assays are unavailable, we augment deployable field laboratories, such as the U.S. Army Theater Area Medical Laboratory (TAML), with a basic capability that includes immunodetection and nucleic acid analysis methods. The current deployment package weighs approximately 400 lb and requires at least 1 year of specialized user training. A number of advanced technologies are emerging that would reduce this logistical and training requirement. Several of these approaches are based upon gene amplification by the polymerase chain reaction (PCR) method.

2. Polymerase Chain Reaction Methods

2.1. STANDARD PCR

The PCR method specifically amplifies gene targets with a combination of oligonucleotide primers [1]. PCR has become a powerful tool to identify infectious diseases [2]. An example of the identification of *Yersinia pestis*, the plague agent, from

P.J. Stopa and M.A. Bartoszcze (eds.),
Rapid Methods for Analysis of Biological Materials in the Environment, 239–249.

other bacteria is shown in Figure 1. Similar methods have been used to address a broad panel of organisms at USAMRIID. However, a number of technical barriers prevent routine use of these methods in the clinical laboratory [3]. Transfer of this technology to field laboratories has been hampered by the lack of rapid and simple methods for specimen processing, product detection, and data analysis. Sample processing with commercial kits can rarely be performed in less than an hour. PCR product detection and analysis methods usually depend upon gel electrophoresis, which requires analytical skills and experience beyond the training of the average Army laboratory technician. In our experience, it may take 6 to 12 months to train a soldier technician in PCR techniques to identify priority biological agents.

2.2. PCR-EIA

PCR product detection approaches that use labeled probes can replace gel electrophoresis. We exploited colorimetric enzyme immunoassay methods (PCR-EIA) for PCR product detection. Figure 2 outlines one approach that incorporates digoxigenin (Dig)-dUTP into the amplified DNA. Dig-labeled PCR products are hybridized in solution to agent-specific biotinylated probes and captured onto streptavidin-coated microtiter plates. With horse radish peroxidase (pod)-labeled anti-Dig antibodies and a chromogenic substrate in a standard EIA format, the captured PCR product can be detected and measured. In contrast to standard PCR methods, PCR-EIA results are reported as optical density (OD), allowing standardization and quality assurance evaluation of the data. Moreover, PCR-EIA is compatible with enzyme-linked immunoassay equipment already available in most clinical laboratories. In some cases a qualitative result can be obtained by eye. The relative specificity of PCR-EIA with standard PCR approaches using *Y. pestis* isolates and hetologous bacterial strains is presented in Figure 1. Similarly, Figure 3 compares the relative sensitivity of PCR-EIA with the standard gel electrophoresis detection method. While PCR-EIA retains the specificity of the standard PCR method, it is at least 10 times more sensitive. PCR-EIA has also been adapted to membrane and chromatographic formats for hand-held qualitative assays.

2.3. FLUOROGENIC 5' NUCLEASE ASSAYS

We simplified agent identification by using fluorogenic 5' nuclease PCR assays [4]. This approach uses oligonucleotide probes labeled with fluorescent dyes (Taqman™; Applied Biosystems Division, Perkin-Elmer, Foster City, CA). The detection probe is labeled with a reporter (R) dye that once released from proximity to the quencher (Q) dye by the 5' to 3' exonuclease activity of the *AmpliTaq* enzyme makes real-time PCR detection possible (Figure 4). Using the Applied Biosystem's sequence detector model 7700, we demonstrated the usefulness of the technology for identifying potential biological agents [5]. We are now extending this assay technology for identifying the principal biological agents. Fluorogenic assays have been demonstrated at USAMRIID for *Bacillus anthracis*, *Y. pestis*, *Francisell. tularensis*, Venezuelan equine encephalitis virus, *Brucella abortus*, and orthopoxviruses. Figure 5 shows the relative sensitivity using *Y. pestis* as a target organism. While the expensive ABI Prism™ 7700 is useful in a fixed facility, it is unsuitable for field use because it's large footprint (over 36 feet^2)

and weight (more than 80 lb). More recently, these assays have been adapted to a miniaturized analytical thermal cycling instrument (MATCI) developed at Lawrence Livermore Laboratory [6]. Figure 6 shows the instrument in a briefcase for field use.

2.4. MATCI

MATCI is an integrated miniaturized analytical system with small-scale reaction chambers integrated with heaters and optical windows, simple inexpensive electronics, and diode-based detection components. Components were constructed with silicon microfabrication technology. Improved software and detection methods allow real-time detection of PCR products by 5' nuclease fluorogenic assays. The current device is lightweight, fits in a briefcase, and is compatible with field laboratory use (Figure 6). We showed that MATCI can duplicate the performance of the commercial ABI Prism^{tm} 7700 but it more rapid, permitting identification in about 25 min (Figure 5). Real time detection is not currently available on the commercial ABI Prism^{tm} 7700 instrument.

3. Comparison of PCR Approaches

Table 1 compares and contrasts three approaches discussed above with *Y. pestis* as a target organism. While PCR-EIA was the most sensitive, immunodetection of PCR products increased performance time. The fluorogenic 5' nuclease PCR format with the ABI 7700 instrument did not shorten performance time or increase sensitivity when compared to standard PCR. If real-time detection was available on this instrument, biological agents could be identified in as little as 50 min (data not shown). The size, weight and price of this instrument make it incompatible with field applications. Performance time was significantly shortened with MATCI, but the current configuration is the least sensitive. We are currently working with a commercial contractor to improve the performance of the MATCI technology and place it in a hand-held format.

4. Summary

Rapid, sensitive, and specific assays to identify biological agents in environmental and clinical samples are the focus of research and development currently being conducted in Department of Defense laboratories. Of four PCR formats, the colorimetric PCR-EIA method was the most sensitive assay compatible with clinical or field applications. Assays based on 5' nuclease fluorogenic assays represent the next generation of nucleic acid detection technology. Maturation of this and other emerging advanced technologies will make rapid, hand-held analysis a reality.

TABLE 1. Comparison of PCR Formats

	Performance Time	Limit of Sensitivity	Specificity
Standard PCR	2.0 - 3.0 hr	300 copies	100%
PCR-EIA	2.5 - 3.5 hr	30 copies	100%
5'Nuclease PCR	2.0 - 3.0 hr	300 copies	100%
MATCI	20 - 60 min	>300 copies	100%

FIGURE 1. Identification of *Yersinia pestis* isolates by Standard PCR and PCR-EIA assays. PCR-EIA results are reported as optical density units at 410 nm. A photograph of a representative ethidium bromide-stained electrophoresis gel is shown for comparison above the graph. *Y. pestis* strains are shown on the left; unrelated isolates of *Y. enterocolitica and Y. pseudotuberculosis* are shown on the right.

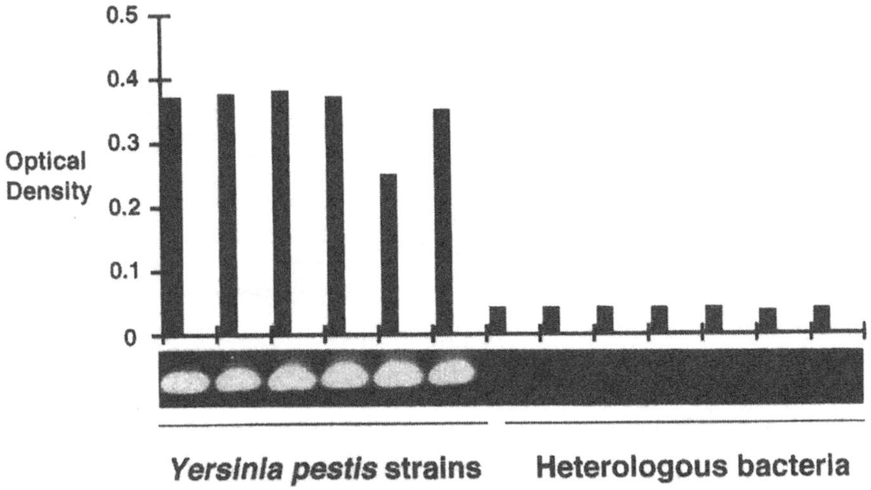

FIGURE 2. PCR-EIA Method

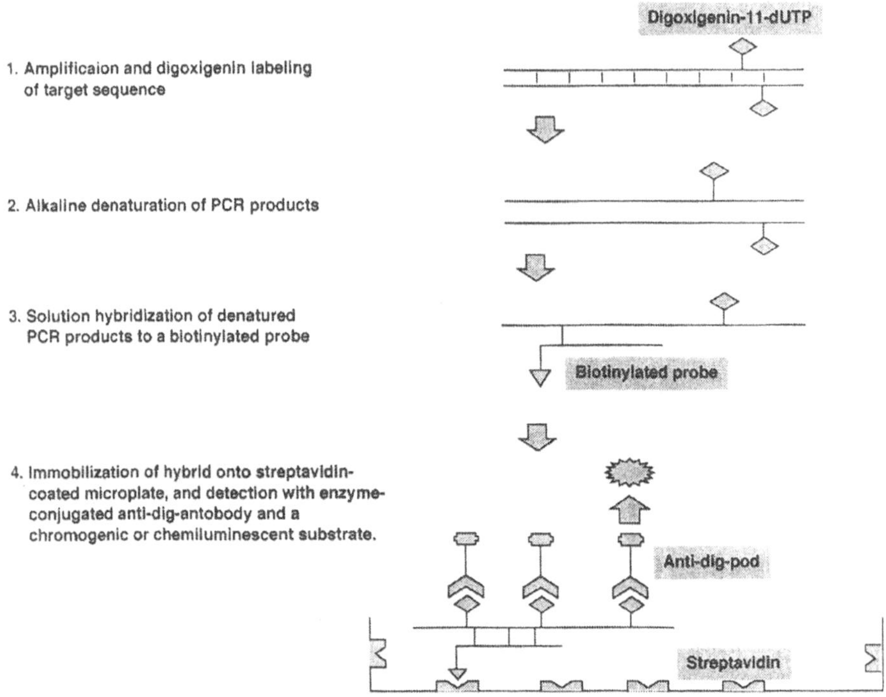

FIGURE 3. Sensitivity of Standard PCR and PCR-EIA assays. Ten-fold dilutions of target *Y. pestis* DNA was evaluated by two different PCR formats. C, negative control.

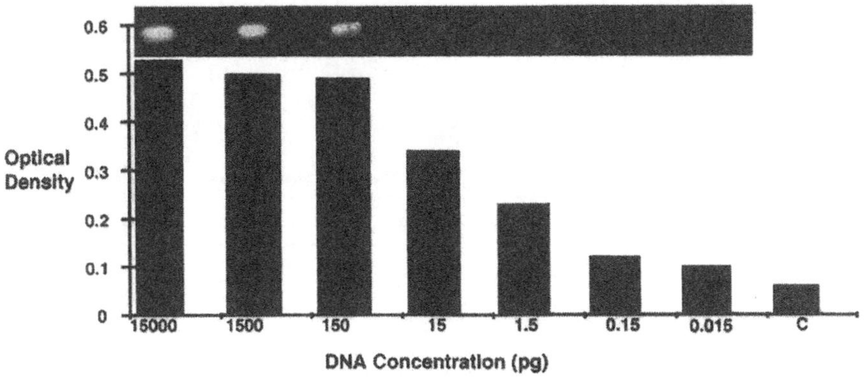

FIGURE 4. Fluorogenic 5' nuclease assay with Taqman™ probes. R, reporter fluorescent dye; Q, quenching fluorescent dye; FP, forward primer; RP, reverse primer. 5' and 3' denote the relative orientation of the nucleic acid strands.

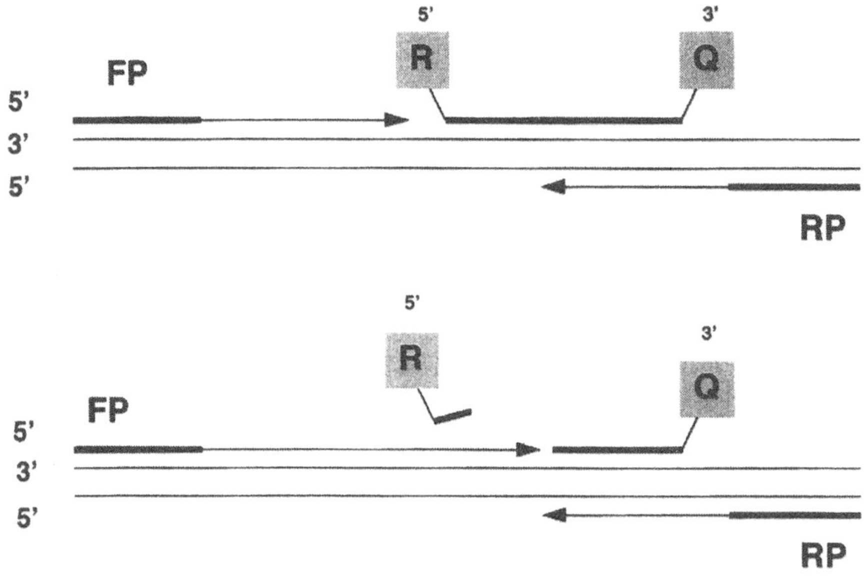

FIGURE 5. Sensitivity of fluorogenic 5' nuclease PCR assays. Ten-fold dilutions of target *Y. pestis* DNA were evaluated by fluorogenic 5' nuclease PCR assays and identical Yersinia pestis-specific Taqman™ probes. threshold values were established using no template controls and non-pestis strains. –Rn represents the relative difference in the intensity of the reporter dye intensity compared to quencher dye intensity.

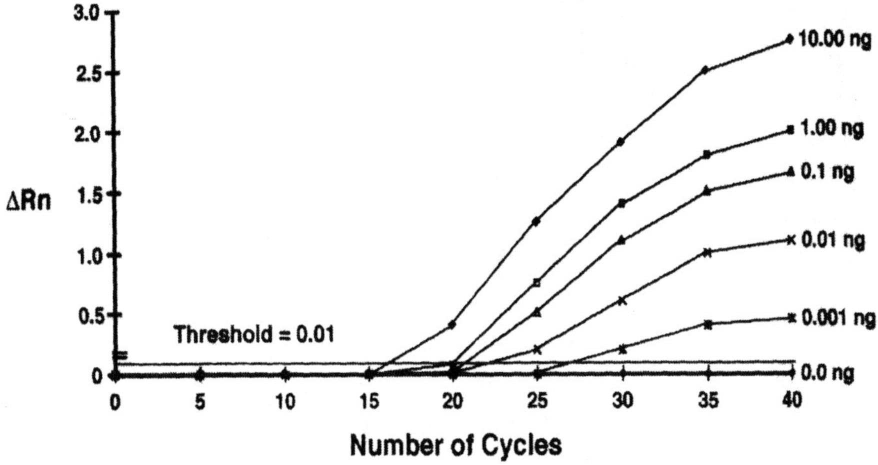

FIGURE 6. Miniaturized Analytical Thermal Cycling Instrument (MATCI)

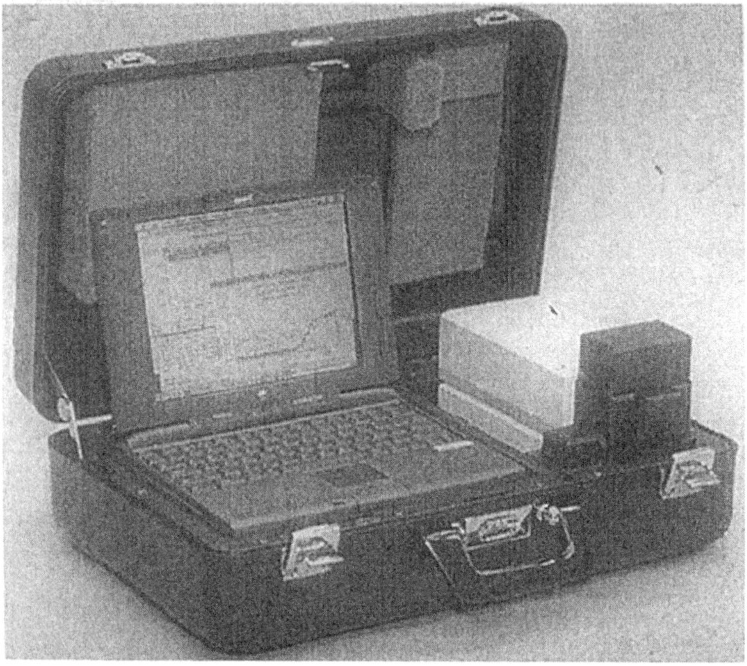

5. References

1. Mullis, K., Faloona, F., Scharf, S., Saiki, R., Horn, G., and Erlich, H. (1986) Specific enzymatic amplification of DNA in vitro: the polymerase chain reaction. Cold Spring Harbor Symp. Quant. Biol. **51**:263.
2. McDade JE; and BE Anderson. (1996) Molecular epidemiology: applications of nucleic acid amplification and sequence analysis. Epidemiol Rev **18**:90-97.
3. Whelen, A.C., and Persing. D.H. (1997) The role of nucleic acid amplification and detection in the clinical microbiology laboratory. Annu Rev Microbiol **50**:349-373.
4. Lee, L.G. , Connell, C.R., and Bloch, W. (1993) Allelic discrimination by nick-translation PCR with fluorogenic probes. Nucleic Acids Res. **21**: 3761-3766.
5. Ibrahim, M.S., Esposito, J., Jahrling, P.B., and Lofts. R.S. (1997) The potential of 5' nuclease PCR for detecting a single-base polymorphism in Orthopoxvirus. Molecular and Cellular Probes **11**: 143-147.
6. Northrup, M.A., Beeman, B., Hills, R.F., Hadley. D.. Landre, P. and Lehew, S. (1996) Miniature DNA-based analytical instrument, p. 153-157. In H M Widmer [ed.]. Analytical Methods and Instrumentation: Special Issue on Micro TAS. Ciba-Geigy. Basel.

IDENTIFICATION OF PATHOGENS USING SINGLE/DOUBLE STRAND CONFORMATION POLYMORPHISM (SSCP/DSCP) ANALYSIS

M.T.C. OFFERMANS[1], H. MEYER[2] and N.D. ZEGERS[1].
[1]Division of Immunological and Infectious Diseases, TNO Prevention and Health, P.O Box 2215, 2301 CE Leiden, The Netherlands. [2]Institut für Mikrobiologie, Sanitätsakademie der Bundeswehr, Neuherbergstrasse 11, 80937 München, Germany.

1. Abstract

Single/double strand conformation polymorphism (SSCP/DSCP) analysis of polymerase chain reaction (PCR) amplified DNA fragments has emerged as a simple and sensitive screening method for the detection of small genomic variations. In this study we investigated the use of a four-color fluorescent-dye-labeled PCR-SSCP/DSCP technique as a screening method to enable the identification of microorganisms. PCR-SSCP/DSCP analysis was performed on 14 kDa fusion protein-encoding gene fragments, derived from strains belonging to different species of the genus *Orthopoxvirus*, amplified with fluorescent-dye-labeled primers. An automated gel analyzer was used for SSCP/DSCP analysis of the products. Preliminary results indicate that the PCR-SSCP/DSCP approach is a powerful method for the detection and identification of strain-specific nucleotide sequence polymorphism within the 14 kDa protein encoding gene. Therefore, it is concluded that this PCR-SSCP/DSCP technique represents a sensitive, reproducible and highly discriminatory procedure for molecular typing of pathogens.

Keywords: microorganism, pathogen, *Orthopoxvirus*, molecular typing, nucleotide sequence polymorphism, fluorescent, PCR-SSCP/DSCP technique

2. General Introduction

2.1. INTRODUCTION

Nowadays DNA-based detection methods become important tools for detection and identification of microorganisms. The polymerase chain reaction (PCR) plays an important role in many detection assays and screening methods to improve the sensitivity and specificity of these detection systems. Several methods for the detection of PCR products have been described previously [2, 9, 12] and the system selected will depend on the sensitivity and specificity of detection that is required in a particular assay. PCR and hybridization methods, however, require an extensive set of pathogen-specific primers and probes for detection and identification of a large variety of

P.J. Stopa and M.A. Bartoszcze (eds.),
Rapid Methods for Analysis of Biological Materials in the Environment, 251–258.

microorganisms. To circumvent this problem one set of a conserved primer pair can be used for PCR amplification of a DNA sequence that encloses genus-, species- or strain-specific sequences within a group of pathogens. Pathogen-specific single base substitutions within a particular sequence of interest can be regarded as nucleotide sequence polymorphism. To screen nucleotide sequence polymorphism rapidly, a strategy based on the SSCP technique [10, 11] has been adapted for this purpose. This PCR-SSCP/DSCP based technology relies on the fact that changes as slight as a single base substitution can alter the conformation of double stranded (ds) as well as single stranded (ss) DNA molecules which can result in a different electrophoretic mobility of these molecules in non-denaturing gel electrophoresis procedures [7, 8].

2.2. PCR-SSCP/DSCP ANALYSIS

Recently, various PCR-based methodologies have been developed for the rapid detection of genetic variability. In PCR-SSCP/DSCP analysis, the DNA sequence of interest is first amplified by PCR, and detection of the amplified product is undertaken using either labeled primers or labeled deoxynucleotides which can be incorporated into the PCR product itself [4, 6]. For SSCP analysis the PCR product is heated to dissociate the DNA strands and subsequently analysed by a non-denaturing polyacrylamide gel electrophoresis [10, 11]. In SSCP combined with DSCP analysis, single-stranded as well as double-stranded (homoduplexes and heteroduplexes) DNA molecules can be analysed simultaneously by a non-denaturing polyacrylamide gel electrophoresis procedure [7, 8]. Under a given condition the single/double stranded DNA molecules will fold into unique conformations determined by their entire nucleotide sequence. Differences in nucleotide sequence result in a different primary and secondary structure which is stabilized by intramolecular interactions. However, there is currently no adequate theoretical model available for predicting the three-dimensional structure of single/double stranded DNA molecules under a given condition. The sensitivity of the SSCP/DSCP analysis depends on the sequence content of the DNA strands, size of the DNA fragment, location of the nucleotide sequence polymorphism within the DNA fragment, temperature and gel electrophoretic conditions [3, 5].

The PCR-SSCP/DCSP analysis can be performed by either a manual or an automated procedure, using radioisotopic-labeled or fluorescent-dye-labeled primers/ deoxynucleotides, respectively. The use of an automated gel analyzer that utilises laser detection of fluorescent-dye-labeled DNA fragments offers many advantages over the conventional technique of autoradiographic detection. Firstly, mobilities can be normalized between lanes by including an internal-lane-standard labeled with a different fluorescent dye in each lane. This system requires an automated detection system that is capable of distinguishing several different colors. The internal-lane-standard eliminates the problems of band-shift artefacts and gel-to-gel variation often encountered with other techniques. Secondly, a system that characterises DNA fragments by automated fluorescence detection, eliminates the need for radioactive labeling of the PCR products and post-run gel handling. Thirdly, maximum information is obtained when complementary DNA strands are labeled with different fluorescent dyes. Dual-labeling facilitates comparison of sense and anti-sense DNA strands and is particularly useful since the reversal of strand position can occur under some electrophoretic conditions [3] and in multiplex-PCR-SSCP/DSCP approaches [13]. In addition, bands resulting from ss- or dsDNA molecules can be easily distinguished.

In this report we demonstrate the applicability of an automated four-color fluorescent-dye-labeled PCR-SSCP/DSCP approach as a screening method for the identification of microorganisms. Identification of different species of the genus *Orthopoxvirus* was selected as a model. These viruses are serologically closely related [1] but display species-specific nucleotide sequence polymorphisms within the 14 kDa fusion protein encoding gene. Therefore, DNA sequence diversity within this gene can be used as a genetic marker for molecular typing of these viruses by SSCP/DSCP analysis. PCR was performed using fluorescent-dye-labeled primers. SSCP/DSCP analysis of PCR products were performed on an automated gel analyzer. Preliminary data indicate that normalised SSCP/DSCP mobility patterns facilitate identification of pathogens.

3. Material and Methods

3.1. AMPLIFICATION

Isolation of genomic DNA of various *Orthopoxviruses* was undertaken as described previously [1]. Fluorescent-dye-labeled 14 kDa fusion protein gene sequences (292 bp) were generated by PCR amplification. The reaction was performed in a final volume of 25 ml of 10 mM Tris-HCl pH 9.0, 50 mM KCl, 1.5 mM $MgCl_2$, 0.1% Triton X100, 0.2 mg/ml BSA; 10 pmol of each primer (FAM (blue)-labeled forward primer; TET (green)-labeled reverse primer (Perkin-Elmer, Nieuwerkerk a.d. IJsel, The Netherlands)), 200 mM of each dNTP (Pharmacia, Uppsala, Sweden) and 1 Unit of *Thermus aquaticus* polymerase (Taq DNA polymerase, Gibco-BRL, Life Technologies Inc., Frederick, MD, USA). Twenty amplification cycles were performed with a 94EC (1 min) denaturation step, a 50EC (1 min) annealing step and a 72EC (1 min) elongation step in a Perkin-Elmer Cetus 9600 thermocycler (Perkin-Elmer Cetus, Norwalk, CT, USA). Negative controls (without a DNA template) were included in each PCR amplification. Samples were analysed by 1% agarose gel electrophoresis (stained with EthBr 0.5mg/ml) and visualised by UV transillumination.

3.2. SAMPLE PREPARATION AND SSCP/DSCP ANALYSIS

Double and single stranded PCR products were prepared directly prior to electrophoresis. To prepare dsDNA samples, 0.5 ml of fluorescent-labeled PCR product was added to 24.5 ml 0.1% SDS, 10 mM EDTA. An equal volume of 5% dextran blue in formamide was added to the dsDNA mixture. Two aliquots of 9 ml were withdrawn from the dsDNA mixture for DSCP and SSCP analysis. For normalisation of SSCP/DSCP mobility patterns an internal-lane standard was added to the PCR mixtures. For DSCP analysis, 1 ml of an internal-lane standard GeneScan-500 TAMRA(red)-labeled (GS500-TAMRA) (Applied Biosystems, Inc., Foster City, CA, USA.) was added to the aliquot and carefully mixed. To prepare ssDNA fragments for SSCP analysis, the dsDNA mixture was heated to 100EC for 5 min and immediately placed on ice to prevent reannealing. Subsequently, 1 ml of the internal-lane standard GS500-TAMRA was added to this ssDNA mixture. After sample preparation, 1.5 ml of the dsDNA mixtures was applied in the even-numbered lanes to a 5% polyacrylamide (19:1 ratio acrylamide to bisacrylamide) sequencing gel containing 90 mM Tris-borate pH 8.4, 2 mM EDTA and 10% glycerol. Electrophoresis was performed at constant power of 15 W at 30EC for 10 min using the temperature controlled Applied Biosystems 377 gel analyzer (Applied Biosystems). Subsequently, 1.5 ml of the ssDNA mixtures was applied onto the gel in the odd-numbered lanes and electrophoresis was prolonged for 19h. Data are automatically collected and subsequently analysed using the 672 GENESCAN Software (Applied Biosystems).

4. Results

4.1. PCR AMPLIFICATION

Fluorescent-labeled PR amplified fusion protein gene fragments were analysed by agarose gel electrophoresis to determine the efficacy of the PCR reaction. Samples obtained in the exponential phase of the amplification reaction were electrophoresed. As illustrated in figure 1 clear PCR products could be observed for all species of the genus *Orthopoxvirus* included in this study (lanes 2-6). Negative controls (without a DNA template) were included in each PCR amplification and no contamination could be detected (Figure 1, lane 1). As shown, discrimination between amplicons derived from different *Orthopoxviruses* was not possible simply by agarose gel electrophoresis.

FIGURE 1. Agarose gel electrophoresis of PCR amplified 14 kDa fusion protein gene sequences derived from five different species of the genus *Orthopoxvirus*. Lanes: 1 = negative control (without DNA-template); 2 = cowpoxvirus strain Brighton; 3 = vacciniavirus strain Elstree; 4 = camelpoxvirus strain CP-1; 5 = mousepoxvirus strain M-1; 6 = monkeypoxvirus strain Copenhagen.

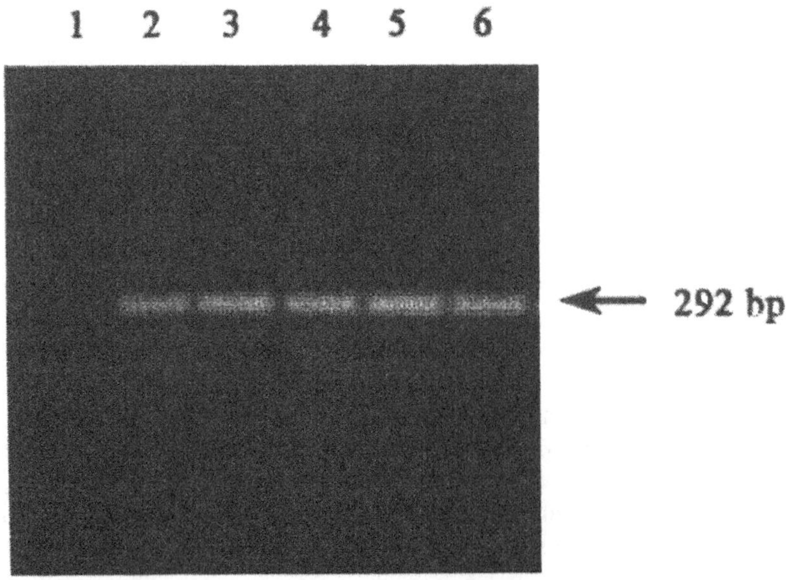

FIGURE 2A. Partial gel view of a representative four-color fluorescent-dye-labeled SSCP/DSCP analysis. Amplified 14 kDa fusion protein gene sequences derived from different species of the genus *Orthopoxvirus* were loaded on a non-denaturing polyacrylamide gel for SSCP and DSCP analyses. Sequences are amplified by PCR using FAM (F=blue) labeled forward and TET (T=green) labeled reverse primer. The red bands are from the GS500-TAMRA internal-lane-standard. Horizontal axis: lane number. Vertical axis: number of scans. DSCP analysis results in dual-labeled dsDNA bands (F+T)(odd numbered lanes). SSCP analysis results in single-labeled ssDNA bands (even numbered lanes). Electrophoresis was performed at constant power of 15 W at 30EC for 19h. Lanes: 1,2 = vacciniavirus strain Elstree; 3,4 = monkeypoxvirus strain Copenhagen; 5,6 = camelpoxvirus strain CP-1; 7.8 = mousepoxvirus strain M-1; 9,10 = cowpoxvirus strain Brighton.

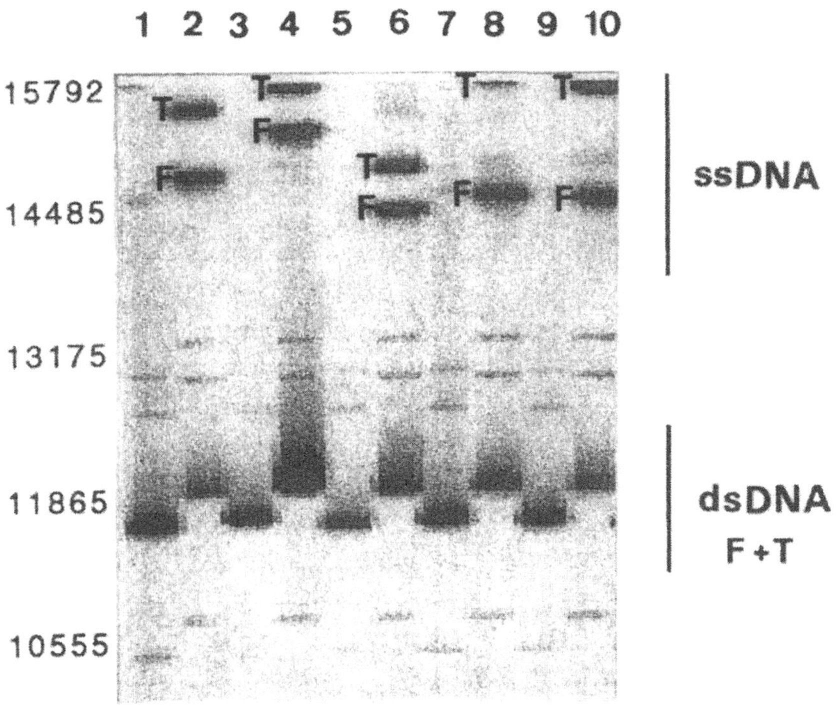

FIGURE 2B. Electropherogram of lane 6 (amplified 14 kDa fusion protein gene fragment derived from camelpoxvirus strain CP-1 analysed by SSCP). Horizontal axis: mobility of single stranded DNA gene fragments presented in terms of arbitrary mobility units calculated by the Genescan 672 software using the internal-lane-standard GS500-TAMRA (red) as a reference. Vertical axis: fluorescence intensity.

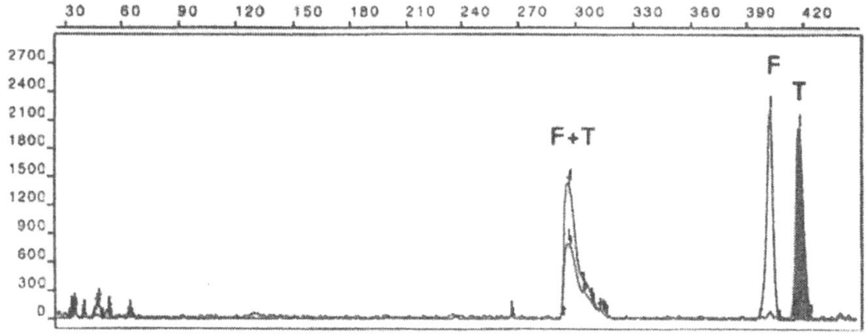

4.2. PCR-SSCP/DSCP ANALYSIS

To investigate the applicability of an automated four-color fluorescent-dye-labeled PCR-SSCP/DSCP approach as a screening method for molecular typing of pathogens, PCR amplicons (see Figure 1) derived from different species of the genus *Orthopoxvirus*, were analysed. After sample preparation, an aliquot of fluorescent-labeled ss- and/or ds-PCR products were loaded on a non-denaturing gel and subsequently electrophoresed.

In Figure 2A, a Genescan gel view of a representative PCR-SSCP/DSCP electrophoresis procedure is shown. The vertical axis indicates the number of scans. The red bands represent the GS500-TAMRA internal-lane-standard. Dual-labeled bands (scan number region 11500-12500) represent complementary dsDNA molecules, whereas single-labeled bands represent ssDNA molecules (scan number region 14000-16000). However, following SSCP analysis dual-labeled DNA bands also appeared at the position of dsDNA molecules. These bands are homoduplex molecules, derived from reannealed denatured ssDNA molecules.

The GeneScan 672 software calculates a calibration curve for each lane of the gel to accurately normalize the mobility of detected DNA bands by using the internal-lane standard. Data aligned by arbitrary mobility units, is shown in the corresponding electropherogram (Figure 2B). As a representative example, an electropherogram of lane 6 is shown. The blue peak represents FAM-labeled ssDNA molecules, whereas the green peak represents TET-labeled ssDNA molecules. Under these electrophoretic conditions no clear mobility differences of dual-labeled dsDNA molecules were detectable by DSCP analysis. However, normalised SSCP mobilities showed very pronounced differences indicating nucleotide sequence polymorphisms within the DNA sequences analysed. These normalized electrophoretic mobilities therefore allow identification of pathogens.

5. Discussion

The most critical element of PCR-SSCP/DSCP analysis is determination of electrophoretic mobility differences between DNA strands differing by as little as one base substitution. With most systems the reliable detection of such electrophoretic mobility differences can be problematic due to lane-to-lane variability that is characteristic of most of commonly used gel electrophoresis procedures. In this study, we describe the use of a multi-color fluorescent-dye-labeled PCR-SSCP/DSCP technique. This technique has the benefit that the complementary DNA strands of the PCR product can be labeled individually with a unique dye label. The data generated using this approach is considerably easier to interpret than data obtained by the single-label systems. For example, more than one stable conformation of the same ssDNA molecule is frequently found and which may result in several electrophoretic distinguishable DNA bands. In single-label systems, these extra DNA bands are difficult to interpret correctly. However, with the unique labeling of each DNA strand it is easier to detect and accurately identify these bands. Another advantage of the multi-color fluorescent-dye-labeled SSCP/DSCP technique is the possibility to use an internal-lane standard as an internal reference for each sample analysed. This internal-lane-standard is loaded in every lane along with the PCR products to enable automated determination of the mobility shifts of the separated ds- and/or ssDNA bands. Automated laser detection of fluorescent-dye-labeled DNA bands is collected by a coupled computer. Subsequent analysis of the laser detected DNA bands by GENESCAN 672 software determines the corresponding electrophoretic mobility and is calculated by means of the internal-lane-standard used as a reference thereby controlling the lane-to-lane and gel-to-gel variability.

In this study the fluorescent-based PCR-SSCP/DSCP technology was investigated as a tool to facilitate molecular typing of pathogens. To determine pathogen-specific SSCP/DSCP mobility patterns due to known pathogen-specific nucleotide sequence polymorphisms within a particular target DNA, PCR amplified 14 kDa fusion protein encoding gene fragments, derived from strains belonging to different species of the genus *Orthopoxvirus*, were studied. Our preliminary results show that in agarose gel electrophoresis and normalised DSCP analyses no clear mobility shifts between the diverse DNA target sequences could be distinguished. However, in SSCP analysis normalised mobility differences between ssDNA molecules of the diverse DNA target sequences could be distinguished. These results indicate that this method permits identification of virus species belonging to a genus using only one conserved primer pair. However, this method requires optimisation. In subsequent studies we will investigate the resolution capacity of the method using a mixture of several different *Orthopoxviruses*.

In conclusion, the results of this experiment indicate that detection and subsequent determination of normalised SSCP mobilities offers the possibility of generating a database of electrophoresis patterns facilitating molecular identification of pathogens. However, characterization of new mutations detected by SSCP/DSCP will require examination of the DNA sequence (by DNA cycle sequencing) to correlate the band mobility shift seen on a non-denaturing gel with the change in nucleotide sequence.

Currently, we are developing a method for identification of a broad variety of bacteria using a multiplex PCR with conserved primer pairs.

6. References

1. Czerny, C.-P., Johann, S., Hölzle, L. and Meyer, H. (1994) Epitope detection in the envelope of intracellular naked *Orthopox* viruses and identification of encoding genes, *Virology* **200**, 764-777.
2. Fujimoto, S., Marshall, B. and Blaser, M.J. (1994) PCR-based restriction fragment length polymorphism typing of *Helicobacter pylori*, *J. Clin. Microbiol.* **32**, 331-334.
3. Glava…, D. and Dean, M. (1993) Optimization of the single-strand conformation polymorphism (SSCP) technique for detection of point mutations, *Hum. Mutat.* **2**, 404-414.
4. Hayashi, K., Orita, M., Suzuki, Y. and Sekiya, T. (1989) Use of labeled primers in polymerase chain reaction (LP-PCR) for a rapid detection of the product, *Nucleic Acids Res.* **17**, 3605.
5. Liu, Q. and Sommer, S.S. (1994) Parameters affecting the sensitivities of dideoxy fingerprinting and SSCP, *PCR Methods Appl.* **4**, 97-108.
6. Makino, R., Yazyu, H., Kishimoto, Y., Sekiya, T. and Hayashi, K. (1992) F-SSCP: Fluorescence-based Polymerase Chain Reaction-Single Strand Conformation Polymorphism (PCR-SSCP) Analysis, *PCR Methods Applic.* **2**, 10-13.
7. Offermans, M.T.C., Sonneveld, R.D., Bakker, E., Deutz-Terlouw, P.P., de Geus, B. and Rozing, J. (1995) Denaturing and non-denaturing gel electrophoresis as methods for the detection of junctional diversity in rearranged T cell receptor sequences, *J. Immunol. Methods* **181**, 101-114.
8. Offermans, M.T.C., Struyk, L., de Geus, B., Breedveld, F.C., van den Elsen, P.J. and Rozing, J. (1996) Direct assessment of junctional diversity in rearranged T cell receptor b chain encoding genes by combined heteroduplex and single strand conformation polymorphism (SSCP) analysis, *J. Immunol. Methods* **191**, 21-31.
9. Olsen, J.E. and Skov, M. (1994) Genomic lineage of *Salmonella enterica* serovar Dublin, *Vet. Microbiol.* **40**, 271-282.
10. Orita, M., Iwahana, H., Kanazawa, H., Hayashi, K. and Sekiya, T. (1989a) Detection of polymorphisms of human DNA by gel electrophoresis as single-strand conformation polymorphisms, *Proc. Natl. Acad. Sci. USA* **86**, 2766-2770.
11. Orita, M., Suzuki, Y., Sekiya, T. and Hayashi, K. (1989b) Rapid and sensitive detection of point mutations and DNA polymorphisms using the polymerase chain reaction, *Genomics* **5**, 874-879.
12. Preston, M.A. and Borczyk A.A. (1994) Genetic variability and molecular typing of *Shigella sonnei* strains isolated in Canada, *J. Clin. Microbiol.* **32**, 1427-1430.
13. Telenti, A., Imboden, P., Marchesi, F., Schmidheini, T. and Bodmer, T. (1993) Direct, automated detection of rifampin-resistant *Mycobacterium tuberculosis* by polymerase chain reaction and single-strand conformation polymorphism analysis, *Antimicrob. Agents Chemother.* **37**, 2054-2058.

USE OF PCR FOR IDENTIFICATION AND DETECTION OF BIOLOGICAL AGENTS

H. GARRIGUE, G. PATRA, AND V. RAMISSE
Biology Department, Centre d'Etudes du Bouchet, 91710 Vert le Petit BP3 – France

1. Introduction

Polymerase Chain Reaction is a very common technic used in most laboratories to identify bacteria, fungi, viruses, or DNA encoding for a toxin, a protein. In this presentation, we will first review some examples of its uses in medical laboratories and then focus on its use for detection and identification in environmental samples.

2. PCR: A Tool for Medical Diagnostic

"The birth of modern biotechnology in the late 1970s depended directly on the DNA replication machinery and synthetic capacity of *Escherichia coli*. Even the Polymerase Chain Reaction (PCR), the method hailed as the greatest development in molecular biology in the last decade, derives much of its power from a DNA polymerase isolated from the thermophilic bacillus "*Thermus aquarticus*." These are two sentences extracted from the preface of a very interesting book of the American Society for Microbiology "Diagnostic Molecular Microbiology" [1]. I extracted them because they recall a very important point, even if PCR is a routine technic nowadays for which commercial kits are widely available, it is very young (the first publication is due to Moseley *et al.* in 1980 [2]) and its progress can be seen as a revolution in diagnostic field.

In the mid 1980s, a lot of companies began to produce commercial probes and kits allowing the technique to spread, there were no more need to develop your own probes for the diagnostic of common bacteria and viruses you were interested in. Technic of detection of target sequences have also evolved during the same time and radioactive probes can be replaced now by reliable non-radioactive techniques and became easy enough to be introduced in many laboratories in developed as well as in developing countries. This technique can support traditional microbiology, as diagnostic tools because they are quick, sensitive reliable and open often the possibility of diagnosis for fastidious or non-cultivable microorganisms.

The field covered by diagnostic PCR is wide, its goes from bacteria, fungi and viruses in clinical samples to food borne diseases where genes encoding for enterotoxins can be detected after cultivation. We will not detail the different techniques possible to use,

P.J. Stopa and M.A. Bartoszcze (eds.),
Rapid Methods for Analysis of Biological Materials in the Environment, 259–277.

just names will be given and we are sure that in this workshop, some will be largely explicited. The basic PCR for identification of agents can be modified to increase its power or allow specific goals. These are the cases of:

- Multiplex PCR in which multiple primer pairs specific for different targets are included in the same amplification reaction [3].

- Competitive oligonucleotide priming which allows detecting point mutations and is used in HIV patients [4].

- Sequencing determination. 16S-23S ISR (Intergenic Spacer Region), *vrr*A [5].

- Nested amplification, which by a two round of amplification is very sensitive but can give contamination and is not really recommended for routine analysis.

- Detection of RNA targets very important for some RNA viruses or r RNA detection.

- Quantitative PCR which consist in amplifying an internal standard, so called lure DNA, and the target DNA of unknown concentration simultaneously under exactly the same conditions allowing the quantification before PCR occurs [6].

The new techniques which are not based on PCR but on different type of analysis of DNA and RNA are not of our concern here.

3. Use of PCR other than for Detection and Identification of Off-Site Environment: Examples in France

CEB depends upon the French ministry of defense and is involved in research for the protection of man against Nuclear, Chemical and Biological agents. The mission goes from detection, identification of therapy. As such it possesses a laboratory for identification of biological agents and has a great concern for some as *B. anthracis* for which have developed various PCR techniques of identification. The most of these techniques are still not fieldable and are used for now on to expertise.

3.1. CONTROL OF CONTAMINATION AND SECURITY

PCR can be used to control laboratory-associated environment. By using multiplex technique, we can detect the plasmids as well as a chromosomal sequence specific (Ba813) [7] of anthrax (Figures 1, 2, and 3). We took surface samples in many places in a laboratory working on attenuated anthrax strains, before monthly decontamination procedures. Results identify centrifuges as the main location where amplifiable and culturable material is present.

We have analyzed different places and laboratories, one of these studies will be presented at the next meeting of "The Molecular Biology of *B. cerus, B. thuringiensis* and *B. anthracis*" Soria Moria Bacillus Workshop at Oslo by 23-25 May 1997. Nucleic acids were extracted directly or after enrichment before PCR detection. The monitoring of labs by PCR in a research or a vaccine production facility could optimize decontamination procedures and prevent the potential risk of spreading microorganisms from the facility to environment. It would allow to diminish contamination during PCR process for analysis of agents; it is always possible that during analysis, a sample is contaminated by agents and lead to a false positive. Of course this can occur mainly when Good Practice Laboratories recommendations cannot be applied. It is recalled that this implies the use of different rooms, one for extraction, one for preparation, one for amplification and finally one for post PCR analyses. More this sequence must be done in a one way direction and it is not allowed to go reverse. Significations of PCR results are discussed later.

3.2. A CASE OF ANTHRAX

In February 1996, we contacted by a Pasteur Institute for confirmation of the identification of *B. anthracis* isolated following a human anthrax case [8]. As soon as we received the samples, a classical PCR was conducted and during the same time an antibiogram was set. Three hours later, the isolate was identified as *B. anthracis* by using a multiplex PCR assay. The next morning, data including antibiogram were given back to Pasteur Institute, which could with its own data confirms the diagnostic of the medical laboratory concerned. A preventive treatment of the family had been done before in order to prevent complications.

This example was given because it illustrates very well how potent PCR can be, when you look for a specific agent and when you deal with culture. Of course, it cannot prejudge of the rapidity and sensitivity with which agents from environmental samples can be identified.

4. Use of PCR for Detection and Identification of Agents from Environmental Samples

PCR directly, after extraction or after cultivation can be applied to various types of samples biological and environmental, we will focus in this para more on environmental than biological samples:

Biological: any classical biological sample can be analyzed, blood, expectorations, sputum, urine, skin, faeces,...to which must be added samples coming from dead animals which can be referred as biological or environmental.

Environmental: water and liquids, soils, food, vegetation, vectors and swabs of any surfaces which could have been contaminated and contain either microorganisms or DNA and of course air samples which can contain aerosols of agents.

The goal of our laboratory is to be able to provide, as quickly as possible with reliable information, authorities that have requested analysis on samples. That can be any French military authority or, through them, any international organization as United

Nation Special Commission (UNSCOM). Some analysis made for UNSCOM will be taken as example later.

What is very important to have clear in mind is that PCR will provide answers on questions addressed. What is looked for must be focussed and known precisely. Probes for anthrax will only be able to identify anthrax. The choice of the probe used is important, its specificity must have been assayed on a lot of close related strains, genes and other different strains.

Direct analysis of the DNA (Southern blot) of microorganisms and its expression in RNA (Northern blot) is possible using methods based on hybridization. Numerous tests using radioactive or non-radioactive probes have been developed to reveal given sequences very specifically. There are the following two possibilities:

(a) direct hybridization with a (labeled) probe, when the target is present in sufficient amounts;

(b) amplification by the Polymerase Chain Reaction (PCR) followed by a detection method, which could be based on hybridization [9].

In vitro enzymatic amplification of DNA or RNA by PCR versus RT-PCR, leads to specific amplification, by several million times, of a precise nucleotide sequence being sought. PCR is derived from the primer extension technique in which short DNA sequences (oligonucleotides) are used as primers by a DNA polymerization enzyme after molecular hybridization on both sides of the area of interest on each of the two DNA strands.

However, the use of the PCR technology is not yet completely controlled, and one of the major difficulties still to be resolved relates to contamination by previously amplified DNA, in view of the method's extreme sensitivity. The problem of the polymerization enzyme inhibitors, particularly in soil samples, is without a doubt a major hurdle. Some soils do not contain any inhibitors, classically sandy soils but this is not always true. For analysis of samples coming from Iraq, which were mainly sandy, PCR was not efficient as proven by the MIMIC technique (Figure 4) used [10]. We had to cultivate them before any analysis and PCR the strains that had grown even if that can introduce a bias due to the competition between the strains existing in the samples. This bias was diminished by cultivation on specific medium as Plet medium.

Recently, Felske and collaborators described a novel method for rRNA isolation from different soil samples. These method that combines an adapted ribosome isolation method and a common RNA extraction step. This study offers some interesting perspectives for the control of the different soils bacterial populations [11].

4.1. MOLECULAR HYBRIDIZATION AND PCR

Direct molecular hybridization using non-radioactive probes can be used to seek a DNA fragment in a sample if the target is present in substantial quantities. If this is not the case, the technique lacks sensitivity, even if the probes are radioactively labeled. By virtue of its great sensitivity, PCR will enable the infectious agent to be detected directly in the sample, even if the concentration is low. In this way, the lack of sensitivity of non-radioactive probes is amply offset by amplification factors of several million achieved by the PCR technique.

4.2. CHOICE OF PROBES OR PRIMERS FOR IDENTIFICATION OF BACTERIA

PCR is based on the principle of hybridization between two DNA primers and the DNA or RNA target, and therefore raises the problem of choice of the target. The same applies for molecular hybridization with a DNA probe. Detection of biological agents by PCR or molecular hybridization may target any available genetic material specific for this agent:

(a) structural genes coding for proteins (surface antigens, toxins, etc.);
(b) the messenger RNA (mRNA) transcribed from those genes;
(c) ribosomal RNA (rRNA) genes.

Independently of the method of analysis (amplification of hybridization) selected, the choice of genetic material targeted is a strategic one.

(a) Targeting of a structural gene usually concerns a virulence factor specific to each pathogen. This approach requires that the presence of the target bacterium in the sample to be analyzed is suspected;

(b) Targeting the mRNA of a structural gene has the same disadvantage as targeting the gene itself. However, this approach may be interesting as it provides an indication of the physiological state of the bacterium. It is thus possible to obtain chronological information on the bacterium, indicating whether it was alive recently or not. Also, there are more mRNAs available, whereas a gene is usually present only as a single copy. On the other hand, mRNA is much more instable than DNA and requires stringent precautions for handling.

(c) Lastly, targeting of rRNA genes is interesting in its universality, as these RNAs represent a very well conserved component of the transnational machinery in the living world. The 16S RNA genes of bacteria are present in one or more copies (here again the number of copies available at the start may be greater) and have a relatively well conserved structure. Often the sequence of these genes can be divergent enough for it to be possible to distinguish between bacteria of closely related species. It is thus possible to determine DNA probes with a variable level of specificity (group, family, genus or species). It is easy to see the advantages of this approach in cases where no specific bacterium is suspected to be present: bacterial rRNA gene from a sample can be amplified, blindly and with a selected level of specificity, and then identified.

In the case of anthrax, ribotyping of 18 *B. anthracis* strains and 5 close related species ([12] with 16S and 23S rRNA probes revealed significant homology in the patterns.

Results showed that the best resolution was obtained with *Acc*I and *Dra*I with the 23S rRNA probe and *Eco*RI and *Dra*I with the 16S rRNA probe (Figures 5 and 6).

Identification may be also carried out either by systematic sequencing followed by comparison of the sequences obtained with a database (the most exhaustive but also the most ponderous method) or by hybridization methods using specific DNA probes.

Nevertheless, designing primers should be done carefully by targeting an appropriate sequence specific to the microorganism to be detected as previously mentioned.

Certain limitations of the approach using the DNA probes coding for rRNAs can be circumvented for bacterial species by exploiting the properties of the transfer RNA gene clusters. Examining one or more intergenic regions like 16S – 23S region would enable all the species belonging to a single bacterial genus to be identified.

4.3. HYBRIDIZATION METHODS

It is not appropriate to measure only the size of the total quantity of DNA synthesized during PCR because non-specific amplification may occur. The analysis must include a more specific detection mean. Molecular hybridization allows specific identification of PCR products on the basis of DNA sequence. Different methods of detecting PCR products can be compared (Figure 7).

Two hybridization systems can be contemplated heterogeneous phase hybridization, in which one of the DNA strands is linked to solid support, and homogenous phase hybridization in a liquid.

In the case of heterogeneous phase hybridization, two options are open depending on whether the DNA fixed is the probe (hybridization in a sandwich with a second probe) or the target. This hybridization can be contemplated only if two different types of probes are available: a capture probe and a detection probe. The former allows specific capture of the hybridization complex on solid phase assay (ELISA plate), and the complex is then revealed by the detection probe. All the homogenous hybridization methods are based on separation of the hybridized probe and the free probe after capture by affinity.

Given the rapidity and simplicity of the test liquid phase hybridization is preferred because an automated protocol can be developed. It can be carried out in 4 to 6 hours, whereas heterogeneous phase hybridization takes 4 to 16 hours. The liquid phase hybridization methods give a result directly in the form of a numerical value (read by spectrophotometry), whereas the heterogeneous phase hybridization methods and electrophoretic analysis of the PCR products necessitate reading by means of a scanner.

4.4. TYPE OF NON-RADIOACTIVE PROBE

Non-radioactive probes can be classified in the three following categories:

(a) signal-emitting probes: this is the most direct system, usually represented by probes coupled to a fluorochrome;

(b) probes coupled direct to the revelation enzyme: this chemical coupling may take place at one end of a synthetic oligonucleotide;

(c) chemically or enzymatically modified probes detected by a molecules with a high affinity for the probe's modification: this is the class to which biotin/avidin systems and hapten (sulphone, digoxigenin, acetyl aminofluorene, etc…)/antigen-hapten antibody systems belong.

The protocol that we classically use in the lab is shown (Figure 8), the specific 152-bp long DNA fragment amplified only when *Bacillus anthracis* DNA was used as target was analyzed by non-radioactive sandwich hybridization in microtiter plate using two oligonucleotides, the capture probe (there the oligonucleotide C1) was covalently linked onto aminated wells of microtiter plates (COVALINK NH® from NUNC). The detection oligonucleotide (here D3) was labeled with biotine. The hybrid molecules were detected by avidine conjugated with alkaline phosphatase and chromogenic substrate (here PNPP).

The specificity of the selected primers derived from the Ba813 sequence were tested with 28 *Bacillus anthracis* strains (9 avirulent strains and 19 fully virulent strains) and 33 strains of heterologous bacteria. No DNA amplification was observed when we used R1 and R2 primers with the other bacillus species or non-bacillus bacteria. Moreover, recently 11 *Bacillus thuringiensis* strains representing the principals serovar was tested, with 3 other strains of *Bacillus cereus* and 2 strains of *Bacillus mycoides*, again a time no DNA amplification was observed [13].

5. Other Considerations

Specific detection of DNA or RNA sequences may be carried out by PCR amplification and/or molecular hybridization. The PCR amplification product is a double-strand DNA fragment usually present in sufficient quantity for it to be directly displayed by ethidium bromide staining after electrophoresis.

In this last approach, the specificity of amplification can be easily verified from certain features of the amplified fragment such as its size and its restriction analysis. The system's reliability can be increased by amplifying various parts of the entire sequence to be detected. Hybridization will increase in fiability of detection of PCR products in future simplification when automation of hybridization and revelation systems render the electrophoresis and ethidium bromide staining technique uncompetitive for development of a fast and simple identification procedure.

Nevertheless, PCR amplification can be entirely automated and, after addition of the probes(s), liquid-phase hybridization would only be one more stage for an apparatus performing PCR. Immunocapture of the hybridization complex, washing and revelation of the probe will be possible in ELISA plates and will thus be fully automated.

The protocol that we classically use in the lab is shown in Figures 9 and 10.

Results given by PCR must be taken carefully and conclusions drawn with care. In some cases, there are no difficulties especially when PCR is not used as a tool for detection but for control or identification but in others it is necessary to consider various parameters. Some practical examples of questions of concern are given:

What is the meaning of a positive result: are you sure of the presence of the agent?, are you sure there is no contamination?, have you controlled first the possible presence of aerosol of agents in your lab? What signification can you give to a sample showing after cultivation the presence of one or two bacteria?

6. Conclusion

Advantages of PCR and non-radioactive probes may be summarized:

(a) the exceptional sensitivity of tests combining the PCR technique and molecular hybridization means that a few DNA or RNA can be detected;

(b) the potential for simplifying the test is a direct consequence of its extreme sensitivity.

As a single stage – amplification by PCR – gives the entire system unlimited sensitivity, it becomes possible to perform the other stages using tools of limited sensitivity that nevertheless satisfy the test's simplicity and rapidity criteria.

We believe important also, in order to have reliable results to control laboratory environment before analysis, to follow GLP, i.e., to separate physically, as much as possible, the rooms in which a sample must be analyzed, step by step in order to diminish the possibility of contamination. The second recommendation is to correlate PCR analysis with classical microbiology and conduct analysis in parallel. That will save time if PCR is inhibited and provide information on antibiotic sensitivity, which in some circumstances can be very useful. Secondly, it allows to detain the strains for their molecular typing and to have reference strains.

Finally, PCR for disarmament analysis seems more and more possible. Most of the techniques and protocols described can be applied and new techniques for differentiation of strains of a same type of bacteria are appearing. It has been possible [12] in the lab by using restriction enzymes to have a tool able to find 5 groups of DNA pattern of anthrax starting from 18 DNA of *Bacillus anthracis* strains digested by *Acc*I restriction enzyme. The aim is to get a tool in some years, which will tell undoubtfully that an anthrax found in a place is natural and endemic or come from an other part of the world and must be considered potentially as the proof that non-authorized studies have been conducted.

FIGURE 1. Results of the amplification of total genomic DNA (~ 100 ng) by 67/68, 25/26, 57/58, 3 /4 and R1/R2 primer pairs. 123 bp ladder (1, 20), negative control (2, 21), *B. anthracis* Sterne 7700 (3), Sterne 7702 (4), pXO1-/2+ ATCC4429 (5), pXO1-/2+ ATCC6602 (6), Cepanzo (7), Davis TE702 (8), *B. subtilis* ATCC6051 (9), *B. pumilus* ATCC7061 (10), *B. licheniformis* ATCC14580 (11), *B. stearothermophilus* ATCC7953(12), *B. thurigiensis* ATCC10792 (13), *B. sphaericus* ATCC14577 (14), *B. alvei* ATCC344 (15), *B. megaterium* ATCC14581 (16), *B. pasteurii* ATCC11859 (17), *B. cereus* ATCC14579 (18), *B. subtilis* ATCC9372 (19), *B. anthracis* 227 (22), 957 (23), 170 (24), 300 (25), 779 (26), 832 (27), 663 (28), 376 (29), 846 (30), 256 (31), 582 (32), 282 (33), 85 (34), 576 (35), 955 (36), 554 (37), 657 (38), 516 (39), 346 (40), 552 (41), 470 (42), 575 (43), 204 (44), 893 (45), and 69 (46).

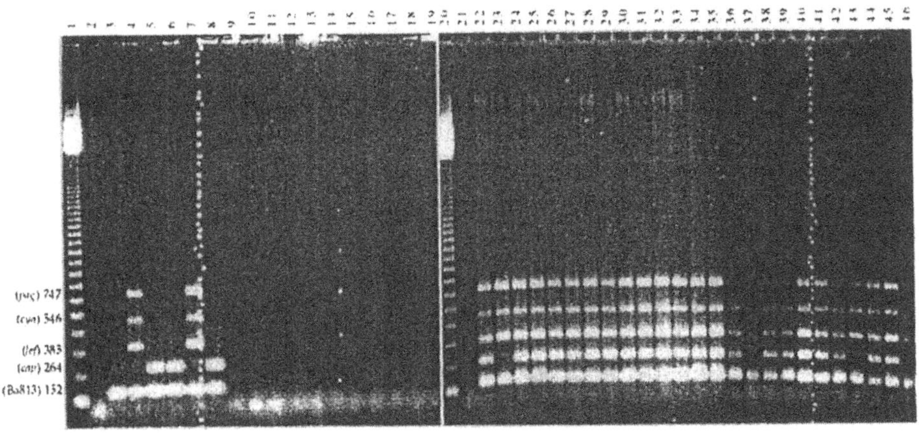

FIGURE 2

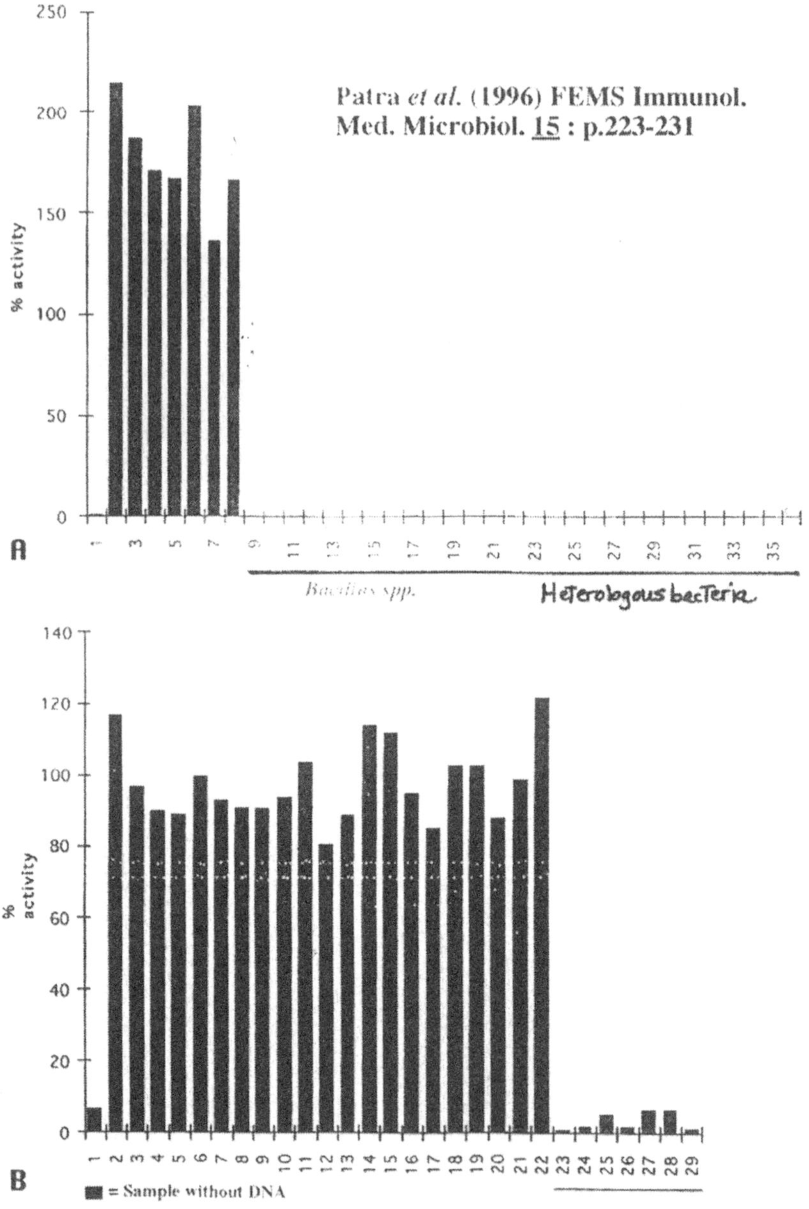
Patra *et al.* (1996) FEMS Immunol. Med. Microbiol. 15 : p.223-231
% activity
A
Bacillus spp.
Heterologous bacteria
B
= Sample without DNA

FIGURE 3

(Patra, G. and J.-L Guesdon, 1997, Unpublished data)

These data complete the article cited below (Table 2, page 225) :

Patra, G., Sylvestre, P., Ramisse, V., Thérasse, J. and Guesdon, J.-L. (1996) Isolation of a specific chromosomic DNA sequence of *Bacillus anthracis* and its possible use in diagnosis FEMS Immunol. Med. Microbiol. **15**, p.223-231

Strains [a]	Ba813 Primers R1/R2	Source/Donor	Derivation [b]
B. cereus	-	U.S.A	CEB 97-08 ← I.E.B.C. 569
B. cereus	-	U.S.A	CEB 97-09 ← I.E.B.C. 738
B. cereus	-	Germany	CEB 97-10 ← I.E.B.C. 001
B. t. israelensis	-	*Culicidae larvae* Israel	CEB 97-11 ← I.E.B.C. T14 001
B. t. guyangiensis	-	Soil China	CEB 97-12 ← I.E.B.C. T43 001
B. t. palmanyolensis	-	Soil Spain	CEB 97-13 ← I.E.B.C. T55 001
B. t. novosibirsk	-	Soil U.S.S.R	CEB 97-14 ← I.E.B.C. T24A 001
B. t. kurstaki	-	*Ephestia kühniella* France	CEB 97-15 ← I.E.B.C. T03A 001
B. t. kenyae	-	Kenya	CEB 97-16 ← I.E.B.C. T04B 001
B. t. kurstaki HD1	-	U.S.A	CEB 97-17 ← I.E.B.C. T HD1
B. t. konkukian	-	South Korea	CEB 97-18 ← I.E.B.C. T34 001
B. t. tolworthi	-	U.K	CEB 97-19 ← I.E.B.C. T09 001
B. t. aizawai	-	*Plodia interpunctella* U.K	CEB 97-20 ← I.E.B.C. 07.29
B. t. morrisoni	-	U.S.A	CEB 97-21 ← I.E.B.C. T08 001
B. mycoides	-	Germany	CEB 97-22 ← I.E.B.C. 001
B. mycoides	-	Colombia	CEB 97-23 ← I.E.B.C. 003

[a] Strains were kindly provided by Dr. M. Lecadet (Director of I.E.B.C)
[b] CEB, Centre d'Etudes du Bouchet ; I.E.B.C, International Entomopathogenic *Bacillus* Centre - Institut Pasteur
T and/or 001 = Type strain

FIGURE 4

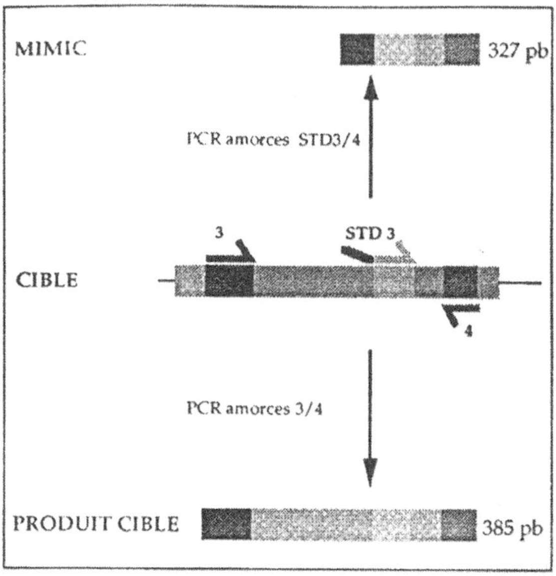
MIMIC
327 pb
PCR amorces STD3/4
3
STD 3
CIBLE
4
PCR amorces 3/4
PRODUIT CIBLE
385 pb

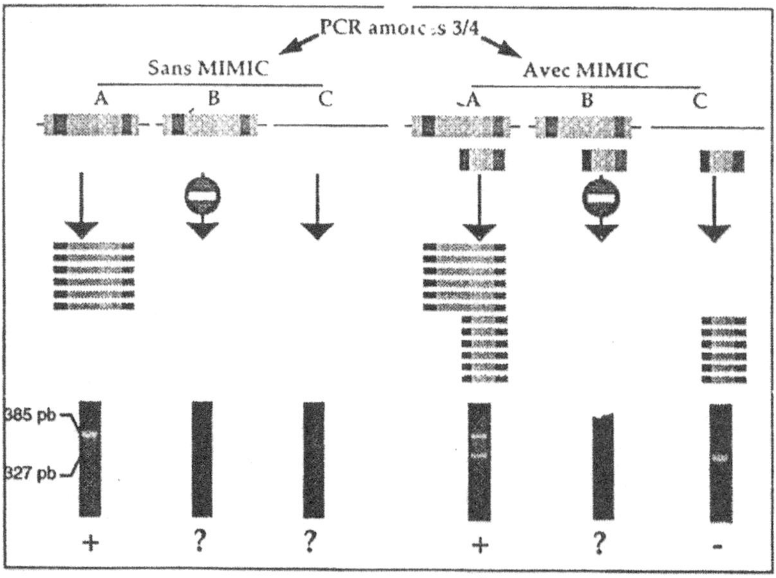
Sans MIMIC
Avec MIMIC
A
B
C
A
B
C
385 pb
327 pb
+
?
?
+
?
-

FIGURE 5

Computer graphics summarising fingerprinting data for *Hin*dIII/16S probe (Block A) and *Hin*dIII/23S probe (Block B), obtained with 18 *Bacillus anthracis* strains.

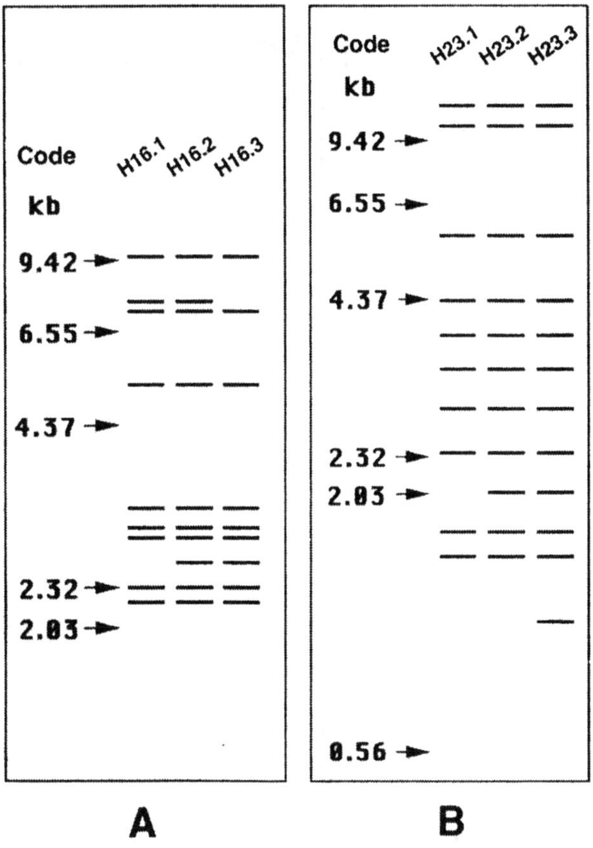

Patra *et al.* (1997), (Manuscript in preparation)

FIGURE 6

Computer graphics summarising fingerprinting data for *Eco*RI/16S probe (Block C) and *Eco*RI/23S probe (Block D), obtained with 18 *Bacillus anthracis* strains.

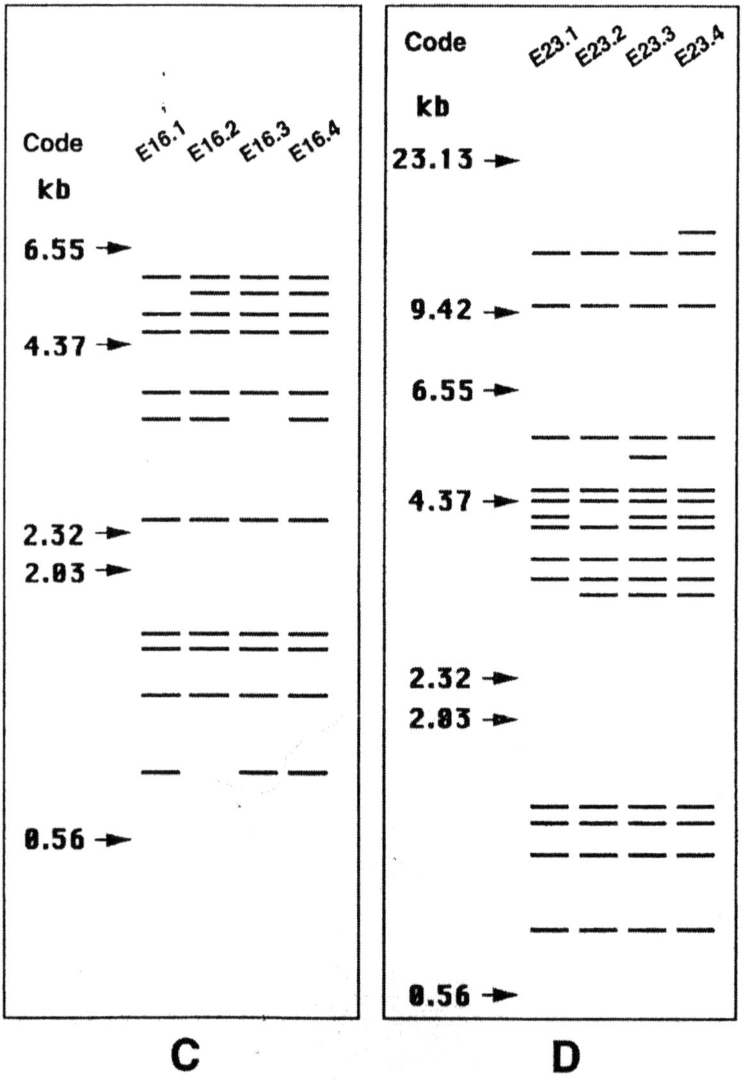

Patra *et al.* (1997), (Manuscript in preparation)

FIGURE 7

COMPARISON OF SOME TECHNIQUES

TECHNIQUES	INFORMATION	TIME
1) LENGHT	+	**1/2 h**
2) ARDRA	+++	**1h30**
3) HYBRIDISATION (Southern blot) (Covalink®)	++++	**≈ 3 days** **≈ 8 h**
4) MULTIPLEX PCR (if more than 3-4 targets)	++++	**2h30**

C.E.B
P.C.C. GARRIGUE
Warsow May 1997

FIGURE 8

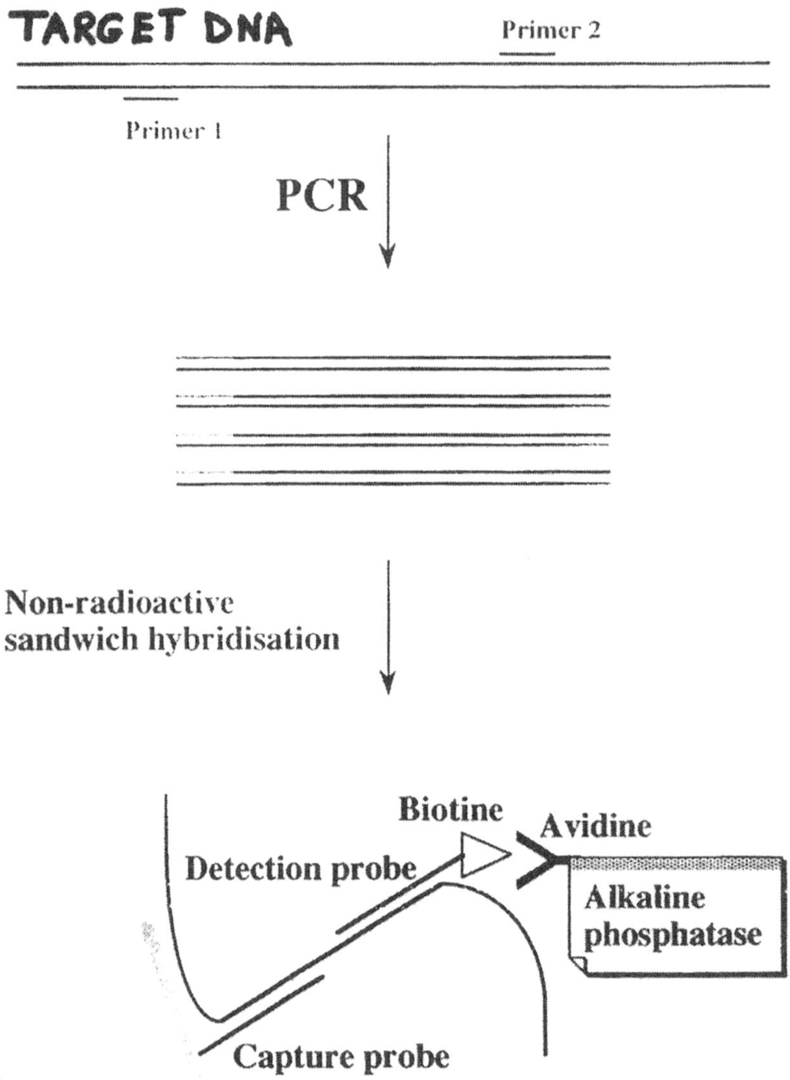
TARGET DNA
Primer 2
Primer 1
PCR
Non-radioactive
sandwich hybridisation
Biotine
Avidine
Detection probe
Alkaline
phosphatase
Capture probe

FIGURE 9

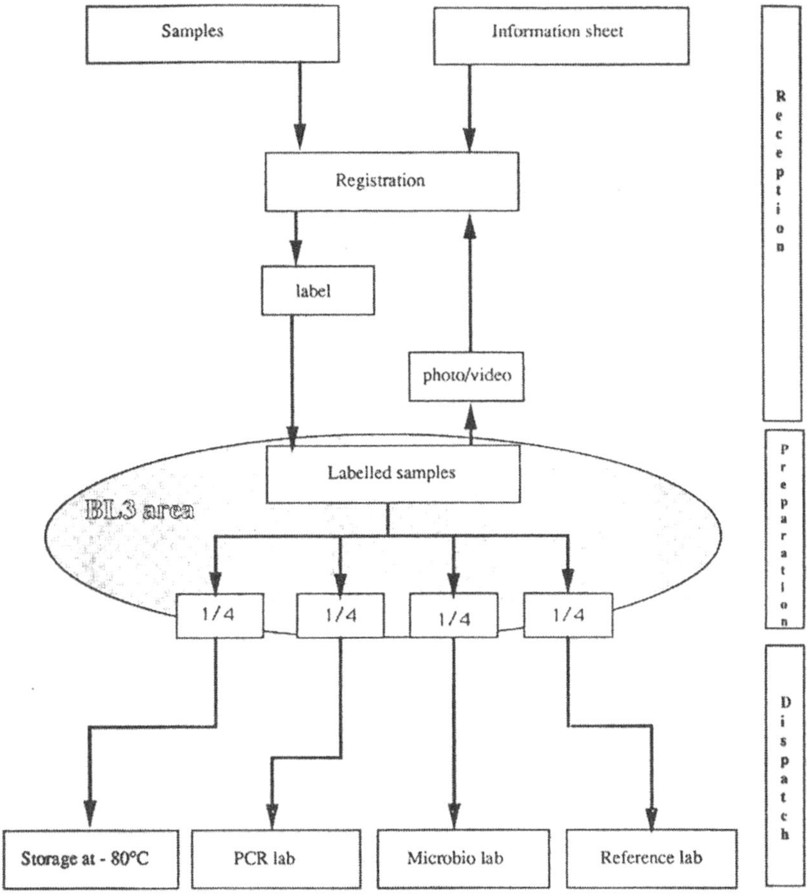

C. E. B.
P.C.C. GARRIGUE
Warsow May 1997

FIGURE 10

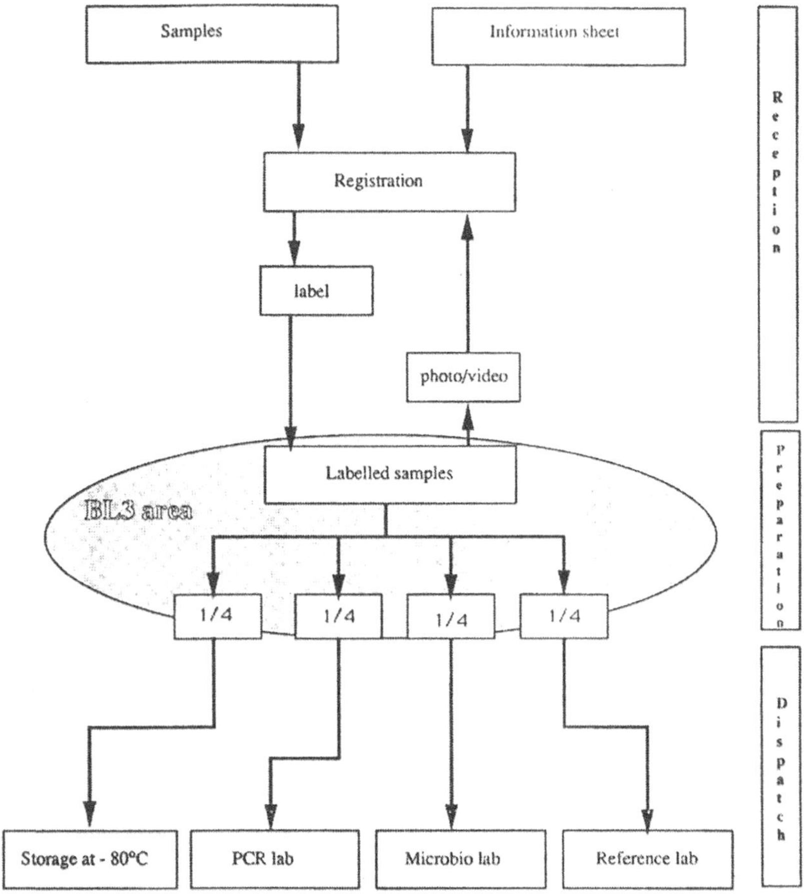

C. E. B.
P.C.C. GARRIGUE
Warsow May 1997

7. References

[1] **Pershing, H.** (1993) Diagnostic molecular microbiology, principles and applications edited by D.H. Pershing, T.F. Smith, F.C. Tenover, and T.J. White, SFM, Mayo Foundation.

[2] **Moseley, S.L., I. Hug, A.R. Alim, M. So, M. Samadpour-Motalebi, and S. Falkow** (1980) Detection of enterotoxigenic *Escherichia coli* by DNA colony hybridization. *J. Infect. Dis.* **142**: 892-898.

[3] **Ramisse, V., G. Patra, H. Garrigue, J.-L. Guesdon, and M. Mock** (1996) Identification and characterization of *Bacillus anthracis* by multiplex PCR analysis of sequences on plasmids pXO1 and pXO2 and chromosomal DNA. *FEMS Microbiol. Lett.* **145**: 9-16.

[4] **Gibbs, R.A., P-N. Nguyen, and C.T. Caskey** (1989) Detection of single DNA base differences by competitive oligonucleotide priming. *Nucleic Acids Res.* **17**: 2437-2448.

[5] **Jackson, P.J., E.A. Walthers, A.S. Kalif, K.L. Richmond, D.M. Adair, K.K. Hill, C.R. Kuske, G.L. Andersen, K.H. Wilson, M.E. Hugh-Jones, and P. Keim** (1997) Characterization of the Variable-Number Tandem Repeats in *vvr*A from Different *Bacillus anthracis* Isolates. *Appli. Environ. Microbiol.* **63** (4): 1400-1405.

[6] **Vincent, U., G. Patra, J. Therasse, and P. Gareil** (1996) Quantitation of polymerase chain reaction – amplified DNA fragments on capillary electrophoresis and laser-induced fluorescence detection. *Electrophoresis* **17**: 512-517.

[7] **Patra, G., P. Sylvestre, V. Ramisse, J. Therasse, and J.-L. Guesdon** (1996) Isolation of a specific chromosomic DNA sequence of *Bacillus anthracis* and its possible use in diagnosis. *FEMS Immunol. Med. Microbiol.* **15**: 223-231.

[8] **Berthier, M., J.-L. Fauchere, J. Perrin, B. Grignon, and D. Oriot** (1996) Fulminant meningitis due to *Bacillus anthracis* in an 11 year old girl during Ramadan. *Lancet* **347** (9004): 828.

[9] **Keim, P., A. Kalif, J. Schupp, K. Hill, S.E. Travis, K. Richmond, D.M. Adair, M. Hugh-Jones, C.R. Kuske, and P. Jackson** (1997) Molecular Evolution and Diversity in *Bacillus anthracis* as Detected by Amplified Fragment Length Polymorphism Markers. *J. Bacteriol.* **179** (3): 818-824.

[10] **Ramisse, V.** (1997) Personnel Communication.

[11] **Felske, A., B. Engelen, U. Nubel, and H. Backhaus** (1996) Direct Ribosome Isolation from Soil to Extract Bacterial rRNA for Community Analysis. *Appl. Environ. Microbiol.* **62** (11): 4162-4167.

[12] **Patra, G., P. Sylvestre, V. Ramisse, J. Therasse , and J.-L. Guesdon** (1997) Characterization of *Bacillus anthracis* strains by rDNA Gene Restriction Patterns (Manuscript in preparation).

[13] **Patra, G. and J.-L. Guesdon** (1997) (Unpublished data).

COMPARATIVE PERFORMANCE OF CONVENTIONAL AND MOLECULAR DOSIMETRY METHODS IN ENVIRONMENTAL BIOMONITORING

Assessment using Bacillus-based commercial biopesticides as models

V.L. SELIGY[1,4], G.R. DOUGLAS[1], J.M. RANCOURT[1], A.F. TAYABALI[1,4], I. OTVOS[2], K. vanFRANKENHUYZEN[2], J. DUGAL[3], G. ROUSSEAU[3], AND A.G. SZABO[5]

[1]Mutagenesis Section, Environmental and Occupational Toxicology, Environmental Health Directorate, Department of Health; [2]Canadian Forest Service, Natural Resources Canada; [3]Societe de Protection des Forets Contre les Insectes et Maladies, Quebec; [4]Carelton-Ottawa Biology Institute, Environmental Chemistry & Toxicology Program; [5]University of Windsor, Department of Chemistry & Biochemistry, Windsor, Ontario, Canada

1. Introduction

Capacity building in environmental biomonitoring is a key component of sustainable development. This paper addresses this subject in relation to: practical methods for general and specific monitoring of environmental biohazards; conductance of compliance monitoring of large-scale releases of microbe-based contaminants, and risk/benefit analysis of biotechnology products such as biopesticides used in "greening" the environment. In addition to providing an environmental toxicology perspective to "*what do we really know about the environmental*"(59), we discuss the safety and utility of *Bacillus thuringiensis*-based, commercial pesticides (BT products) for developing and validating environmental dosimetry methods and strategies. A performance evaluation of "conventional" and "advanced" molecular dosimetry methods is made from a large data set based on an aerial spray calibration. The latter methods are by no means as "advanced" as those under development for climate control and defense research (42; see authors in this NATO ASI edition), but they are often regarded at the limit of capacity (molecular skills and specialty equipment) in most field work conducted in developing and developed countries.

2. Global Environmental Surveillance

The closing of this century has witnessed the quadrupling of the human population, the appearance of several revolutionary technologies, and > 30 new diseases (21,41,49,71). The current revolutions in "food production" and "healthcare" are based on "platform" biotechnologies, supported by remarkable molecular genetic and cell biology discoveries within the last 25 years. The growth in new biological knowledge and bioanalytical tools is spectacular. Biotechnology is unveiling the secrets of biodiversity and human well being, the biosphere's fragility, and the solutions for "greening"

P.J. Stopa and M.A. Bartoszcze (eds.),
***Rapid Methods for Analysis of Biological Materials in the Environment*, 279–297.**

industry and sustaining *global health; its surveillance and management* (18,23,24,46,55,57,65,75). These biorevolutions are our quintessential extra-genetic adaptations to counter the negative impact of our global invasion; the indiscriminate energy demands, spread of diseases and pollution.

In this context, the infant field of environmental biomonitoring (EBM) (Fig. 1), is a core, and route activity for identifying and instituting "common sense" regulation of environmental contaminants, confidence building for public acceptance of biotechnology applications, and validation of safety/efficacy measures for managing resources such as forests and biodiversity (15,18,27,55,58). Global EBM capacity-building(29,44) is needed to address *traditional risks*, related to "*natural*" xenobiotics (plant/animal/microbial chemicals and particulates)(12,34,39) and the numerous *modern risks* (chemicals/biologicals) derived from the last 200 years of industrial activity which overlap *traditional risks*, and create *transition risks* and *synergisms*(25,57). This ongoing legacy of contamination requires a global strategy to ensure quality of life as we know it today and survival beyond the next century (19,21,23,29,55,64,65,73,75).

As a generality, life histories (size, longevity, reproductive and recombinant rates) of higher organisms make them unpredictably reactive to microbes (existing or newly mutated) and novel chemicals(12,31,35). Striking examples are the variable responses we have after exposure to serious pathogens and our own body micro-flora, whose biocomplexity potential (benevolence or pathogenesis), in relation to our biochemistry and exposure to xenobiotics, is poorly understood (Fig. 1)(12,35,64,65). As local populations grow, so do these and other microbes that are involved in zoonosis. The sharp increase in survival rate since 1950 is linked to epigenetic defense measures, like antibiotics which are presently not always available or adequate, depending on risk group, geographical location and risk synergism. Unless release is deliberate, most disease threats to humans, wildlife and domestic animals would come from known xenobiotics whose ranges have been extended by changes in our population dynamics, drug usage, geoclimate and regulation or public health(19,21,24,44). Currently, we have a semi-detailed hazard profile for < 5% of the priority pollutants, a genomic description for a handful of microbes and their probable life histories, and scanty details for most viruses ~1%($5x10^4$), bacteria <1%($4x10^5$), fungi <5%(10^5) and protozoans <20%(10^5)(12,13,73). In a world context, we have *only a limited knowledge of the environment,* and our EBM capacity, while growing, is inadequate to manage *traditional/modern risks* and their synergisms, and *traditional* risks created by initiatives to remedy existing environmental problems.

3. Candidates for Environmental Biomonitoring (EBM)

EBM provides a detailed knowledge currently lacking about exposures and dose-responses pertaining to the complex nexus of cause-and-effect relationships between chemicals and biologicals, and us, and other organisms (Fig.1)(48,57). The subject list for EBM is large: environmental pollutants in waste waters and landfill leachates, whose toxicity may be potent or may be present at sufficient levels to pose threats to biota (14,26,43,55); biohazards that cause current serious world health problems, and can act as agents for biowarfare/bioterrorism (31,35,75), and man-induced outdoor (ambient) and indoor, air/water contaminants of common opportunistic microbes which arise from scaling-up agri-forest industries, and building dynamics for energy and air conservation (39,46,50,51,54). The latter require monitoring as *transition* risks to human health, livestock and damage to material goods (12,18,39,50,51,54,75). An emerging area for EBM use is in the R&D and regulation of microbe-based biotechnology products (MBPs)(9,15) such as those used in biowarfare (31,63,75), bioremediation and pest control (1,4-11).

Our interest is in assessing environmental health benefits from new technologies, which are essentially *transition risks*. Pest control through use of biopesticides is a global experiment in progress, involving a marriage of old and new concepts about crop management as well as health and safety. Present commercial biopesticides are basically concentrates of crude bacterial or viral fermentations (1,11,52,54). The most popular are BT products (Fig.2) whose short-term release by aircraft or ground devices, contributes a high *transient*, bioparticulate burden within target sites and any area where drift occurs. Compared to chemical pesticides, their overall potential for *hazard*, in balance with other biological threats (Fig.1), is likely to be transient and favorable for environmental health (1,36,56,68). However, evaluation is needed for these experimental products to be fully developed, and to be used in any environment without expert qualification or concern over high frequency, repetitive use, and use in high population density areas (4,6,17,56).

In terms of Bt literature (>6000 publication since 1901), there are shortfalls relating to technical gaps between laboratory and field/product research, and the perpetuation of weak or outdated studies relating to health risks (52,56). The majority of the molecular work and literature reviews focus on δ-endotoxin R&D (1,8,11,60) with little or no detail on actual BT products and their biochemical genetics and pathogenesis (52,61,62). In a global context there are concerns over quality of unregulated production and misuse (9,75). BT products are not *natural* as often stated. They are very complex (Fig.2) and when initially released, the spore component (Fig.3) exceeds any 'indigenous' population, if at all present (1,52,53). The biohazard concerns focus on the strong genetic identity to *Bacillus cereus- anthracis* pathogens (see details Fig.6), and the bioparticulate-*synergisms* that could unfold with frequent high exposures by high risk groups (Fig.1). This pathogenic potential is underscored by discoveries of functional cytolysins and enterotoxin-like factors produced by vegetative cells (3,5,16,32,33,61,62,74). The 'novel' cytolysin (VIP or vegetative insecticidal protein) strains for future insecticides (40) share their genes with *Btk* and *Bti* commercial strains (53;see Fig.6). The tendency to generalize experimental and theoretical work on δ-endotoxin mechanisms of toxicity and species specificity (7,8,11,66) confounds research aimed at precisely understanding what is the *active (toxification) ingredient(s)*of commercial BT products, depending on ecozone, climate, dose and target (1,52,61,62,67).

4. Exposure Markers and Environmental Sampling

We have examined existing BT products and have monitored their use in agri-food and forest management operations (6,52,53,62), in order to collect data on utility of EBM methods for modeling release, exposure and persistence of microbe-containing biotechnology products (MBPs), and biohazard of *B. cereus-B. anthracis*-related genes. These efforts were directed at product quality control/assurance and calibration methods (Fig. 2) and investigations into rapid *in vitro* bioassays (61,62). These efforts gave for the first time, baseline data on most relevant subcomponent(s) for toxification of target and non-target cells (efficacy *vs.* safety), and use of these markers in field dosimetry assays (Fig. 3D) (52,61). The use of scanning microscopy (S.E.M) and biophysical subfractionation methods (centrifugation, electrophoresis, filtration) were key to devoluting the complex, sludge-like properties of these products. All products tested so far had at least five candidate markers that could be exploited to varying degrees in molecular assays (Figs. 3 and 4). Most of these markers have been ruled out for mid- to long-term persistence monitoring by rigorous testing of their availability in dilutions of BT products (Fig. 4) and samples recovered from field sites revisited at intervals over consecutive years after product release(53). In conductance of short-term mapping, most markers could be used for the cadre of sampling activities defined by objectives, product application mode, and properties of the field site (see Fig. 3B and C).

Our experience has been that with the appropriate marker assay (Figs. 3 and 4), it is relatively easy to establish and quantify BT product release, using even the simplest, time-resolved, passive samplers such as synthetic and natural fomens (Fig. 3B and C). The methods for recovering BT product are comparable to those recently cited as useful for EBM of worker and bystander exposures to airborne microorganism contaminants (15, 18). Most flat or 3-D surface collectors can provide meaningful data on droplet capture /cm^2 over controlled exposure time intervals. In short-term quantitative tests, recovery of BT subcomponents in samples was generally high, depending on surface transfer mode (dry swab < wet swab < submersion washing) (see Fig. 3E). However, unlike our experience with indoor air monitoring, dose recovery outdoors is highly dependent on deposit-site (sheltered vs. unprotected) and weather conditions. After 2-3 days, losses due to dispersion (movement by wind and moisture to other environmental compartments) and/or decay (sun/UV damage) ranged from ~10% to >99.999% (53). Most subcomponents are unaltered by dispersing in water or wash solutions (0.01% SDS-0.1mM EDTA, pH 8.0) and vigorous mixing to ensure uniform aliquots when product additives (oil) are present. Detection of trace levels (~ 0.01%) of extracellular subcomponents (the 'free'NA, protein and natural or exogenously added colorants) can be done after concentration of washes by lyophilization or adsorption to hydrophobic or charged membranes (0.2μm pore size) (52). However, the analyses summarized in Fig. 4 show that only spores can be detected at any dose, largely because of their inherent stability and amplification potential that can be monitored by growth-induced turbidity and bioreduction, and PCR. Scanning E.M. analysis indicates that quantities of intact PIBs in micro-dissected droplets are highly variable relative to spore numbers (Fig. 2).

In practical terms, the choice of marker (e.g., presence of PIB-proteins, spore or specific gene) depends on whether a *general* or a *specific* (target *vs.* product component) *field assessment* is needed. In either case, the standard for dose calibration should be the actual BT product used in the operation. Except for morphology-based quantification, quantitative analysis of droplets by various markers can be based on equivalent product dilution concentration. However, marker access in droplets and in solubilized BT product is not always equivalent. The biosludge-like properties (particulate density and viscosity) of most biopesticides increases with dehydration and requires vigorous mixing before sampling of pre and post application stocks. Therefore an important reference to have in profiling exposure would be BT product droplets harvested by an array of collection systems, positioned (horizontally and/or vertically if needed) along a spray pathway representing a maximum exposure. These reference samples are important for QC/QA analyses and also testing effects of additives (if any are used).

5. Performance of Conventional/Molecular Detection Assays

For BT product applicator industries across North America and the world, there are no official monitoring methods to track release, and off-target drift(4). The approaches described in Figs. 3D-F and 4 generally go beyond what has been used in field work so far.

Repetitive analysis of each assay method against BT product dilutions show that the practical range for most methods is no greater than about 4 logs of product dilution. Only those methods that track the spore viability (or DNA amplification) are useful over almost the entire dilution range (~ 10^{10}). A comparison of these same assays in the quantification of BT product deposited on standard collector plates, obtained from a staged aerial application (Fig. 5), showed that all methods give similar results when compared against the exogenous tracer dye. The exception is the lack of quantification using *in vitro* gene amplification (PCR) even with a reference template such as λbacteriophage (Fig. 6). The use of PCR to access the 'free' nucleic acid (FNA) of BT products as well as the DNA in spores has been systematically analyzed by us and will be the subject of a separate report (53). The sensitivity of this detection method is ~ 10 spores, but reliable detection is 200 to 200,000 spores / reaction which agrees with others (30,63). From coamplification tests using defined amounts of λbacteriophage as template (and commercial primers), the FNA serves as a very poor template probably because of size and degree of nicking, and also interferes with the amplification reaction at high product concentrations. Additives such as dyes were also found to decease template priming efficiency.

A summary of performance and cost of most assays likely to be used in EBM monitoring of microbe-based biotechnology products is given in Fig. 7. Spray droplet detection methods are evolving and will in time prove superior for most purposes. At present, general detection (droplet scanning and tracking dye additives) is probably the simplest and most economical approach, but it is not nearly as informative as manual quantification, using markers such as spray droplets, spores and intact PIB crystalloids (with/without stains). The latter assays are labor-intensive, and like droplet/dye scanning do not give any information on 'condition' of putative *active ingredients* (*i.e.*, the integrity of PIB crystalloids and endotoxins and spore viability). Droplets <30μm usually remain airborne and can not be collected without air sampling (aggressive) equipment (2,54). They can be visualized with magnification and adhering fluorescent dyes (e.g., rhodamines) improve their detectability down to 1-5μm. Current droplet scanning devices give only coarse information on number and size as captured by passive samplers (e.g., pre-sized paper or kromkote® cards (60). The droplets detected (>50 - 500 μm) are biased to those that settle before collectors are retrieved within the usual 0.75 - 1 hour post-release period.

The plating method gives reliable information on colony morphology (identity), most probable number of viable spores (cfu), and extrapolated dose based on serial dilutions of BT product. Based on 5 years of working with *Bti* and *Btk* based BT products, we have found that the batch to batch variation in spore content was the least variable of any of the endogenous markers (52). When performed properly, the plating method is dependable and sensitive, but it is tedious to perform and limited to plating densities < 8-10/cm^2 (Fig. 4B). New methods based on real time monitoring (growth and bioreduction assays Fig. 4C and D) permit quantification over a broad range of BT product concentrations (> 8 logs serial dilution) with high sample throughout which reduces cost. Along with these new quantification tools that we have introduced, there are others already in use for environmental microbiology and ecology (12,17,30,69). These include: methods based on nucleic acid (DNA hybridization and PCR), protein and carbohydrate-lipid components by direct analysis (HPLC gas chromatography and immunodetection)(1,9). Few of these methods have been used on any routine basis for BT pesticides except by us. There is considerable value in monitoring by immunoassay the putative functional domain(s) of δ-endotoxins as a substitute for cytotoxicity assays. However, our investigations into the antigenicity of whole BT products has revealed two serious problems. First, much of the pro-δ-endotoxins of PIB structures was inaccessible or non-reactive (non-antigenic), but cell wall debris and membrane proteins were very reactive. Second, antigenicity was altered by weathering of whole BT products. In general, monoclonal detections systems are preferred for proprietary control, but in our experience polyclonal detection systems (or monoclonal cocktails) offer the broadest epitope scanning capacity for environmental monitoring. Scanning electron microscopy has become a routine tool for us, and has proven to be highly informative for product QC (52). It shows that spray deposits are not always predictable in their composition and can reveal semi-quantitative information about active ingredients. The cost for S.E.M. assays through use of contracted services is in the range of the sophisticated molecular genetic assays (≥$30US) but is faster to perform. Most of the tools used here would be standard for molecular biology institutions, and only applicable for non-field use such as in devolving contents of microbe-based biotechnology products (insect viral products included, V. Seligy in preparation). Essentially an analysis of 50 products or more could be accomplished within just a few days, at costs under $40 (US) per product replica. Direct costs for analyzing field samples will be less depending on marker/assay system chosen.

6. Conclusions

a) Performance of markers and assays. The use of EBM is important for evaluating biopesticide product efficacy through formulation changes such as concentration of subcomponents, (e.g., vegetative cells and/or their novel cytolytic byproducts, and/or chemical or bio-organic additives, and/or novel subspecies), frequency in re-application, and use in highly populated, urban environments that could generate potentially serious transition risks. Here we have covered some of the practical aspects of microbe-based bioparticulate monitoring by data based on our work on several BT products that were released in different ecozones across Canada. From 5 years of laboratory and field analyses, we conclude that BT products are complex in composition, yet remarkably similar to each other (Fig. 2; see also 52). The most stable and metabolically active component is the spore whose abundance is ~ 1 to 7 times that of the intact PIB crystalloid. Except for highly irregular, amorphic structures, no other crystal forms were ever seen. Along with proteins (~75% are related to δ-endotoxins; 44), BT products contain significant amounts of "free nucleic acid" and also variable levels of hydrolytic enzymatic activities (9, 45,60, 61,73). From an *in vitro* toxicology perspective, spores, specifically second generation vegetative cells, and their extracellular byproducts, are the most powerful source of cytotoxin in BT products for either insect cells or human cells (40,61,62).

The data presented in Figs. 4, 5 and 7 demonstrate that most of the subcomponents of current BT products (*Btk* and *Bti*) can be successfully used as dosimetry markers in conventional and molecular assays with varying reliability, sensitivity and cost. Our side by side comparison of seven different dosimetry assays, and their validation through use of common dilution standards and samples from a demonstration aerial field application (Figs. 4-6), has been an important undertaking which gives credence to emerging views on monitoring expectations for process micro-organisms in biotechnology (15), biological contaminants (12,18) and pathogens (30,47). We expect that further improvements to existing biopesticide dosimetry and standardization methods, along with a deeper understanding of the life cycles of these organisms, their species specificity and pathogenesis, will enable successful exploitation of new products derived from the over 40 subspecies of *B. thuringiensis* now identified (8,11).

b) Environmental Health Risk Perspective. At present it is not clear why BT products are relatively free of hazard (infection and allergic responses) for users and general public that presumably may have been exposed so far. The relatively good safety record of BT products has less to do with the spore's hazard potential than the product formulation attributes (GMP in the elimination of contaminating organisms and vegetative cells) and excellent application practices that control product drifting, particle size and dose, and management in the selection of target locations in relatively unpopulated areas (9). From a contemporary immunotoxicology view, the current BT products are significantly rich in allergen (6), spores and extracellular organic matter such as protein, nucleic acid and carbohydrate (Fig. 2). The spores in BT products exhibit a high stability to temperature, and also to extremes in pH (52). They survive pseudo gastric pH conditions and at alkaline pH (a target species property) are activated along with pro-δ-endotoxins(52,72). Spores germinate in a cooperative manner, exhibiting a growth optimum ($\geq$ 37°C, pH 7-8) and nutrient usage similar to *B. cereus* and animal cells (52). Spores and the exogenous nucleic acid in BT products carry virulence gene information that is closely related to that of *B. cereus* pathogens (10,16,17,28). The potential for mobilization and re-expression of these sequences is likely remote (see Fig. 6 and ref. 45,53). Recent *in vitro* bioassays of ours, the only tests conducted on whole BT products so far, indicate that vegetative cells derived during incubation exhibit a general and powerful cytotoxic response to both insect and human cells alike *in vitro* (61,62,74). This hazard information must be considered within the platform of current risks involving exposure to the most severe of pathogens (Fig.1).

The physical methods required for partitioning the components of BT products place these products into the broad category of xenobiotics described as industrial powders or particulates (2,39). The analytical methods and data compel us to consider that MBPs are bonafide members of the ' powder technology products' grouping, whose formulation and application modes (4,11) are key to their efficacy and safety (2). Safety of bioaerosols (12,15,23,39,42,54,56) and toxic substances within industrial powders depends on formulation properties and modes of exposure(2). The refinement and precision of current technologies used by aerial applicators for release of BT products significantly governs the particulate properties which appear to fall within the category of industrial dusts (5,60).

The latter are generally easy to manage, but this depends on frequency and extent of exposure, along with the "health" status of an individual immune response system and other life-style factors (12,16,17,19,21,28,56). The addition of tracer dyes to track the release and estimate deposit of BT products (60), depending on dye type, potentially adds undefined risk synergisms, such as in the case of rhodamines whose toxicology is only partly defined (37). In this regard the use of a dye accepted in foods, drugs and cosmetics (FD&C brilliant blue#1; Figs. 4 and 5) has been introduced by us (SOPFIM, Quebec) to improve environmental safety of BT products.

c) Global Environmental biomonitoring capability. The current status of global EBM reflects technology deficits between developed and developing countries, and the more serious gap most countries have, between awareness of new technology and its actual applied use (44,55,57). The distribution of world expertise in EBM and regulatory surveillance capability is fragmented, largely uncoordinated, and certainly not advanced overall to combat any widespread, major crisis. Real-time and long-term EBM is still largely an experimental activity. Also, standardized methodologies for EBM and access to specimen banks that profile individual Nations particular pollution/ contamination biases, and use of certified reference materials (CRMs), for monitoring toxic chemicals and biologicals are largely non-existent or under development (23,24,26,27). The role of NATO might well be to assist in establishing the necessary infrastructures as outlined in Fig. 1. This would include using existing NATO linkages for coordinating networks to supply expertise, certified references materials (CRMs), baseline data (for harmonization of surveillance criteria), and advance ultra sensitive detection/diagnostic technologies (see other authors of this NATO SI series) to assist developed and developing countries combat natural and deliberate spread of chemical and biologically induced diseases.

FIGURE 1. Environmental biomonitoring (EBM) relationship to xenobiotic exposure.

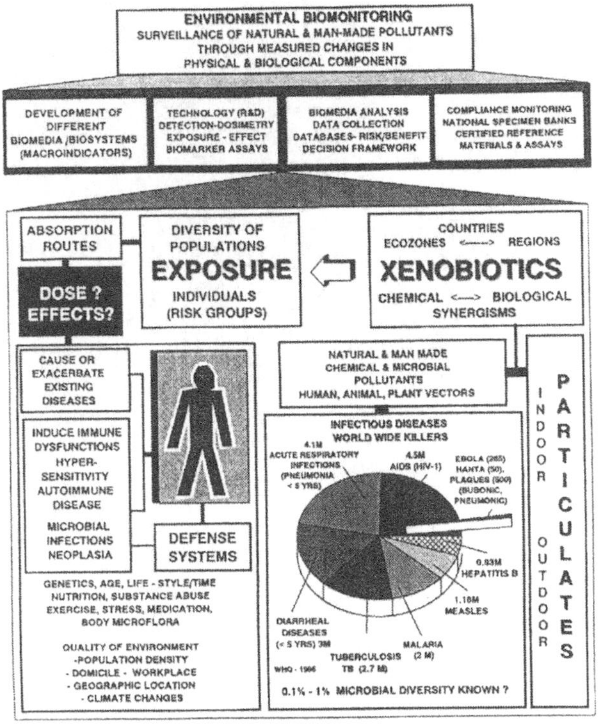

For most xenobiotics there is only a broad (not unique) knowledge about how life-time/life-style exposures affect health. Effects of exposure to bioparticulates can vary, depending on dose, internalization pathway and risk group affected (fetus, infant, child, senior, various immune impairments- SCID, AIDS, cystic fibrosis, asthmatics and workers)(12,39,48,54). For bioparticulates, entry pathways include: skin/mucosal lesions (doses $< 10^3$), respiratory tract (lung access > as bioaerosol size $\leq 10\mu m$ and dose $>10^3$), ingestion and urogenital tract (doses $>10^4$-10^9). Mucosal integrity and interactions with "normal" resident body flora (>500 species at 10^4 to 10^9 /mL faecal fluid) 'prime' the immune system daily. Body flora also scavenge 'excess' nutrients and minerals and occupy cell receptor sites required for invaders to multiply and facilitate transmembrane movement.

FIGURE 2. Properties of most currently used commercial BT products.

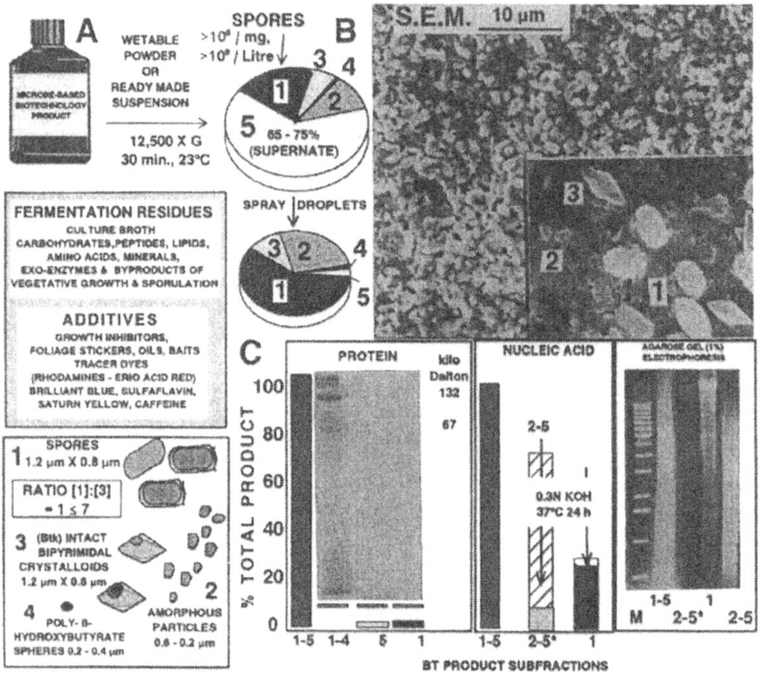

A. BT products are complex substances that originate from sporulation phase cultures of . *thuringiensis* (11). Subspecies such as *kurstaki* [*Btk*] and *israelensis* [*Bti*] are used to control defoliating and blood sucking insects, respectively (8,11). As used here, 'BT' signifies 'commercial'; it is not to be confused with 'Bt' which is often used to mean: 'toxin'. the 'inactive' pro-δ-endotoxin (132kDa in size) or truncated 'active' δ-endotoxin (60-65kDa; the organism, *B. thuringiensis*, or a culture of a subspecies producing "toxin", and commercial products which, in fact, are rarely studied, but often stated as consisting of "toxin". **B.** BT products examined by biophysical methods (38,52,62) have similar biocomponents: spores[1]; amorphous particles[2] related to parasporal inclusion bodies (PIB)[3] and PHB spheres[4]. The liquid phase[5] varies with extent of dehydration. The PIB [3] is referred to as 'crystal', 'crystal toxin',' crystalline' or 'crystalloid'; it is heterogeneous and arises as a cytoplasmic condensate, rich in many proteins and nucleic acids (NA), membrane and secondary metabolites like PHB. PIB size/composition, and temporal formation with spores (usually in 1:1 copy), depends on subspecies and other poorly defined factors. The spore / PIB ratio of specific BT products is relatively stable (e.g. F48 is ~7). **C.** Electrophoresis of components shows related proteins (~75% related to proendotoxin) and NA contents. "Free NA"(FNA $\simeq$ 80% in [2-5] defined by A260nm = 45µg; λ260/λ230 = 1.9-2.1) has a KOH (RNA) sensitivity [2-5*] ≈ 50%. FNA is mostly degraded; it is a spore-harvest component of *B. anthracis* and *cereus* (28,30,47,63), and related to spore mother cell degradation, but not necessarily critical to PIB formation (7,53).

FIGURE 3. Sampling priorities and detection assays for biopesticide monitoring.

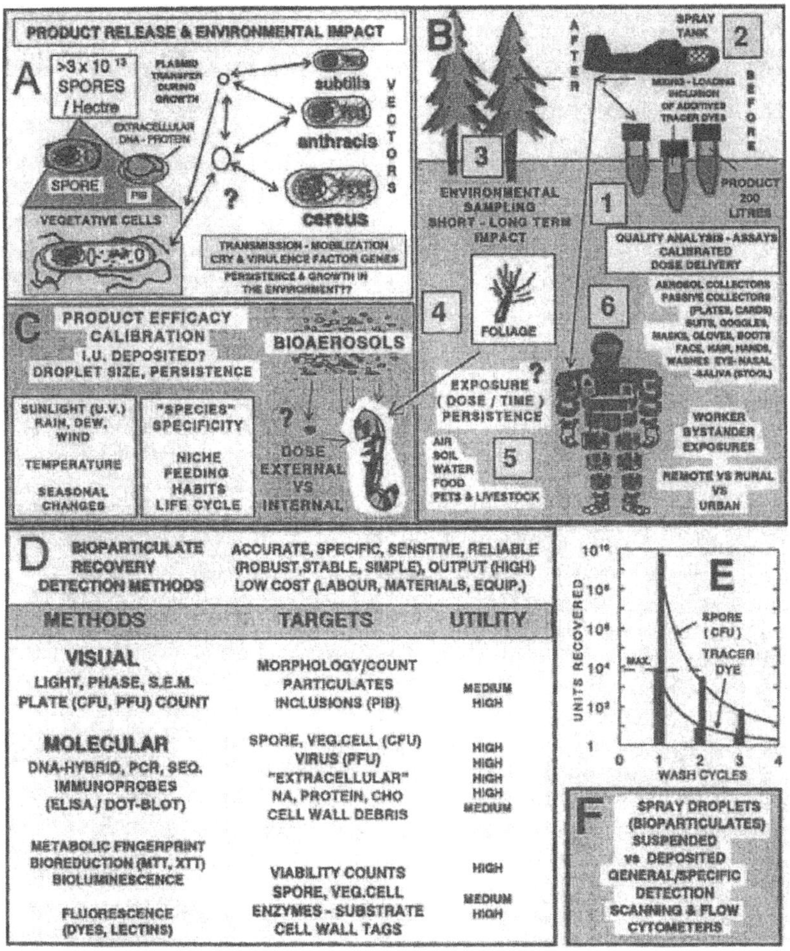

A. Mid/long-term studies (months/ years) must track the regenerative potential of spores (or DNA), but short-term mapping (days/weeks) **(B,C)** can use other markers (Fig.2). Sampling strategies in **B[1-6]** (also 15,18) must include standards **[1,2]** to calibrate delivered dose to intended target **[4]** and non-targets **[5,6]**. Human mapping **[6]** can use both passive and aggressive sampling methods (39). In most cases criteria for recovery and detection **(D)** can be generally met. Almost any surface can serve as a reliable sampler (receptor) for BT product deposit register. Elution of deposits with known wash volumes (0.1 to 5X surface area) gives a transfer pattern similar to that shown for a cotton glove **(E)** tested for rhodamine and spore release. Bacterial culture plates can capture relatively large quantities of droplets for use as standards in droplet size and ingredient concentration analysis (Figs 5, 6). Methods used by the BT industry vary from practical, manual (visual) droplet inspections to bioassays, droplet and tracer dye scans **(F)**, cfu count/morphology, and 'toxin' ELISA assays (4, 60, 66,67,69).

FIGURE 4. Subcomponent dosimetry related to BT product dilution standards.

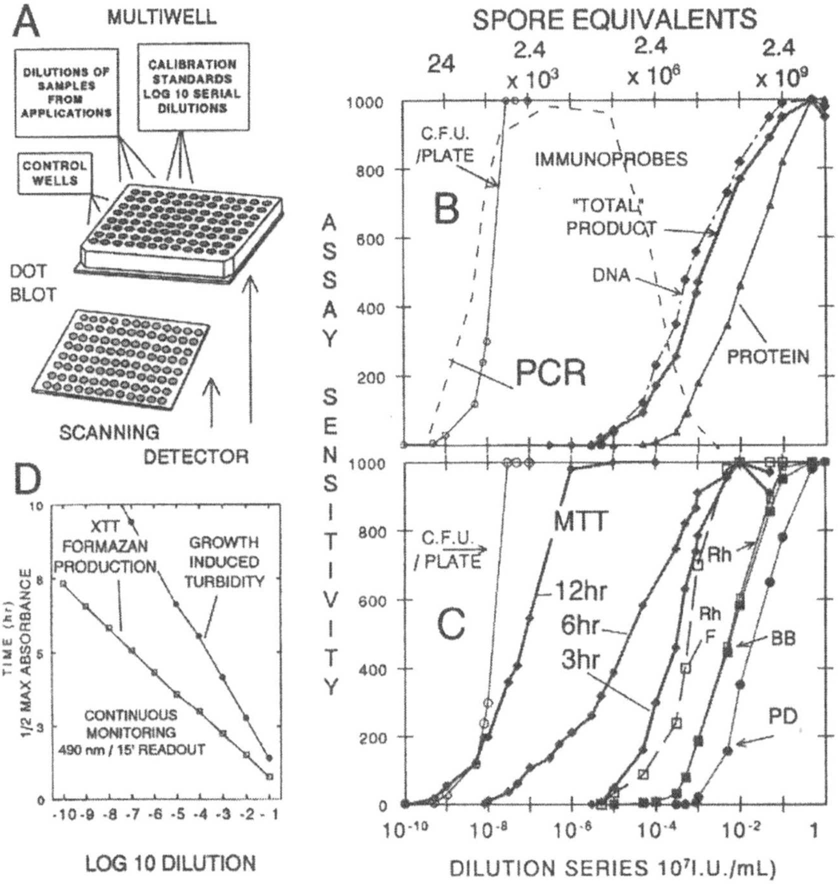

A. Product dilutions (F48B in sterile water ref. 52) were aliquoted in microwells, tubes or arrays on nylon or PVDF membranes before assaying as in **B-D**. **B.** Assays- plating (cfu), PCR (*cry1*Ac gene primers, - -), DNA hybridization dot blot (*cry1*Ac), immunoblot with polyclonal antibody against whole product/protein (30, 52,62). **C.** Assays- bioreduction (MTT added after 3, 6,12hr incubation in LB broth) (52); absorbance of dyes, brilliant blue (BB, FD&C #1at 630nm and natural colorant (PD) and rhodamine (Rh) at 535nm or fluorescence (RhF at 553nm). The min-max of most detection methods spans ~ 3 $\log_{10}$ units and covers the product concentrations from 100% to 0.05%. For all BT products tested so far, the viable spore was the most robust and sensitive indicator. Conventional plating methods, based on most probable number of organisms/sample, are quantitative up to ~ 8 - 10 colonies/cm^2. Bioreduction monitored by step (**C**) or continuous growth in liquid or membranes (**D**) at 37°C with/without XTT (45, 52) are able to estimate the entire dilution range, but the latter while fully automatic are limited to 96 samples / scanner / 12 hr run.

FIGURE 5. Comparison of assays using samples derived from aerial spray applications.

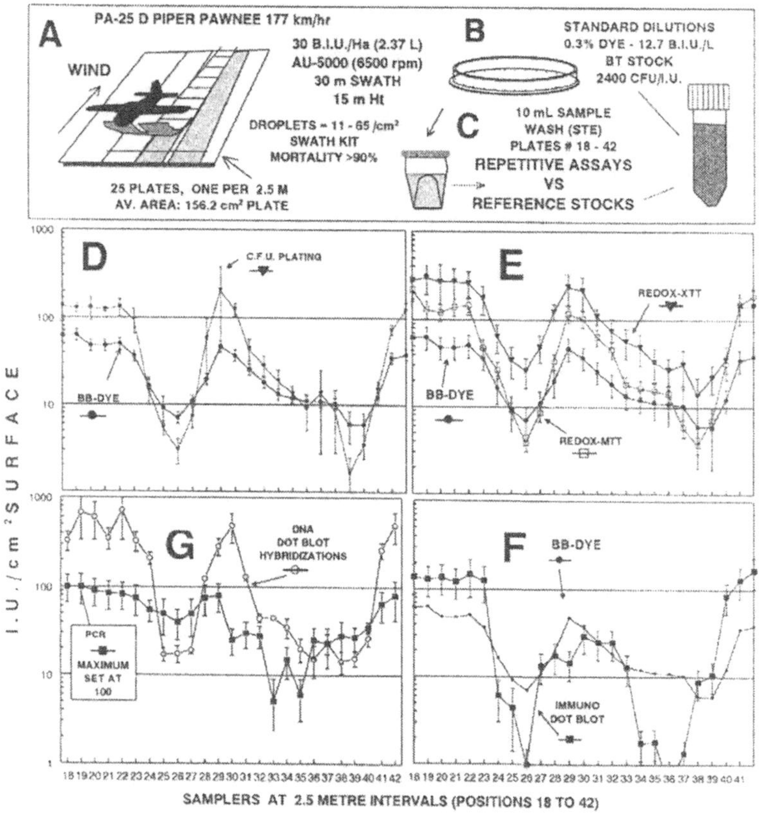

A-C, Production of spray samples using a flat-terrain model was carried out in ecozone Atlantic Maritime R-117 (20). This generated 25 collector plates which were eluted (**B**) and assayed as in Figs 3.4,6 along with dilution standards ($\log_{10}$ -3 to -8) of the product. Data were expressed in I.U. (international units of potency) based on one unit = 2400 cfu (52). Replications were N$\geq$ 5. **D.** Plate assay data based on extrapolations of replicas of three dilution series and optical measurements of the FD&C brilliant blue dye #1 additive. **E.** Bioreduction assays were continuous read (XTT for 16 hrs) or step (MTT added after 3, 6 or 9 hrs). **F.** Immunoblot of samples trapped on PVDF membranes and probed with murine polyclonal antibody to whole BT product. Dilution was 1:1000 and visualized by goat-antimurine-horse radish peroxidase assay, and quantified by scanning (61). **G.** Comparison of DNA detection by dot blot hybridization and PCR using replicas of different genes which included the endotoxins (cry1Ac, cry1Ab) and genes common to *B. cereus* (4,10,32) labeled by *in vitro* amplification (PCR) in separate assays (Fig. 6). Maximum relative PCR product was set to equal 100 I.U.

FIGURE 6. Monitoring the DNA component of BT products.

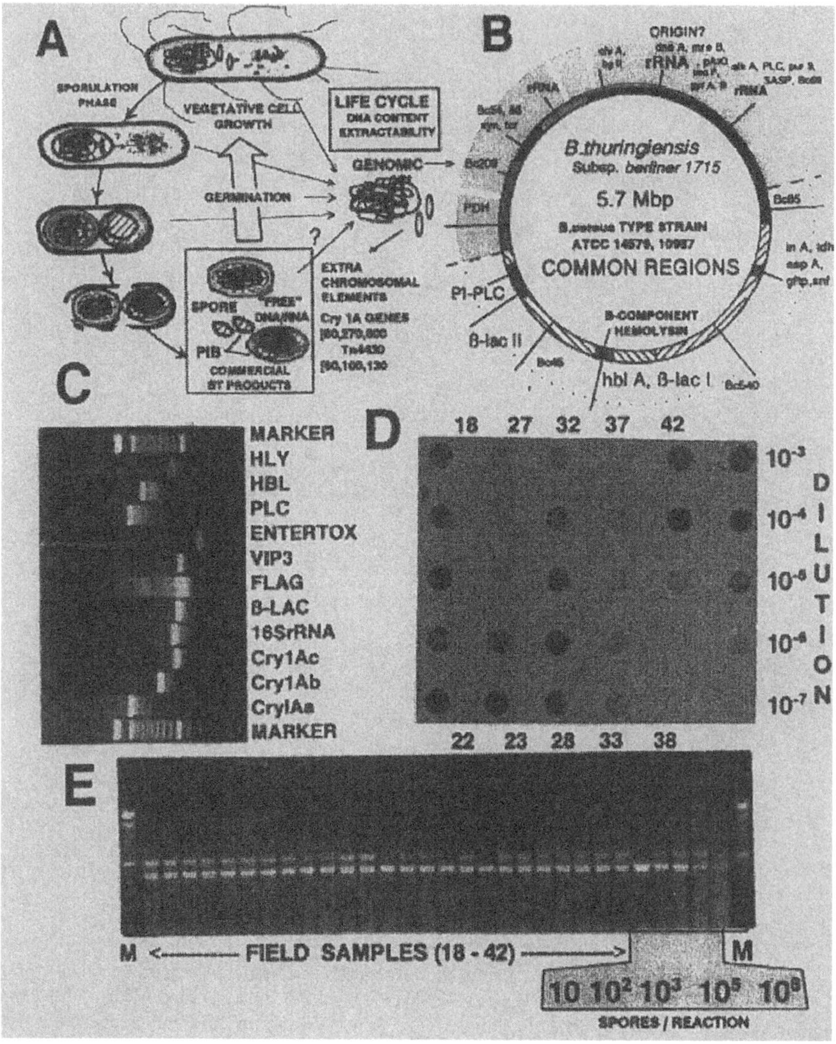

A. DNA extraction changes with life cycle stage; spore-DNA of BT products is the most difficult to access. **B.** Emerging views on genomic organization indicate a close identity to pathogens (3,5,7,10.16,17,28,75). Sequence of plasmid-borne *cry* genes which characterize the various *Bt* subspecies, and common *B. cereus* genes and can serve as diagnostic probes using *in vitro* amplification with polymerase chain reaction (PCR) methods (8,10, 32,52,62) (**C**). **D** and **E** are examples of quantifications of BT product samples (see also Fig.5) by DNA (dot blot) hybridization (**D**) and PCR (**E**) methods using *cry*1Ac probe (52) and BT product dilution standards. **E** (right side) shows that PCR of spore genes is semi-quantitative within a narrow range relative to an exogenously added template (λphage at 10^4). Our coamplification method establishes relative copy and production efficiency of amplicons, and demonstrates potential detection of virus particles, if present in commercial BT products and in field samples. The primer set for λ is routinely used as a commercial QC standard (Perkin-Elmer).

FIGURE 7. 9-block performance plot of various detection assays.

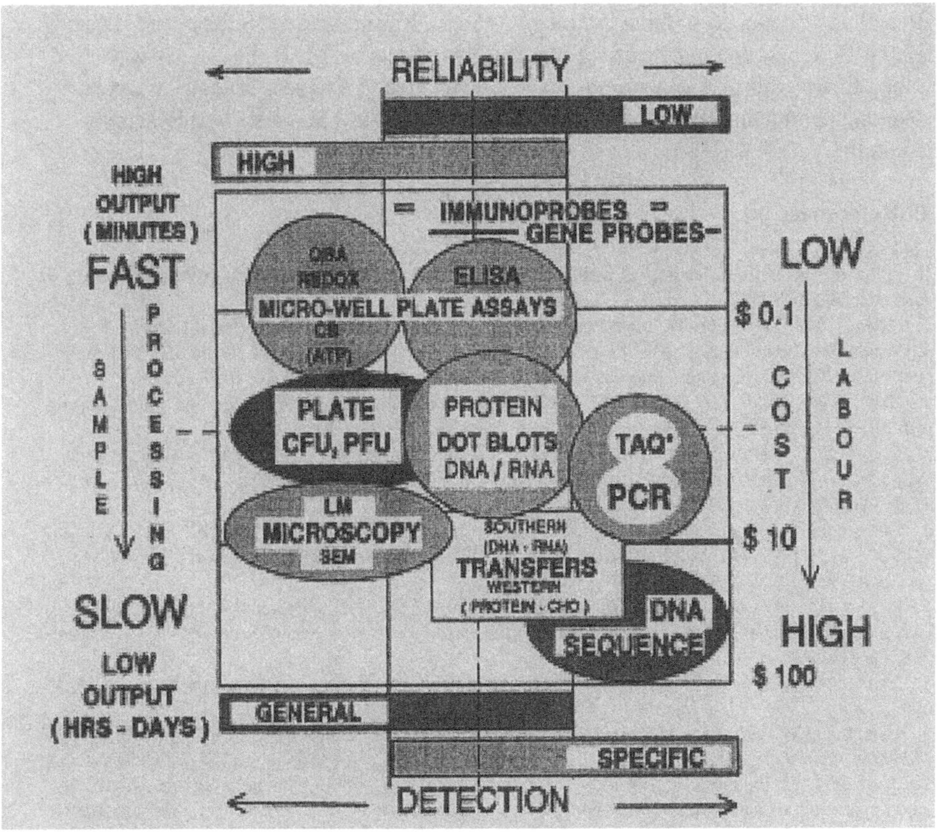

The most laborious part of any environmental monitoring activity centers on gathering and inventory of samples. After samples have been secured they can be dispensed into replica formats (master stocks) for access by any number of assays using semi-automated approaches (multi-channel pipetting, automatic read and data transfer, etc.). Material and labor costs (for skills) increase with specific monitoring capability, the ultimate being DNA sequence that is used in mutation screening (22). The tracer dyes are lowest in cost but they are coupled to important BT products subcomponents. Bioreduction and ELISA type assays are attractive but only the former can scan the entire product dilution range and track the spore, which is regarded as the most important for health/safety. Probes for antigen and genes offer specificity. Immunoassays can be low in cost, but choice of probe (mono vs. polyclonal system) and target (protein vs. CHO) are critical. DNA (colony) blot systems are semi-quantitative and low in cost if performed in large batches. Least quantitative is PCR (70), but they are the most critical for toxin/species diagnostics and production of gene probes. Versions which use single or combinations (e.g., TAQ©, Perkin-Elmer) of fluorescently tagged primers offer sensitivity and quantisation, but they fail due to BT product complexity and inhibitors such as formulation additives (53).

7. Acknowledgments

For the excellent NATO workshop arrangements and selection of topic by Peter Stopa and Michael Bartoszcze, M.D.; the review of this paper by co-authors, and Drs. D. Blakey and A. Godfrey (Biotechnology Section, Environmental Substance Division), and G. Coleman; the generous supply of BT products by Dr. R. Fusco, Abbott Laboratories, and research support to V.L.S. by Health Canada, Industry Canada (National Biotechnology Strategy Program) and Natural Sciences and Research Council.

8. References

1. Addison, J.A. (1993). Persistence and nontarget effects of *Bacillus thuringiensis* in soil: a review. *Can. J. For. Res*.**23**, 2329-2342.
2. Allen, T. (1997) Particle size measurement. Vol. I and II. 5th Ed. Chapman & Hall, London.
3. Asano, S-I., Nukumizu, Y., Bando, H., Iizuka, T. and Yamamoto, T. (1997) Cloning of novel enterotoxin genes from *Bacillus cereus* and *Bacillus thuringiensis*. *Appl. Env. Microbiol.* **63**, 1054-1057.
4. Barry, J.W., Skyler, P.J., Teske, M.E., Rafferty, J.A. and Grim, B.S. (1993) Predicting and measuring drift of *Bacillus thuringiensis* sprays. *Env. Tox. & Chemistry* **12**, 1977-1989.
5. Beggs, R.W., Tayabali, A.F., Rancourt, J.M. and Seligy, V.L. (1996) Virulence enhancing potential of extracellular metabolites produced by commercially-formulated strains of *Bacillus thuringiensis*. *Cellular and Molecular Biology* **42**, S63 - S64.
6. Bernstein, I.L., Berstein, J.A., Miller, M.E., Terzieva, S., Lummus, Z.., Berstein, D.I., Seligy, V.L. and Selgrade, M.J. (1997) Health effects research on microbiologic insecticides in farm workers. *J. Allergy Clin. Immunol.* **99**, s81.
7. Bietlot, H.P., Schernthaner, J.P., Milne, R.E., Clairmont, F.R., Bhella, R.S. and Kaplan, H. (1993) Evidence that the cryIA crystal protein from *Bacillus thuringiensis* is associated with DNA. *J.Biol. Chem.* **268**, 8240- 8245.
8. Cannon, R.J. (1996) *Bacillus thuringiensis* use in agriculture: molecular perspective. *Biol. Rev.***71**, 561-636.
9. Capalbo, D.M. (1995) *Bacillus thuringiensis* fermentation process and risk assessment. A short review. *Mem. Inst. Oswaldo Cruz*, **90**, 135-138.
10. Carlson, C.R., Johansen, T., Lecadet, M.-M. and Kolsto, A.-B. (1996) Genomic organization of entomopathogenic bacterium *Bacillus thuringiensis* subsp. berliner 1715. *Microbiology* **142**,1625-1634.
11. Cibulsky, R.J., Devistetty, B.N., Melchior, G.L. and Melin, B.E. (1993) Formulation and application technologies for microbial pesticides: review of progress and future trends. *J. Testing and Evaluation* **21**, 500-503.
12. Collins, C.H., Aw, T.C. and Grange, J.M. (1997) *Microbial diseases of occupations, sports and recreations.* Butterworth-Heinman. University Press, Cambridge. pp 1-174.
13. Colwell, R.R. (1997) Microbial diversity: the importance of exploration and conservation *J. Industria Microbiol. & Biotechnology* **18**,302 - 307.
14. Crathorne, B., Dobbs, A.J. and Rees, Y. (1996) Chemical pollution of the aquatic environment by priority pollutants and its control, in *Pollution: Causes, Effects and Control.* R.M. Harison (ed.) 3rd Ed. Royal Society of Chemistry. pp.1-25.
15. Crook, B. (1996) Review: Methods of monitoring for process micro-organisms in biotechnology. *Ann.Occup. Hyg.* **40**, 245-260.
16. Damgaard, P.H., Larsen, H.D., Hansen, B.M., Bresciani, J. And Jorgensen, K. (1996) Enterotoxin-producing strains of *B. thuringiensis* isolated from food. *Lett. Applied Microbiol.* **23**, 146-150.
17. Damgaard, P.H., Granum, P.E., Bresciani, J., Torregrossa, Eilenberg, J. and Valentino, L. (1997) Characterization of *B. thuringiensis* isolated from infections in burn wounds. *FEMS Immun. Med. Microbiol.***23**, 47-53.

18. Dillion, K.H., Heinsohn. P.A. and Miller. .J.D. (1996) Field guide for the determination of biological contaminants in environmental samples. AIHA Biosafety Committee, *Am. Industrial Hygiene Assoc.*,Virginia USA., pp167.
19. Dobson, A., Campbell. M.S. and Bell. J. (1997) Fatal Synergisms: Interactions between infectious diseases, Human Populations Growth. and Loss of Biodiversity. F. Grifo and J. Rosenthal (eds.) In, *Biodiversity and Human Health.* Island Press Wash.DC.pp.87-110.
20. Ecological Stratification Working Group (1995) *A national ecological framework for Canada.* Agri-Food Canada and Environment Canada. Ecozone Analysis Branch, Ottawa/Hull.
21. Engelman, R.(1997) Earthly dominion: population growth, biodiversity, and health. F. Grifo and J. Rosenthal (eds.) In, *Biodiversity and Human Health.* Island Press Wash., DC pp 39-59.
22. Eng, C. and Vijg, J. (1997) Genetic testing: the problems and the promise. *Nature Biotech.***15**, 422-426.
23. Gould, R. and Connell, N.D. (1997) The public health effects of biological weapons. In B.S. Levy and V. Sidel (eds.) *War and Public Health.* Oxford University Press and Am. Public Health Assoc. pp 98-116.
24. Groce, N.E. and Reeve. M.E. (1996) Traditional healers and global surveillance strategies for emerging diseases: closing the gap. *Emerging and Infectious Diseases.* **2,** 351- 353.
25. Framework for environmental health risk management. (1997) Final report *Presidential/Congressional Commission on risk assessment and risk management.* Vol.1 pp.1-64.
26. Hee, Q. (1997) Mutagenesis and acute toxicity studies on saliva-leached components of chewing tobacco and simulated urine using bioluminescent bacteria. in K.S. Subramanian and G.V. Iyengar (eds.) Environmental Biomonitoring. Exposure Assessment and Specimen Banking. ACS Symp. Series 654.Wash., DC. pp.77-82.
27. Iyengar, G.V. and Subramanian. K.S. (1997) Environmental biomonitoring and specimen banking: Analytical perspectives, in K.S. Subramanian and G.V. Iyengar (eds.) *Environmental Biomonitoring. Exposure Assessment and Specimen Banking.* ACS Symposium Series 654.Wash., DC. pp.1- 16.
28. Jackson, S.G., Goodbrand. R.B.. Ahmed, R. and Kasatiya, S. (1995) *Bacillus cereus* and *Bacillus thuringiensis* isolated in a gastroenteritis outbreak investigation. *Lett. Appl. Microbiol.* **21**,103-105.
29. Jorgens, H. (1997) *National Environmental Policies: A comparative study of capacity-building.* M. Janicke and H. Weidner (eds.) Spriner UN University, Finland.
30. Johns, M., Harrington, L.. Titball. R.W. and Leslie, D.L. (1994) Improved methods for the detection of *Bacillus anthracis* spores by the polymerase chain reaction. *Lett. Appl. Microbiol.* **18**, 236-238.
31. Kaufman, A.F., Meltzer, M.I. and Schmid. G.P. (1997) The economic impact of a bioterrorist attack: are prevention and post attack intervention programs justifiable? *Emerging Infectious Diseases* **3,** 83-94.
32. Lechner, M., Kupke, T.. Stefanovic, S.. and Gotz, F. (1989) Molecular characterization and sequence of PIPLC of *Bacillus thuringiensis. Mol Microbiol.* **3**, 621-626.
33. Li, R. S., Jarret, P., and Burges. H.D. (1987) Importance of spores, crystals and δ-endotoxins in the pathogenicity of different varieties of *Bacillus thuringiensis* in *Galleria mellonella* and *Pieris brassicae. J. Invertebr. Pathol.* **50**, 277-284.
34. Markert, B., Oehlmann, J. and Roth, M. (1997) General aspects of heavy metal monitoring by plants and animals, in K.S. Subramanian and G.V. Iyengar (eds.) *Environmental Biomonitoring. Exposure Assessment and Specimen Banking.*, ACS Symp. Series 654.Wash., DC.pp.19- 29.
35. Marrs, T.C., Maynard, R.L. and Sidell, F.R. (1996) *Chemical warfare agents. Toxicology and Treatment.* John Wiley & Sons, UK.
36. McClintock, J.T., Scaffer, C.R., and Sjoblad, R.D. (1995) A comparative review of the mammalian toxicity of *Bacillus thuringiensis*-based pesticides. *Pestic. Sci.*, **45**, 95-105.
37. Nestman, E. R., Kowbel, D.J. and Ellenton, J. (1980) Mutagenicity in Salmonella of fluorescent dye tablets used in water tracing. *Water Research* **14**, 901-902.
38. Nickerson, K.W., Zarnick, W.J. and Kramer, V.C. (1981) Poly-β-hyroxybutyrate parasporal bodies in *Bacillus thuringiensis. FEMS Microbiol Lett.* **12**, 327-331.
39. NIOSH Case Studies in Bioaerosols. Ed. Ness, S.A. Government Inst., Inc., Rockville, Maryland 1996.
40. Nunez-Valdez, E. (1997) *Bacillus thuringiensis* conference in Thailand: a widening "umbrella". *Nature Biotech.***15**, 225-226.
41. OECD (1991) *Technology in a changing world.* Paris France.
42. Peter, T. (1996) Airborne particle analysis for climate studies. *Science* **273**, 1352-1353.
43. Preston, M.R. and Chester, R. 1996. Chemistry and pollution of the marine environment, in *Pollution: Causes, Effects and Control.* R.M. Harison, ed., Third Edition. Royal Society of Chemistry. pp 26-51.
44. Plotkin, B.J. and Kimball, A.M. (1997) Designing an international policy and legal framework for the control of emerging infectious diseases: first steps. *Emerging Infectious Diseases* **3,**1-9.

45. Rancourt, J.M., Beggs, R.W., Coleman, G., Tayabali, A.F. and Seligy, V.L. (1996). Quantitative redox assay applications for assessing antibiotics against the microbe component of commercial *Bacillus thuringiensis* biopesticides. *Cell and Mol. Biology* **42**, S63 - S64.
46. Reid, W.R. (1997) Opportunities for collaboration between the biomedical and conservation communities, in F. Grifo and J. Rosenthal (eds.) *Biodiversity and Human Health* Island Press Wash., DC. pp 334-351.
47. Reif, T.C., Johns, M., Pillai, S.D. and Carl, M. (1994) Identification of capsule-forming *Bacillus anthracis* spores with the PCR and a novel dual-probe hybridization format. *Appl. Env. Microbiol.* **60,** 1622-1625.
48. Rhomberg, L. (1993) Use of biomarkers in quantitative risk assessment, in Travis, C.C. (ed.), *Use of biomarkers in assessing health and environmental impacts of chemical pollutants.* Plenum Press, N.Y., pp. 331-345.
49. Roizman, B. and Huges, J.M. (1994) Effects of changes in human ecology and behavior on infectious diseases: an introduction. *Proc. Nat'l Acad. Sci. USA* **91**, 2377.
50. Schmidt Etkin, D. (1994) *Biocontaminants in indoor environments.* Cutter Information Corp. 1-155.
51. Seligy, V., Dove, M., Yaguchi, M., Willick, G. and Moranelli, F. (1986) Genetic engineering of cellulase genes from a fungus, in M. Moo–Young, S. Hasnain and J. Lamptey (eds.) *Biotechnology and Renewable Energy.* Elsevier Applied Science Publication Ltd. Essex, pp.118–124.
52. Seligy, V.L., R.W. Beggs, J.M. Rancourt, and A.F. Tayabali. (1997) Quantitative bioreduction assays for calibrating spore content and viability of commercial *Bacillus thuringiensis* insecticides. *J. Ind. Microbiol. and Biotech.* **18**, 370-378.
53. Seligy,V.L., Beggs, R.W., Coleman, G.J., Rancourt, J.M. (1997) Environmental monitoring of the DNA component of commercial products derived from *Bacillus thuringiensis* subspecies *kurstaki* and *israelensis* (manuscript in preparation)
54. Seltzer, J.M. (1997) Sources, concentrations, and assessment of indoor pollution, in E.J. Bardana and A. Montanaro (eds.) *Indoor air pollution and health.* A. Marcel Dekker. N.Y., pp.11-60.
55. Shugart, L.R. (1993) State of the art- Ecological biomarkers, in Travis, C.C. (ed.) *Use of biomarkers in assessing health and environmental impact of chemical pollutants.* Plenum Press, N.Y., pp.237-245.
56. Siegel, J.P. and Shadduck, J.A. (1982) Safety of microbial insecticides to vertebrates-humans, in M. Laird, L.A. Lacey and E.W. Davidson (eds.). *Safety of microbial insecticides.* CRC Press, Inc., Boca Raton, Florida. pp.101-113
57. Smith, K.R. (1995) Environmental hazards during economic development: the risk transition and overlap. Assessing and managing health risks from drinking water contamination: approaches and applications. *IAH Publ.* **233**, 3-13.
58. Solbrig, O.T., vanEmden, H.M. and van Ordt, P.G. (eds.) (1992) *Biodiversity and Global Change* Monograph **8**. Internat'l Union of Biol. Sciences, Paris, pp 221.
59. Stopa, P.J. (1997) Opening remarks, "*What do we really know about our environment*", posed as a theme and challenge to participants of the NATO Advanced Research Workshop on *Molecular Markers for Monitoring Environmental Biohazards*, Warsaw, Poland.
60. Sunderham, K.M.S., Sunderham, A. and Sloane, L. (1996) Effect of tracer dyes on initial deposits and persistence of *Bacillus thuringiensis* subsp. *kurstaki* toxin after application of two commercial formulations onto spruce trees. *J. Environ. Sci. Health,* **B31**, 1341-1362.
61. Tayabali, A.F., Beggs, R.W., Rancourt, J.M. and Seligy, V.L. (1996). PI-PLC-like activity in cultures of human and insect cells exposed to commercial *Bacillus thuringiensis*-based products. *Cell. & Mol. Biol.* **42,** 63-64.
62. Tayabali, A.F. and Seligy, V.L. (1997). Cell integrity markers for in vivo evaluation of cytotoxic responses to bacteria-containing commercial insecticides. *Ecotox. & Environ. Safety* **37,** 152-162.
63. Titball, R.W., Turnbull, P.C.B. and Hutson, R.A. (1991). The monitoring and detection of *Bacillus anthracis* in the environment. *J. Appl. Bact. Symp. Suppl.* **70**, 9S-18S.
64. Turnberg, W.L. (1996) *Biohazardous waste: Risk assessment, policy, and management.* Wiley-Interscience Publication, N.Y., pp. 422.
65. UNEP. (1991) The world conservation union, United Nations Environment Programs and world wide fund for nature. Caring for the Earth: a strategy for sustainable living. Gland, Switzerland. See also, 1987 GEM System. *Global Pollution and health.* Yale Press Ltd., UK.
66. Vachon, V., Paradis, M.-J., Marsolais, M., Schwartz, J.-L., and Laprade, R. (1995). Ionic permeabilities induced by *Bacillus thuringiensis* in Sf9 cells. *J. Membrane Biol.,* **148**, 57-63.
67. van Frankenhuyzen, K. (1994) Effect of temperature on the pathogenesis of *Bacillus thuringiensis* Berliner in larvae of the spruce budworm, *Choristoneura fumiferana* Cem. (Lepidoptera: Tortricidae). *Can. Entomol.* **126**, 1061-1065.

68. Vlak, J.M. (1995) Developments in engineering baculovirus insecticides and concepts of biosafety, in J. Landsmann and R. Casper (eds.) *Key biosafety aspects of genetically modified organisms*. Blackwell Wissenschafts-Verlag GmbH Berlin pp. 54-60.
69. Watkinson, I. (1994) *Bacillus thuringiensis*- product standardization. *Agri. Ecosys. Environ.* **49**, 37 -38.
70. White, T.J. The future of PCR technology. *TIB TECH* **14,** 478-84.
71. WHO. (1996) World health report: fighting disease, fostering development. Geneva.
72. Wilson, G.R., and Benoit, T.G. (1993). Alkaline pH activates *Bacillus thuringiensis* spores. *J. Invert. Pathol.* **62**, 87-89.
73. Zakrzewski, S.F. (1997) *Environmental Toxicology*. 2nd ed. American Chemical Society Mono. 190 pp 320.
74. Zhang, M-Y., Lovgren, A., and Landen, R.(1995) Adhesion and cytotoxicity of *Bacillus thuringiensis* to cultured *Spodoptera* and *Drosophila* cells. *J. Invert. Pathol.* **66**. 46-51.
75. Zilinskas, R.A. (1997) Iraq's biological weapons. Past as future? *JAMA* **278**. 418-424.

WORKSHOP CONCLUSIONS

At the conclusion of the workshop, the participants discussed the various aspects of the workshops. From their discussion, the following conclusions were drawn:

1. The monitoring of the environment for hazards posed by biological materials is just beginning to be appreciated.

2. Not much is known about the biology of the environment.

3. Tools need to be developed and refined to allow study of the microbiology of the environment.

4. More than one technique needs to be utilized to confirm the presence or absence of a particular biological entity in the environment.

5. Microbiological monitoring of the environment has implications in many wide ranging applications.

APPENDIX

FIRST DAY'S PANEL DISCUSSION ON THE STATUS OF THE ENVIRONMENT:

Peter Stopa: It's kind of late. I think that we had a very good session this afternoon. I see a lot of implications in terms of sampling, in terms of detection equipment, and analysis equipment.

Dr. Ivanov, one question which I would like to ask you is that with the thin film, the bacteria do stay on the top of the water. Do we really need to sample deeply to find out if there are any pathogenic organisms in the water. Or is it better to skim the sample along the top of the lake? Or to go very deep and come up with a sample?

Dr. Ivanov: It's very difficult for the natural aquatic system to say what is the concentration of these cells. The bacterial cells are distributed in a very thick field in the first phase.What is the concentration is this ..... It is very difficult to say what it is. Another is that the bacterial cells which are distributed as free cells in the bulk of water. Some -- a great quantity --of the bacterial cells adhere to the surface -- these bacterial cells which are swimming in the bulk of the water to the bottom. The term which is usually used in literature on the concentration of bacterial cells in the water is a very discussible matter. I propose to use different terms or different parameters for the different biotopes. For the bottle of water, okay, we can say that the concentration of bacterial cells, for example, some carcinogenic cells is this -- that number. But for the biofield you need to determine what you are talking about. Regarding the bulk of water, we have a standard method -- how to take this sample, this bacterial number from here, but from the.. Where sometimes 100 percent of the bacteria can be accumulated if you know any standard. We have a lot of methods, techniques, methodics but no standard for these biotopes, but I think that for some cases, for some specific cases -- not only bacterial, but for some viruses monitoring by the standard methods may be very difficult.

Peter Stopa: How would you count the bacteria?

Dr. Ivanov: Its possible to count the bacteria per square meter in an aquatic system. Maybe it would be better, or the simplest way, if you numerate the bacteria in the biofield and in the bulk of water. To characterize ..this dilution -- this share of bacteria, well they are there and the consideration of the bacteria in the bottle of water if you have the standard determination.

Peter Stopa: One more question from our colleague from Rumania. You will have several posters here on monitoring the water in lakes and in the river and that. Do you see any variations in counts or anything from where you take the sample of the water? Or do you just take a plunge with it? Would any of you care to comment on that?

P.J. Stopa and M.A. Bartoszcze (eds.),
Rapid Methods for Analysis of Biological Materials in the Environment, 303–308.

Michaela Lazarescu: We use standard sampling methods.

Peter Stopa: So you do see differences in levels. So it kind of does matter --there's really no good way to do the sampling on that.

Michaela Lazarescu: Yes, yes.

Peter Stopa: Anybody else have any questions or comments?

Dr. Jim Ho: I was taking to Doug Kell's group and we thought it would be useful if we could introduce the flowcytometry to people who are now just getting into looking at microbial populations in diverse locations like soil, permafrost and exploit some of the useful techniques that Hazel (Davey) and some of the people in Doug's group have worked up. You can actually do a lot of good microbial physiology using available probes that are currently available right off the shelf. Some of which were mentioned earlier.. I guess the comment that was given to me was that flowcytometers are hard to come by. And this is true, but Shapiro has actually written books on how to put together home-built flowcytometers in the last seven or eight years.

Stopa: Actually, the third version of his book was about two years ago and for about twenty-five thousand dollars he will sell the pieces and his expertise on how to put these together. Now Howard's rationale is, and I've had this discussion with him, is that he'll help you build a small system so you can measure the data, and then go to the appropriate official and show them the data, and maybe they'll give you the money to buy a big one.

Dr. Ho: The point that I'm trying to make here is that using flow cytometry in the appropriate guise can actually answer a lot of those fuzzy questions as regards to non-culturable, unculturable, viable, non-viable -- and you can actually define these terms. And I'd like to see that happen, starting from this conference. Thank you.

Dave Parrish: Actually, being the only commercial voice, I'll address the Howard Shapiro topic. I would like to see somebody learn flow cytometry and build flow cytometry all at once. That's a lot to write off. It can be done if you have somebody in house, but that's a lot.

Dr. Henry Meier: I have to agree with Dave and our commercial voice. I think it would be much better if a country bought a single machine and you could have multiple users. I know that in some places and universities they run their flow cytometers around the clock and you can easily pay for it if you can justify enough samples in a single day. It becomes economical and very do-able To give you some data on this, I tell you some of the things we've done. As you can here everyone is talking about flow cytometery. So if you can cell a center where something like ten universities can use it and companies and things like that can come in you can really make it financially accessible.

Dave Parrish: Additionally, if you take a company's -- any company's older equipment -- that's a pretty good write-off for that company. You can use it as a tax write off.

Dr. Henchal: I have a question. It actually makes sense on me that we can make devices to go to the field. The idea to have centralized laboratories to do this testing... I think its difficult to really get accurate analysis. If you set up laboratories that are far away from the sight to do the testing, you have a lot of problems with retaining integrity of the sample during transport. It might be better if we could concentrate on some devices that could at least give us a minimal amount of data that could give us information on sight.

Michaela Lazarescu: I'd like to ask our colleagues at this forum if you have made or have a monitoring network also from a biological point of view. for water and air, because from a chemical point of view -- I am thinking of a network of.. . and barometric transmission of parameters . .. I don't know anything about firm, if there is anything like that.

Peter Stopa: You're absolutely right there. I've been to several meetings -- this is strictly from a military-- on to start looking into air and water monitoring for biologicals and we have also done some work on occasion with the Environmental Protection Agency within the United States, and they are now starting to talk about getting into this business. The way the laws in the US are written, if there is not a law then you don't need to do it. So lately they are revising the laws so that they are looking for a mandate from the government to be able to monitor the air and monitor the water.

Dr. Kaprylants: About this flow cytometry. I would like to voice some skepticism on this approach in general because you know that in terms of viability in microbiology hey have only one discrimination: If they can grow -- if they can multiply -- this is just a bench mark for the microbiologist. From this point of view, any technique is good, including flow cytometery, also the concentration of ATP or I don't know NADH or these others. But you could make a mistake if you try to find a correlation between viability say of a concentration of a substance say.. which could be a correlation, but not obviously in all cases, and this makes it significantly difficult for us.

Peter Stopa: One comment that I have ... one kind of problem in using flow cytometry to answer these questions right now is that Hazel hasn't done enough work yet to figure out the best way to actually use flow cytometry to measure these parameters. Every year in Edgewood we have a research conference in which we can ask about that in the meeting. Last year I had a panel of Dr. Hazel Davey and Dr. Kell, Jim mentioned

Howard Shapiro, and Dr. Steen. These folks are probably the best in their field and we had some good discussions on the right way to use a flow cytometer. Two to four years from now -- once we get a better handle on the dyes and what's going on in the physiology, also with BioRad and Aber Instruments you are starting to see smaller, field portable instruments that are relatively inexpensive, and I think this is going to become a powerful technique.

Dave Parrish: Flow cytometry is a technology, not an instrument. And it might be referred to as two instruments or a few instruments. But it is a technology and it is certainly evolving.

Dr. Henchal: Peter you've had experience with both small devices and large devices. Now are there tradeoffs?

Peter Stopa: For the most part, if we take an instrument that has an argon laser, if the instrument is this big or that big, it behaves about the same -- more or less.

H. Davey: Okay, I've worked with about three flow cytometers in the last seven years or so, starting off with the Skatron then moving on to one with three lasers in it, and now using the *Micro-cyte*which is a small, portable, diode-based instrument and broadly speaking the bigger the instrument, the more complicated it is to operate. When you've got three light sources to align, its going to be a lot more difficult than when you just have an arc lamp and the optics to draw to it...and you have to bear in mind that when you are using more than one fluorescent dye, you are going to need compensation or filters to separate the different signals. So really the choice of instruments comes down to what question you are trying to answer at any one time. For research purposes where you've got skilled operators -- people who have had a lot of training in flow cytometry, you can go straight to it -- a big research instrument with three lasers or whatever it is that you want and get good results. But if you actually want to put an instrument in the field, I think the considerations that you've got to make are very different. For each thing it is a different answer. You can't just say this flow cytometer is better than another.

Dr. Meier: I know Peter has come up with a small system to use in the field, and it is very practical for use but it is a very limited use. It won't have, I assume, a UV laser on it. There are lots of interchangeable things that you can do with a high power machine. If you just want to run -- if you can afford to be specific. The US Army can afford to put thousands in the field, but in small developing countries where you have to justify every penny you are going to spend, I think you are better off sticking with one machine that is as versatile as possible without compromising the multiple uses.

Peter Stopa: In the last 15 minutes that we are going to have, I would like to redirect our thinking a little bit and talk about the environment in a little more detail. We've seen sources of bacteria and naturally occurring biological aerosols from different cell cycles. What implications does this have, for example, for monitoring programs for bacteria -- monitoring bacteria in the air and in the water. We've seen these elegant studies done by Dr. Ivanov and Dr. Vorobyova. The question is, if we were to monitor, what should we monitor for? And what influences the flux? Any comments on that? Would anyone like to say anything? And also... here's another question I would like to put out. We've seen the work by Dr. Kaprylants, and also mentioned by others about the viable and the non culturable issue. If we take an air sample and we grow it up on the plate, because right now the petri dish is the gold standard, in terms of measuring microbiological activity -- does everyone tend to agree with that right now? The gold standard. What if you got nothing on the plate, does that mean that nothing is there? Any parameters, you know, along that line? Any comments?

Dr. Goodacre: I was at a systematics, which a sort of monitorial, conference at the beginning of this year, and that particular area of science has taken a huge impact from methods to measure the small subunits, 16s, where you have very stringent primers, where you can do PCR and single organisms and get very good and very reliable data and if you want to monitor an environment that is going to be the route to take. As you said, if you can't culture it -- you know, conservative estimates are that you can only culture 0.1 percent of what's in the soil. If you are ignoring 99.9 percent of what is by trying to culture it, well that's a big mistake. And this small subunit, 16s, is throwing up a lot of interesting things, a lot of new sequences that aren't in the data base, which means a lot of new organisms that basically haven't been described yet. And if there is 99.9 percent of the organisms still out there --left to be described -- then the bio-diversity question is really huge.

Peter Stopa: So I would like to summarize this up. I think the first point is that we have a lot to learn about the environment. Would you agree on that? That's one factor. We still have to develop technology. We heard about 16s. They still have to develop flow cytometers. Dr. Loomis here has been talking about ATP and some other methods to measure different things where we still need to look at technology. We have a ways to go, I think. At least now in the perspective that I see, we are starting to understand what's going on in the environment. We are starting to be in a position that we have the tools to go in those directions.

Dr. Kaprelyants: I would like to add that the environmental system is even more complex than what we know now. I mean, for example, nowadays, we have the very strong suspicion that that bacteria can persist in so-called miniforms -- which is less than 0.2 microns. While I mentioned that -- this is standard talk on the separation of bacteria and filtration of bacteria. So it means that above 0.2 micropns, there is no bacteria, but in reality it is not the same. So we have examples from several places and very likely from tuberculosis All the bacteria can form small, tiny forms and we don't know exactly what they are. So nobody has even shown the conversion of these small forms. You have some experiments, but they are not very reliable from my point of view. So you have to also keep in mind these possibilities.

Dr. Loomis: Just as a continuation of that concept, we have some other things that we certainly haven't explained like what happens in mad cow disease in prions, and the role of prions in terms of their expression in terms of diseases and what have you, and what regulates those things. So I think there is an awful lot of that we have yet to find.

Dr. Bruce Harper: We've talked quite about bacteria on the interface of liquids, and bacteria in the soil, but I think we failed to scratch the surface on the different kinds of bacteria and environmental factors in aerosols.

Peter Stopa: I've kind of been leaving that to our aerosol session to be held later in the week. So I'm sure Jim can give you a very spirited discussion on that area. We'll have Dr. Grinshpun too. We'll have a whole session devoted to that.

DISCUSSION ON MASS SPECTOMETRY:

Dr. Del Vecchio: Wayne, you mentioned that Bob Carter was doing sequencing. Is he doing any probing with specific sequencing or is he just identifying segments of DNA by their sequence.

Dr. Bryden: What they are doing.. they are working with gene chips. They get a mass spectrometer readout for gene chips. They put down a target sequence on the chip and read it out on the mass spectrometer. They'll need labels and things like that to do it. There is some controversy actually in the literature right now. I guess the main issue in Analytical Chemistry -- there was a critique by the editor on mass spectral techniques for doing sequencing which was actually pretty interesting. It stirred up a lot of controversy in the mass spectral community. So you might want to clue into that A lot of people say, it is very promising because of the multiplexing and things like that that you can do, and a lot of people . .They can do about 80, I believe.

Dr. Del Vecchio: But the chip would actually act as the probe by hybridization.

Dr. Bryden: Yes.

Peter Stopa: Actually -- this is not so much of a question as a comment -- on the pyrolysis system. You talked about the MSMS mode. The system which is out there now which is produced in Germany does have the MSMS mode. One thing that we had some problems with it -- you can sort of classify maybe down to the genus level. You get a little past that and it's sort of hit or miss. It primarily looks at the **(unclear)** of fatty acids and small molecular weight material. So for the moment it is the best that we have. Many of us are really looking forward to the ... There is also an electro-spray technique which we really didn't discuss here, which Pete would have gone over, that's a little bit different. You do a direct injection into it.. Are you going to explain this? Maybe you can explain it a little bit better than I can...

Dr. Bryden: I'm not aware of it so... Maybe it will come to me (laughs).

Peter Stopa: Uh, let's see. I know it's atmospheric pressure ionization. You really need to watch the ions that you stick in there with the sample, but we've done work with a number of toxins and peptides on that and it looks good for proteins -- direct protein injection. And what happens is the solvent -- the water -- boils off of the protein molecule and causes it to ionize, and you wind up with different charges, and you have to watch the counter ions that you have in there. What we look at are the harmonics, so to speak, of different pieces of different charge ratios. And actually you get pretty good mass to charge ratio separations .. and the other advantage to it is you can do big things with it. For example you can take staph enterotoxin B, molecular weight 35,000, put it directly into this instrument and come out with those answers. But the problem we have now is that it is a big machine. These MAULDI type instruments look like they are going to get smaller.

P.J. Stopa and M.A. Bartoszcze (eds.),
Rapid Methods for Analysis of Biological Materials in the Environment, 309–313.

Dr. Henchal: You kind of went past your developmental item fairly quickly. Could we look at that a little bit? You had a box up early on...

Dr. Bryden: Right. This thing? I must say that that's an incarnation drawn by an artist.

Dr. Henchal: Yeah. We're dedicated at USAMRID towards developing point of care medical devices. .. I was concerned because it looked like to me like the laser that you are using was starting to exceed in size in your chamber. Are there developments in the laser area that would advance this procedure? By this I mean a smaller item.

Dr. Bryden: Yeah. We're working with a non-disclosure with someone -- on what is known as a microchip laser. The laser itself is a chip of ??? It's a millimeter on the side. It's a tube, actually It's a real nice developmental item. So actually, the laser is being handled quite nicely. It's a ... laser, an infrared laser, so when you put in doubling and tripling and quadrupling crystals, you can change it into the ultraviolet invisible wavelength ... .. Very high efficiency. The laser itself is actually.. We're definitely working on that and we are also working with other companies on smaller components. The laser on that is actually the least of my worries.

Dr. Garrigue: Do you have any information the kind of results you could obtain with what you call RELATIONS? Tape 9. I mean either the virus or the bacteria? .

Dr. Bryden: On laboratory systems. I don't have them with me, but they are being developed by our collaborators. We know what it looks like, and we know what it is going to look like, and right now it is being transitioned out, and will start to be run on the smallest group that we have.

Q (unknown): Do you expect that to be available soon. I mean this kind of data? This is very interesting. For electospray, we know very well. Everything can be done and we have had very good results from this. But okay, with the electrospray technique, we cannot use bacteria. This is something that...

Dr. Bryden: We're workng very hard on getting that done. Did you want to say something, Millie?

Dr. Donlon: I just wanted to say that Catherine Fenselau who is working for us has a book from the American Chemical Society that defines many of these criteria that she had done since 1976 -- on defining bacteria with different methods. And I would like to say that, until we were certain that we could miniaturize and regularize a small mass spectrometer, the libraries for it were not done. So our first approach was too handle the engineering and make certain that we could get pumps small enough and get lasers small enough to do this reliably, and now you are right that we have to go back and work very hard to build the library that goes on with this technology. So remember as I said, DARPA works in the future. So you have to wait for it.

Dr. Bryden: The library building takes two approaches. One is, use what's available in the literature to know what you are looking for. But the library has to be developed on the instrumentation that you are using. These laser-based techniques and so forth have data bases and things out there that you can lean on.

Speaker: Royston Goodacre UK
University of Wales Intelligent Systems for the Characteriazation of Microorganisms from Hyperspectral Data.

Questions:

Dr. Ho: You obviously were quite concerned with the time of training is it true that you really only train it once, and then you can keep running the training set afterwards so that the absolute time of training is not really critical?

Dr. Goodacre: In an ideal world, yes. But this is not generally the case because anyone whose worked with a mass spec and infrared knows that the signal does change with time. So it's useful to recalibrate the system quite frequently. A little side note which I haven't mentioned at all is that as far as prime S is concerned, I'm pretty sure that we've cracked that one. I have a method now to correct the drift over time. And if that's the case, then maybe the training time isn't such a problem. But waiting thirty hours for a net to train without knowing whether it's going to be right or wrong is a little bit different than thirty seconds.

Dr. Ho: How critical is it for you to grow the organisms consistently and how would that affect the final results.

Dr. Goodacre: At the moment the approach is such that any of these spectroscopic methods are measuring Phenotype. They are measuring expressed gene types. So if the organism changes slightly, then yes, because you are measuring the Phenotype, you will encounter that. The thing is that if you can use these methods that will allow the instruments to remain stable, then you just collect the information from your bacteria in as many states as you can. Then that information is encompassed into the model. So that you can train your network with organisms grown in three or four different media and it will identify those.

Dr. Ho: Now, I know that I am asking a lot, but what are the chances of collecting the organisms direct from your raw clinical samples and then running them through your PYMS and seeing what you can see from there?

Dr. Goodacre: With the current I don’t think that would be really realistic, because the sample size is large. And I think that you would do developments on that. And we heard earlier that you can get decent signals on a mass spec, so I don't see that as being a problem.

Dr. Ho: Now, I spoke to Horizons some five years ago, and at that point they indicated that the instrument wasn't all that sensitive and required a lot of sample material. Have they actually improved the system since then?

Dr. Goodacre: Well they've actually just gone out with pyrolysis, so I don't know. At the moment the system is set up for automated presentation of samples. You can load a carousel to 150 and press "go". What this means, because you have a high throughput of samples, what you don't do is insource pyrolysis You have a calumniating tube with an expansion chamber. Basically, it means that you don't put false rubbish into the on the source and alter its performance in that way. So, no. Unless you can concentrate your sample up, then you wouldn't use that instrument. You would go and use an in-source pyrolysis.

Dr. Ho: Last question. Are you using standard off-the-shelf neural network programs?

Dr. Goodacre: Yep. 32bit.

Dr. Ho: So any person could buy them.

Dr. Goodacre: Yep. I mean the learning curve isn't that great. The thing is its just like another... It's not quite an analytical tool, but it's just a tool provided you do the recipe ABCD you will get the right answers and I see that as an advantage, rather than a disadvantage.

Ivanov: Did you compare the dendrogram obtained by you, with the dendrogram obtained by traditional chemical taxonomy, or by molecular methods?

Dr. Goodacre: Not for this example. We have a poster out there where we have done ... where we have looked at Candida and with PyMS and FT-IR, and actually, it probably doesn't say on the poster, but we have done and are doing the genetics on there. And if you look at the literature, that maps up quite well. Broadly speaking, the sorts of things you get out of these do mirror the other methods. But there are some discrepancies.

Question (unknown): I saw this data on the poster. But I mean, I'm interested in the quotient of correlation between the relatedness of different bacterial specie From the end of this paper and the data obtained from traditional methods. How much will be the correctional quotient between your methods and traditional methods

.

Dr. Goodacre: Well, if I understand the question correctly, the scale on these dendograms is a relative scale, okay. If we are looking at this group of organisms from bacteriuria, if are looking at that plot. If you went out and sampled something, say a lump of cheese, or a dog, whatever, something completely outside our universe. The similarity scale would just be crushed right up to 100 percent. So it's relative. You don't get on the scale. You don't get...Oh well, rot... Basically, it's a relative scale. If you put organisms that you know their relatedness in there, then you can use that relative scale to get the information out. This isn't generally used for that. It's used for rapid identification. It's not used for chemotaxonomic studies.

DISCUSSION ON CLASSICAL TECHNIQUES:

Peter Stopa: In the remaining 20 minutes or so that we have. I think we got a little overview today of some of the classical techniques. We started drift a little into emerging.. and this last paper shows that maybe there are some other tools available to look at the data, rather than just some pretty lines or pretty dots on a piece of paper. So I guess one assessment. I'd like to ask -- I know we have some people who work in the field -- is 'Well what do you think'. Are we there yet with field portable equipment. Are we close? Do we need to go. Are there any users that would like to comment on that?

Dr. Ho: I don't really have any answers. But I just wanted to puts things into perspective. Ever since the mid 70's -- before I got into this business -- the military has laid down the very strict guidelines and rules at to what they envisioned the bio-detector was going to be. It was going to be a simple box, run by batteries, detect less than 100 micro-organisms and cost more than 5000 dollars or so -- 1970 dollars..

Peter Stopa: Jim, could you generally say that it does everything, costs nothing and weighs nothing.

Dr. Ho: Yeah. And we have sort of followed that for a while. We had various programs. And it didn't get us very far, did it? So we can sit back and take a look and think that well, maybe that really wasn't the approach that we should have taken in the first place. Maybe, identification and confirmation really doesn't have to be out there in the field. All we really needed to do was to go and collect a sample when you knew when to collect a sample and bring it someplace where you could use more efficient technology to do your confirmation for you. I'm not saying that that should be the way to do it, but maybe we can accept other ideas and other technology so that a field portable black box that's only the size of a shoe box might be chasing too much of a rainbow.

Dr. Henchal: I think that one of the things that we've learned -- and I'm kind of a program manager too -- is that there is no magic single technology that we're going to be able to use out on the battlefield for threat agents or contaminants. I think that we have to spend a lot more time thinking about what other systems of technology that we can employ to do accurate identification.

Peter Stopa: Also, there was the point that you raised yesterday about trying to samples -- as much as you can -- in the field. .. Sample integrity is an issue.

Dr. Donlon: But speaking from a doctor's perspective. I think you have to solve the problem now with the current technology. You have to establish your god standards. I think that developing current and future in parallel, learning from each other as we go is an essential component of the development of the biological detectors.

P.J. Stopa and M.A. Bartoszcze (eds.),
Rapid Methods for Analysis of Biological Materials in the Environment, 315–319.

Dr. Meier: I don' have a perspective. Basically, I'm the odd man out in this organization because I don't deal with bacteria. But it strikes me that there are some things that have become very obvious. If you collect samples in the field and they have to return to base your units will already be decimated by the time you get back to them. There are two different things It's what you want to assay. It is different if you are looking at pollutants in the water and trying to protect troops and civilian populations from bacterial exposure. The other thing is I think that one of the things that we are missing is that if I were going to be exposed -- or expose someone -- I would look for very common classical bacteria that I could modify. And the thing that I'll have to -- that you'll have to look for -- mechanisms that will detect modifications of everyday bacteria that will make them more virile. Also, one of the things that I would concentrate on would be to look for something that doesn't show all the cattle dying in the field, but look for something that will make people sick. One thing, if you have a bacteria or any kind of agent that kills people right away, people know to stay away from it. If you get them sick and you tie up there medical facility, then you have a thing that will do much more damage than something that is toxic.

Peter Stopa: One comment on that. We are looking at the BW scenario, and what we are basically saying that we need to identify up front as quickly as possible in order to minimize the damage. Go ahead Vito.

Dr. Del Vecchio: Along with what you said, and you sort of asked me a question about that this morning. You are probably aware of the whole new theory of pathogenicity islands where you have almost these cassettes that can go in an horizontal fashion, and they are formed by insertional sequences. And these have just sort of been defined over a year and half that I became aware of this. You know, a lot of people are defining these and I think that's something that you have to keep in mind. I don't know if you will be looking for organisms as much as collections of genes that giver rise to pathogenic mechanisms. ...You know, think of the nucleotide level. .

Dr. Garrigue: Maybe more what seems to be evident at the moment is that not any single technique, and this is what you said, is able to give us identification. We need a combination of techniques. For the moment hope that one day we will be able to go further than that. But at the moment we need a combination. The second is that we are dealing with local detection, local identification. If were able to do any remote identification, the problem will be seen differently. So in a more operational point of view, what is important is what kind of strategy do we have to use. The first is on the field to be able to detect some kind of particles, something strange which will be is called the chemical.. that will tell all soldiers take care now. Use your protection suits and the work for biologists is beginning. So, time is the key in the whole process. Really, if we had time like in a lab where we could spend days and days to be sure to get the exact type of bacteria -- to get all the information -- we wouldn't have any problems. The problem is that we need time, and we do not have any time. So time is a key factor also at the moment.

Peter Stopa: So what we are saying is that from the BW scenario, we need to get it up front as far as possible --probably more than one technique involved and of course bring it back for confirmation.

I'd also like to go in a different direction, for example Dr. Michaela Lazarescu, you and your colleagues are going to give us a couple of posters on water. What type of things do you take out with you in the field. Do you go out and sample an go back to the laboratory? What sort of things would you like to have?

Michaela Lazarescu: For biological monitoring we use a classical method. We measure the cultures from water samples. We wait and see what cultures grow and then we look at the water the kind of the bacterias we found in the water. It's a very old method, but unfortunately we are not so rich to use new technologies and from what I saw today, almost all of these are laboratory instruments, not... on the market.

Peter Stopa: Dr.Ivanov, with some of the work that you've described... you go out and you study in the field. Is that correct?

Dr. Ivanov: I think it is only a part of the system. The new methods -- as you can see in the development of the field of genetics systematics -- very often gives a new systematics or new data. Therefore, my opinion for all data and ideas which were demonstrated here is that it would be interesting not only to show that impossible to identify some microbes or some species, but it is possible to compare the obtained data with known taxonomy, or with the molecular taxonomy because there are our greatest preferences between medical taxonomy and traditional taxonomy. It would be very interesting to compare this data also because many feel that identification is a part of a big gap in science.

Dr. Donlon: I'd like to respond to that because, if you saw the March 31st Science issue, you can see that in the 16sRNA, is the complete hierarchical philogenicity of the microorganism, so you could, and Dr. Woesse, has studied this for many years. And the environmentalists now know that you can trace based on the 16sRibosome the entire lineage of the bacteria, so you could imagine a chip that had the Bacillus species and those strains that were capable of doing that. So that on your identification you would have the complete philogenecity of the microorganism from its origin. And you could see on that -- from that same chip -- that a virulence factor, for instance, that doesn't belong there, was now an E coli organism, being read completely on this hybrid chip.

So that I think that's where we are headed five of ten years from now, so that you can have this complete history of what this is by reading not the DNA, but by reading the 16sRNA...... And by making that a designer chip, so that you have put in your own factors and the antibiotic sensitivity -- that is the vision that the doctor had of producing something like that.

Dr. Ivanov: I have a dream .. to have something , something similar to this. Similar to a map of stars. Every star has its own name and some coordinates. I'm sure that in principle it is possible to have for identification purposes not only the traditional names but some coordinates in this dimensional area.

Dr. Donlon: But all the history of the microorganism is contained within that RNA molecule. So you know it's origin and where it comes from. If you start looking at the 16sRNA you have that pattern. And in addition to that, what we are finding out by looking at 16sRNA is that if you look at just culturable organisms we don't have even a tenth of a percent of what is in the total environment that we cannot culture.

Dr. Bartoszcze: I would like to return to the earlier discussion. We are looking at the ideal method, and we would be very happy with results that would take five, ten or 15 minutes, but we must remember that we will lose a lot of time on transportation. So the future system must be integrated, centering on detection and identification.

Dr. Seligy: I have two points. One of them is addressing the 16s. Not everybody who is the field of molecular genetics agrees entirely with that. And particularly, whatever you stabilize on that chip, we already know that organisms, like viruses, can drift off. And this is a serious point. The other one -- I really enjoyed the point -- even bacteria in particular can drift off. But the other thing that I thought was an important issue, is that most bacteria that can operate around 37 degrees can be opportunistic. They can exploit *us.* We live with organisms. We have 100 trillion cells ourselves and there are as many organisms in us. And we have evolved basically to protect ourselves against those organisms. Most of our sophistication is in that -- in our immune response -- and I was thinking that in biological warfare -- coming from a health side, in which our mandate is to look at public health -- in biological warfare I would be very concerned about what are doing in terms of being able to detect immuno suppression molecules, etc. because those are the ones that could have a very, very strong impact. And I am speaking more now like a pharmacologist than I am a geneticist at this point.

Dr. Donlon: I didn't address the other program, but their is a whole other aspect in looking at what our institute is doing which is finding ways to prevent and infection of any kind in a human being by going after domains. Responses. There are common responses that are given within the host organism. So not studying the bacteria -- which is what I do, but studying the host immune system which is another whole approach. Obviously, if we can prevent anyone from get an infection this is an enormous pharmacological investment there. And that's another big area that DARPA is entering into. So I agree with you and I think that the best we can do in detection, we should do, and pursue again both directions so that we can serve mankind by not having them be infected.

Dr. Meier: Concerning the idea of suping up the immunological response. Well what happens if you get a bug that's part of your normal flora and you modify it so that it's now making botulism toxin. So you wind up incorporating that in and it winds up killing you. Whether rattlesnake venom or influenza virus, the ability to change organisms is at a point where we are going to really need to have -- from my opinion, from being an outsider -- we will need to have a tremendous capability to modify and really be on alert for what's going to happen.

Peter Stopa: Well, in the remaining minute or two, I guess, in summary on this it looks like: the DNA/RNA markers –16S RNA or some combinations of that plus some other nucleic acids, fingerprinting -- may be definitely the way to go on that The question about chemotaxonomy versus biotaxonomy isn't... You know one thing, my classical training is as a chemist, and actually, in biochemistry if you wanted to show that you had a pure protein, you needed to show it by a physical method and a biological method, and maybe... well this seems to be what's emerging out of here. Maybe, we know about very powerful immunoassays, and I think we are starting to get an understanding of the genetics such that, that's going to tell us something so that we'll probably need that and some of the mass spec work that you saw described today, and on the infrared, we will be bringing in the chemical piece.

DISCUSSIONS ON NUCLEIC ACID PROBE TECHNIQUES:

E. Henchal, USA AMRIID -- "Evaluation of Nucleaese Flourogenic Polymerase Chain Reaction Assays for Identification of Biological Agents."

Questions:

Dr. Meier: How large a library of PCR probes are you going to have to carry into the field.

Dr. Henchal: That's a good question. Right now we are focusing pretty much on ten different biological agents, in addition to that we are folding in endemic infectious diseases also into this pattern. Your right, Henry. The ultimate target for about 60 agents. A strategy that we are using -- that we are developing is one that is based on geography. We imagine that we'd be able to -- once we identified where a particular laboratory was going -- we could set up panel that would be specific for a particular place... For example, it doesn't make much sense to survey for Dengue viruses in Bosnia, but it would make sense to have access to primers that were to amplify, let's say, tick-born encephalitis virus in that region. We can pretty much tailor assays and assay sets based on a particular need in a region.

Unknown: I have a question and I know that you didn't talk about it, but I know that the instrument -- ABI claims -- that you also can do quantitative analysis with these instruments. Did you try this method?

Dr. Henchal: We haven't done that yet.. and that requires the development of some control probes and that the next step. My priority has been to make sure that we can at least do the qualitative identification that you need -- this is of first importance - the medical careperson very often doesn't care how much, but what. So that was my first priority to be able to tell a medical caregiver what is the agent. The next step is maybe to fold in more quantitative aspects. These sometimes aren't very important to the medical community, but they could be important for environmental detection.

Dr. Meier: Besides detecting the presence, will this machine be able to at all detect how virile the strain is.

Dr. Henchal: That's a good question. One of the things at USAMRID is that we concentrated pretty much on the virulence. So we are pretty much targeting those genes that are important for causing disease. One of the things that we are going to be doing, is that we are going to be working with DARPA to look at other segments like the 16sRNA, so that we will be able to do more specieation. We consider the fact that because its so easy to modify, we consider looking and targeting the virulence gene as being the first priority.

P.J. Stopa and M.A. Bartoszcze (eds.),
Rapid Methods for Analysis of Biological Materials in the Environment, 321–324.

V.G. Del Vecchio, USA, University of Scranton, -- "Use of a Taqman Light Cycler, and Confocal Microscopy to Detect Specific PCR Products."

Questions:

Dr. Garrigue: The light cycler is apparently able to give you an answer very quickly. By quickly I mean, you said in between 15 or 30 minutes. Do you think that this kind of equipment will be available soon enough to be feasible. Because one of the problems that we in military face is that we want to identify the agents as soon as possible, and classical PCR normally takes something like three hours. If this equipment, in your point of view, is fieldable, will you give us an idea as to when it will be done.

Dr. Del Vecchio: I think it's fieldable. From what I understand, they are working on a system that is going to be much smaller.

Peter Stopa: On the light cycler, how many different things could you screen for simultaneously. Could you kind of take a sample and say I think this might have plague in it? Could you put in a battery of 10 or 12 different suspected materials and come up with an answer? Maybe this or that.

Dr. Del Vecchio: You can multiplex with the Taqman for instance,. if you wanted, you could have universal primers, or part of the RNA, and then have your probe which would be in different parts of the variable region... But if you want, you can also run 24 simultaneously and it only takes 15 minutes. But if you want to multiplex in one capillary tube -- you know, several reactions -- you probably could do it. It is hard to multiplex though because you really compromise the optimum conditions.

Peter Stopa: So you would know... (spoken without a microphone)

Dr. Del Vecchio: Exactly.

Dr. Henchal: I just wanted to make a comment on the fielding question . We are anticipating a fieldable device that is based on the Taqman technology by 1999. We're going to some of our first field trials next year with our prototype devices. There will be a second confirmation trial in the year 2000. Because we are working in medical diagnostics we are going to have to become involved in mult (?) clinical trials or clinically-based trials in order to validate some of the assays, especially for FDA. Some of these technologies could be available for environmental detection without the need for licensing probably in the year 2000.

Dr. Harper: On your last slide, showing the scheme for your identification program -- what sort to of time frame are you looking at for completing the whole thing including the amplification by growing it and constructing your libraries? ..

Dr. Del Vecchio: Its a very long time scheme. Don't forget what we are trying to do is to come up with a specific probe for each of the organisms that are found in a sample. So really we consider that part of the development process. Once you get the probe you want, then you know, it's a matter of putting in the light cycler. That part is very labor intensive, but once you've done it -- once you have the probe -- you have the organism.

Dr. Ho: I'm going to ask you a very irreverent question. In the last few months I've seen people over each other to buy the Idaho machine. Did you actually buy some stuff from the company?

Dr. Del Vecchio: No I didn't. Why? Am I that enthusiastic about it? It's a good instrument, but I didn't buy... in the company.

H. Garrigue, France, CEB -- "Use PCR for Identification and Detection of Biological Agents."

Questions:

Dr. Henchal: Yes, thank you for your excellent discussion. I think that I have one question. In the human case of anthrax that you investigated, did that start as a continuos infection, or was that an inhalation infection.

Dr. Garrigue: No, that was anterior anthrax. And we had a local problem because they have never been able to understand how she got it. In fact, the family -- the grandmother also was ill -- and one relative. The family and one relative. At the beginning we thought that they had eaten some sheep that were contaminated. They used to eat (a lot of sheep), and in France we have some fields -- devil fields. This means that in these places spores of anthrax live. So normally you would never breed any animals there. They could have come from one of these fields, but we have never been able to prove that.

Dr. Henchal: Your last slide also reminded me that our knowledge about variations between these strains is very small. I think you might agree with me that we need to broaden overall our analysis of variations going on in places by doing some world-wide collections on some things. I think that maybe we should think about establishing some kind of central library to evaluate the world-wide evaluation that may be going on in some of these places.

Dr. Garrigue: I agree. It would be very useful.

Dr. Seligy: I noticed that you had one turingensis strain in there. Do you know what subspecies it is?

Dr. Garrigue: Oh. I can show you -- well I have a slide -- where we passed a lot of different Bacillus turen.. and we were able to show that up to now the BA813 works only for anthrax.

Dr. Seligy: No. No, that's fine. The only point that I wanted to make for the audience is that when you actually look at the chromosome map, and not the plasma map, there is a very high degree of mols between these two. In fact well over half of the genomes almost identically align so far. So you have really done something quite nice here.

Dr. Garrigue: We have tested something like ten to twelve different thuringensis some coming from Korea and some coming from all over the world.

Dr. Seligy: But wouldn't it be wise, because in the nature of the use of thuringensis - as a bio pesticide -- that it would be better to make a bank -- I don't mean to criticize -- I just mean to use a bank of products in which you have identified in there to make sure that your identification kit for anthracis is not going to cross over?

Dr. Garrigue: Yeah, you are right. This is why we are trying to do this. In the beginning we were focusing on anthrax and we are really eager to get some strains and if some lab here... well at the moment we are having some problems to get the strains themselves but the DNA is fine.

PANEL DISCUSSION ON DNA TECHNIQUES:

Dr. Del Vecchio: I think what we saw this morning is that there is a whole variety of different techniques that are available to for analyzing your PCR products. Some are high tech and we still have to resort cultural diagnosis to be sure that -- at this point we have to resort to cultural diagnosis to parallel all the work that we are doing, but Henry brought up the point that you have to analyze the PCR products either by SSCP, or by restriction enzyme analysis to be certain that -- even if you are getting something of a certain type -- that actually is the ... and the organisms that you are involved with -- that you are looking for. I guess, we can say that we have a technology that has evolved, and presently we are at the threshold of more rapid detection. So maybe if we start in the present, then branch out to the future and open up the discussion. Maybe somebody wants to say something along these lines.

Dr. Henchal: One thing I noticed in the field is that there is a tendency for USAMRIID to have primers and probes, for the French to have primers and probes and for the Brits to have primers and probes... and right now there doesn't seem to be any concerted effort to understand whether these different systems have equivalency. I was wondering if maybe some future discussion should be concerned in maybe being able to put together evaluation panels of target organisms so there could be an exchange between the laboratories so that we would know which primer and probes are really the most efficient and which work the best in any particular system.

Dr. Garrigue: You are perfectly right. And that was one of the things that we discussed last week in Edgewood. As you in NATO we have a group which is called SICA. One of the problems that we had is that we wanted to do exactly what you wanted, but this is a very political issue. Because some countries think that to let the others know what is a primer, we give them the capability to react. They think that -- one of the things that is put forward -- to say that an enemy could be able to modify the sequence based on the primer and avoid any detection. This is a matter of policy. For us, in France, what we have decided -- and what we do -- is that we publish everything because we think that this risk is very small. I mean, by changing the genome of an agent --in two or three different places -- because it would not use only one type of primer --- You could finally obtain an which is no longer a threat. This is a political issue and we have a lot of problems in doing that. Also we are organizing what is called a round-robin test. And we will have some samples to be analyzed in March '98. We will receive five samples and we will have fourteen days to give that up. We can use any kind of technique that we want to use, but the only thing is to be able to give the answer as soon as possible -- in 15 days.

Dr. Henchal: This is a follow up comment. I think that we have the mechanisms to be able to detect your information in regards to a particular primer or probe set. But what is probably most important is that we come to some kind of agreement on what would be a standardized panel of organisms so we can talk about some common language for the standard.

P.J. Stopa and M.A. Bartoszcze (eds.),
Rapid Methods for Analysis of Biological Materials in the Environment, 325–330.

Peter Stopa: I actually have more of a comment than an question. I'm a dyed-in-the-wool diagnostics kind of guy. And in the early eighties, some of you may know monoclonal antibodies. They hit the field . Those of us who were chemists by training... The microbiologists were running around saying these things are specific, they are not going to cross react to anything. We are going to prove our detection limits by so much. And the world was a nice place. But those of us who suffered through the pangs of physical chemistry and molecular orbital theory, all these things that not only you can learn some science from, but it helps you get through some philosophy courses and you can start relating this to the existence of God and angels, and does truth bend around, black body radiation cells, anyway. The point being that we know that electron configurations, if they look alike, they are going to bind to a receptor. So what that meant was that these monoclonal antibodies are going to cross react. And for about the past eight to ten years, we immunologists guys have been taking it in the shorts because all the DNA guys are going to us saying your stuff cross reacts and ours doesn't. And what I am hearing now is maybe more of a sober approach to that saying, 'Yes, they can cross react. They can have shared etymologies different areas. And the point that Henri is making that we need more than one method to identify and confirm is probably the right way to go.

Dr. Donlon: One of the projects that (we have undertaken) is to eliminate anthrax as a threat... to prevent spore germination. At the university of Alabama we have identified four or five different enzymes that are germination enzymes for spores, and we have purified and crystallized them to make small molecular weight inhibitors, which will inhibit the germination of spores, and therefore they cannot produce the disease. And we are beginning a... chemistry approach to actually prepare a cocktail of these small weight substances that probably has very low toxicity to the body because we are not spore germinating. And I think this is a very exciting possibility. We have USAMRIID involved to be able to test these. Of course there is a long line after that for FDA approval. But I wanted to you to understand some of the leap-ahead technology that DARPA is doing to prevent the disease anthrax.

Dr. Del Vecchio: You know along with that, the recent report is in "Science" on the neuropathogenic E. coli strains. They found a sequence -- there was gene there and they used that to get a vaccine in rats and, hopefully they are going to be trying this in humans. So we are going to have that transition from the pure geno studies -- you know, once again going back to pathogenicity genes -- and then using and applying these in vaccine form. And of you have the exciting area in DNA vaccine and nucleic acid vaccines. And that's another why I think we are on the threshold. Well, maybe we are not on the threshold but maybe we are seeing the threshold at this point.

Dr. Zegers: I have a question for Mrs. Donlon. The method which you just mentioned. You mentioned that the spores were germinated. Is this method used as a profolaxis method or just after...

Dr. Donlon: It could be use for both because the spores as I understand it stick around a little while in the body. The secret also to the success is to choose some enzymes just out, after germination, that will be involved in this process. As far as a breakthrough, we have that available. This technology was actually based on technology developed at Alabama and other places where the flu virus.. specific inhibitors made against their immunities which allows the virus to exit the cell and infect other cells have been now developed by this method. They are being tested in phase three trials by the FDA, so that when you feel the flu symptoms around the flu season, you take this cocktail of inhibitors and prevent the flu virus from continuing its disease. I mean its a way in which we may go in the future in order to prevent infectious diseases. I would see it that this particular cocktail could be used as both, particularly if in the cocktail we have the inhibitors that could prevent not only sporulation but continued germination. I see it as particularly useful if you are going into a threat situation in which one of the detectors which for instance have dectected anthrax, and regardless of the strain or what's in it, you could take this pill and prevent it from germinating.

Peter Stopa: One thing about the library of genomes and DNA probes and sequences... I'd like to throw something out onto the floor. I think it would make an excellent topic for another workshop -- to get everybody together and discuss that matter and bring the sequences with you and throw it up there -- but I do realize some of the sensitivities that Henri mentioned. Some of the countries that we have worked with in the past at times were difficult too. But it's something to think about.

Dr. Garrigue: About what you said, you must know that in some countries the level of classification or confidentiality of these probes are secret, and country eyes which makes things more difficult to discuss. The second thing is about what you were saying what you were saying about these enzymes. Is there no risk that you are only delaying... are they able to block indefinitely, or when the concentration begins to slow down, that finally, as in the case of the flu, the virus that are in the cell burst out. Can you explain more?

Dr. Donlon: We are only in the beginning phases of this technology. I need to make sure that you understand high risk means that payoff would be high. So we have nothing to test yet. But one could imagine that because of the non-toxicity of this drug, that you could take it for a fairly long time until the body responded by clearing out these spores which still would be considered as toxic, and... (eliminated). That's hypothetical. I think the approach is the leap ahead approach that DARPA likes to take -- the high risk -- And it may not work. We'll certainly know that in a couple of years. But I did want you to know the approach is coming.

Dr. Del Vecchio: I think one thing. You know we are talking about fingerprinting and genotyping. Probably what is going to happen in the next few years is the ultimate genotype. This is going to be the big one. We're getting there slowly but surely. And of course, as you get easier and more accessible sequencing technologies that's being made available to people, that's what's going to happen. So I mean, you can't get discouraged by seeing the future. It's sometimes very discouraging.

Dr. Zegers: I also want to make the point on the PCR method. USAMRID and also DARPA are now developing the TIP technology and they are also using the PCR, but there are also other methods for amplification, for displacement amplification which works isothermal, so I think this is also a good approach for developing a chip.

Peter Stopa: Actually, we proposed to do strand displacement on the cytometer. I think that will work quite well. We have a plan to look at that next year. All we need to do is punch a hole in the bug and get the enzyme and the primer in there, but we have to take a look at the best ways to do that. Uh, I hear the bells ringing. I'm sure that they are probably anxiously awaiting us for lunch upstairs. Just a couple of overall summary points. Number one, the level of technology to bring a practical PCR or DNA probe -- Nucleic acid probe technology is pretty close at hand to get it out into the field. We are probably two to three years away. There are still many approaches to do the technique, and it's going to be dependent upon different scenarios, laboratory capabilities and what level you want to get to. I think now there is also an understanding that DNA or nucleic acid probes is not the ultimate definitive technology, but it's a major piece of it. And last, but not least, there are problems and can be problems with samples in terms of inhibitors and other things so we must be careful with sample preparation.

V. Seligy, Canada, -- "Comparative Performance of Conventional and Advanced Molecular Methods for Environmental Exposure Mapping of Biological Aerosols: Validation using Large-Scale Aerial Application of B. thuringiensis Kurstaki-based Commercial Biopesticides.

Peter Stopa: I don't know. I guess there are a lot of questions. Any comments from the audience? What I find particularly intriguing is the kind of an overlap between what you are doing for health prevention and we of course go out in the field and do on the VW practice scenario. I really think it's going to take us a minute or two to really sit down and try to digest this. The amount of information we just got was incredible. Bruce? or Jim? Or.. put you on the spot a little bit. One thing.... What other techniques do you think would be helpful to you in terms of doing this. Anything you've seen here today that you think might be useful?

Dr. Seligy: I think all of the flowcytometry -- the way its going I think is absolutely gorgeous. And there's no question that that is an important tool. I think what -- Jim Ho made reference to it -- and there were a couple of other people early in the meeting that... there's a general and you have a general probing formulation and I think it's very clear from what Henri said too that a battery of backups And I think most of the designs that I've seen this week are thinking about that. And certainly flow cytometry is one of those that can incorporate a number of those specific diagnostic probes. But that one for sure, and it would be nice -- I just loved that one, in fact lets go back to what was missed maybe, that was on your shoe. There it is (points to a slide cartoon). There it is. That's what I want. Maybe not that small, but I do feel that it looks like to us that anything in the direction which it is going would have immediate impact on the public. Including indoor air quality. Because this is becoming an extremely important issue. You know in Canada 80 percent of our population. Eighty percent of the average Canadian's time is spent indoors. You know, we go from one place to another, but we are indoors. And it is now being taken seriously that there are a lot things like sick children with building syndrome, crazy forms of allergies, etc., that we know virtually nothing about.. So this type of technology would be important to have. And certainly high quality sampling, high quality calibration, would be important.

Dr. Ho: The data shows us the that the preparations that we are looking at are nowhere close to pure. Back in the late 70's I came across the reference that the...

Dr. Seligy: I'm sorry. Can you say that again?

Dr. Ho: The preparations, the BT preparations that you were showing us...

Dr. Seligy: It's not the preparations, it's the product that's not pure.

Dr. Ho: Right. Back in the 70's or the 80's, I came across references to the fact that the Israelis came out with a way to purify the protein fragments where it was becoming economically feasible. What ever happened with that correction.

Dr. Seligy: Well, in the attempts, where only a pure protein was used, outside of a transgenic, these have given many times disappointing results. And actually, if you took what I showed you -- that product -- and you tried to do a purification of the protoxin as well as the activated endotoxin, which is a cleavage, the efficiency is something like 0.01 or 0.001 percent. It is not very efficient. It's very labor intensive. Unless, like sure, if it is totally mechanized. The way it normally is, is that you take a simple.

Dr. Ho: Can you explain in a little bit of detail, how did the industry manage to get approval for using this material in public spaces?

Dr. Seligy: Well, that's probably before either of our time. Your talking about the early 50s, when the standards of understanding, just like in this room, you know, some of us would say gee, compared to the public, we know a lot. But as Peter said, we still have an enormous amount to learn. And in the 50 they could only act on what it was that they had. I don't think that the entire structure of understanding of the protection of society, of the cohesiveness of society, was there. The point I wanted to make at the very beginning with the cell wars, is that while we have all of these top diseases in there, BT doesn't come close to that. The question really is, when you talk to people in the public who are upset that they might have had the spray drift on them, or they think that's happened, you don't tell them well, there's no one else, or nobody has keeled over or gotten sick, or maybe some sporadic diarrhea reports or irritation, what you have to do is try to understand, why hasn't it. Because it does have the capability. Where as before we thought that it never had. Most of the literature, they don't bother drawing this thing -- to test it rigourously at 37 degrees. It is grown at 25 degrees or 27 or 30 degress. And that's the other part of it. So when you are factoring in all of these other components... and you are spraying it on crown land, you know, in Canada, way out there -- maybe where you are -- I don't know -- but it's way out there and it's not bothering anybody. So it's acceptable now. When it comes into our community we, then we have to have to have better tools to deal with the prevention of it.

V. Seligy, Canada, -- "Comparative Performance of Conventional and Advanced Molecular Methods for Environmental Exposure Mapping of Biological Aerosols: Validation using Large-Scale Aerial Application of B. thuringiensis Kurstaki-based Commercial Biopesticides.

Peter Stopa: I don't know. I guess there are a lot of questions. Any comments from the audience? What I find particularly intriguing is the kind of an overlap between what you are doing for health prevention and we of course go out in the field and do on the VW practice scenario. I really think it's going to take us a minute or two to really sit down and try to digest this. The amount of information we just got was incredible. Bruce? or Jim? Or.. put you on the spot a little bit. One thing.... What other techniques do you think would be helpful to you in terms of doing this. Anything you've seen here today that you think might be useful?

Dr. Seligy: I think all of the flowcytometry -- the way its going I think is absolutely gorgeous. And there's no question that that is an important tool. I think what -- Jim Ho made reference to it -- and there were a couple of other people early in the meeting that... there's a general and you have a general probing formulation and I think it's very clear from what Henri said too that a battery of backups And I think most of the designs that I've seen this week are thinking about that. And certainly flow cytometry is one of those that can incorporate a number of those specific diagnostic probes. But that one for sure, and it would be nice -- I just loved that one, in fact lets go back to what was missed maybe, that was on your shoe. There it is (points to a slide cartoon). There it is. That's what I want. Maybe not that small, but I do feel that it looks like to us that anything in the direction which it is going would have immediate impact on the public. Including indoor air quality. Because this is becoming an extremely important issue. You know in Canada 80 percent of our population. Eighty percent of the average Canadian's time is spent indoors. You know, we go from one place to another, but we are indoors. And it is now being taken seriously that there are a lot things like sick children with building syndrome, crazy forms of allergies, etc., that we know virtually nothing about.. So this type of technology would be important to have. And certainly high quality sampling, high quality calibration, would be important.

Dr. Ho: The data shows us the that the preparations that we are looking at are nowhere close to pure. Back in the late 70's I came across the reference that the...

Dr. Seligy: I'm sorry. Can you say that again?

Dr. Ho: The preparations, the BT preparations that you were showing us...

Dr. Seligy: It's not the preparations, it's the product that's not pure.

P.J. Stopa and M.A. Bartoszcze (eds.),
Rapid Methods for Analysis of Biological Materials in the Environment, 331–332.

Dr. Ho: Right. Back in the 70's or the 80's, I came across references to the fact that the Israelis came out with a way to purify the protein fragments where it was becoming economically feasible. What ever happened with that correction.

Dr. Seligy: Well, in the attempts, where only a pure protein was used, outside of a transgenic, these have given many times disappointing results. And actually, if you took what I showed you -- that product -- and you tried to do a purification of the protoxin as well as the activated endotoxin, which is a cleavage, the efficiency is something like 0.01 or 0.001 percent. It is not very efficient. It's very labor intensive. Unless, like sure, if it is totally mechanized. The way it normally is, is that you take a simple.

Dr. Ho: Can you explain in a little bit of detail, how did the industry manage to get approval for using this material in public spaces?

Dr. Seligy: Well, that's probably before either of our time. Your talking about the early 50s, when the standards of understanding, just like in this room, you know, some of us would say gee, compared to the public, we know a lot. But as Peter said, we still have an enormous amount to learn. And in the 50 they could only act on what it was that they had. I don't think that the entire structure of understanding of the protection of society, of the cohesiveness of society, was there. The point I wanted to make at the very beginning with the cell wars, is that while we have all of these top diseases in there, BT doesn't come close to that. The question really is, when you talk to people in the public who are upset that they might have had the spray drift on them, or they think that's happened, you don't tell them well, there's no one else, or nobody has keeled over or gotten sick, or maybe some sporadic diarrhea reports or irritation, what you have to do is try to understand, why hasn't it. Because it does have the capability. Where as before we thought that it never had. Most of the literature, they don't bother drawing this thing -- to test it rigourously at 37 degrees. It is grown at 25 degrees or 27 or 30 degress. And that's the other part of it. So when you are factoring in all of these other components... and you are spraying it on crown land, you know, in Canada, way out there -- maybe where you are -- I don't know -- but it's way out there and it's not bothering anybody. So it's acceptable now. When it comes into our community we, then we have to have to have better tools to deal with the prevention of it.

PANEL DISCUSSION: SUMMARY OF THE MEETING

Peter Stopa: Okay, I would like for us to think about what we got out of this in terms of importance, and what we should suggest to NATO or suggest in the book that will come out of this. Does anybody have any comments?

Dr Donlon: We need to say something relevant. That is the key. And everything has been said here that needs to be said about the possibility of systems, the summaries that people have given at the end of there talks. I think it should be extracted by you, as the manager of this, to make some statements on some of the summaries that have been presented by the people here. And for me to stand here and restate what they have said so eloquently, I think this is...

Peter Stopa: Okay, let's make a few statements then. I think No. 1 is that many of us that are on the military side, I think we could certainly make the first statement that thing is there is a lot more to a biological insult than somebody setting off a bomb or intentionally releasing a cloud. The first thing is, that there is a lot in the environment that we don't know about and we need to study the environment a lot closer. A good example of this is the good work that was presented in the first session by all of our good friends from abroad -- by Dr. Kaprylants, Ivanov and all the rest. In terms of equipment, I think, yes, there is the need for rapid screening in order to have an idea that if something here is dangerous to quickly monitor food, the environment, soil, whatever. In terms of rapidly identification we need more than one way to definitely say that something is there.

I guess one question I wanted to ask you, Dr. Zakharov. In terms of taking this technology and sticking it on Mars, what do you have to say about that.

Dr. Zakharov: I would say that I have not spent much time at these kind of conferences -- I mean from the investigation of the environment from a pathological point of view -- and it has been demonstrated to me that we have a number of methods for such an investigation, and I believe that they will be useful for the Mars environment too. Indeed, we use some of these methods for exploration, for example, I mean mass spectrometers and some others. And I believe it is very important to have a shared experience in each of our fields. Of course, many of the techniques which were demonstrated are very difficult to use for space exploration because, you know we have to have very small instruments -- automated instruments but nevertheless, the methods are very useful, and not only for the investigation of Mars, but for example, in the space activities, the decontamination of spacecraft is a very important problem. And from this point of view, we can use the same instrument for sterilization. And from this point of view it was very useful for me and for Viola to participate in this workshop. I thank you Peter and our hosts, and I wish you success.

P.J. Stopa and M.A. Bartoszcze (eds.),
Rapid Methods for Analysis of Biological Materials in the Environment, 333.

GPSR Compliance

The European Union's (EU) General Product Safety Regulation (GPSR) is a set of rules that requires consumer products to be safe and our obligations to ensure this.

If you have any concerns about our products, you can contact us on ProductSafety@springernature.com

In case Publisher is established outside the EU, the EU authorized representative is:

Springer Nature Customer Service Center GmbH
Europaplatz 3
69115 Heidelberg, Germany

Batch number: 10823957

Printed by Printforce, the Netherlands